American Hegemo
Reconstruction of Science in Europe

Transformations: Studies in the History of Science and
Technology
Jed Z. Buchwald, general editor

American Hegemony and the Postwar Reconstruction of Science in Europe

John Krige

The MIT Press
Cambridge, Massachusetts
London, England

First MIT Press paperback edition, 2008
© 2006 Massachusetts Institute of Technology

This book was set in Sabon by Binghamton Valley Composition and was printed and bound in the United States of America.

Library of Congress Cataloging-in-Publication Data

Krige, John.
    American hegemony and the postwar reconstruction of science in Europe / John Krige.
        p. cm. — (Transformations)
        ISBN 978-0-262-11297-0 (hc. : alk. paper)—978-0-262-61225-8 (pb. : alk. paper)
        1. Science—Europe—History—20th century. 2. Technology—Europe—History—20th century. 3. United States—Foreign relations—Europe. 4. Europe—Foreign relations—United States. 5. United States—Foreign relations—20th century. 6. Europe—Foreign relations—1945–   I. Title. II. Series: Transformations (M.I.T. Press)

Q127.E8K75 2006
509.4'09045—dc22
                                                                2006044420

10 9 8 7 6 5 4 3 2

Contents

Acknowledgments

This book lies at the intersection of my current preoccupations with the nature of American power and how it is projected abroad, and my long-standing interest in the relationships between science, technology, and foreign policy—going back to my work on the history of CERN (European Laboratory for Particle Physics) and of ESA (European Space Agency). That interest was enriched by five years at the European University Institute in Florence, Italy, where I was immersed in a world dedicated to understanding the postwar reconstruction of Europe in all its economic, political, cultural, and ideological richness. My subsequent move to the United States has brought me closer to American sources and provided me with added insights into the nature and functioning of American society.

The approach that I have adopted has required my consulting archives in both Europe and America. My first debt of gratitude is to those many archivists who were so helpful to me, notably Ghislaine Bidault (French National Archives), Anita Hollier (CERN), Jonathan Green (Ford Foundation), Alain Paul (French National Archives), Ann Marie Smith (NATO), and Tom Rosenbaum (Rockefeller Foundation). Other colleagues helped by generously providing me with some of their own documents or by finding documents for me: Cathryn Carson, Allan Needell, Johan Schot, Tim Stoneman, and Helmut Trischler. This book was completed while I was the Charles A. Lindbergh Professor in Aerospace History in the Space History Division at the National Air and Space Museum in Washington, D.C. The hospitality of my colleagues there and the intellectual interaction with them regarding my next project (on space technology and American hegemony) have added immensely to whatever merits this book may have.

Some parts of this book have been published before. My thanks to John Heilbron, editor of *Historical Studies in the Physical and Biological Sciences*, and to Roy MacLeod, the editor of *Minerva*, for allowing me to make free use of the articles of mine they published.

I must also thank the Friends of the Center for History of Physics of the American Institute of Physics for a Grant-in-Aid, as well as the Cold War History Project of the Rockefeller Archives Center. The Kranzberg Support Fund, generously made available by the Stern Foundation, provided a stable source of travel money over four years without which this book would not have been possible.

Many people have made comments and criticisms of the arguments developed here or have contributed in other important ways to the book. Special thanks to Ken Alder, Mitchell Ash, Volker Berghahn, Jed Buchwald, Robert Bud, Angela Creager, David Edgerton, David Ellwood, Jean-Paul Gaudillière, Giuliana Gemelli, Gabrielle Hecht, Dan Kevles, Howard Kushner, Grégoire Mallard, Lydie Mepham, Joanna Ploeger, Viviane Quirke, Federico Romero, and Bruno Strasser.

Writing this book would not have been possible without the support and encouragement of those whom I love, and who love me. I dedicate it to them.

List of Archives

AIP	Center for History of Physics, American Institute of Physics, College Park, Maryland
AmPhilSoc	American Philosophical Society, Philadelphia
CACF	Centre des Archives Contemporaines, Fontainebleau, France
CERN	European Organization for Nuclear Research/European Laboratory for Particle Physics, Geneva, Switzerland
FFA	Ford Foundation Archives, New York City
LoC	Library of Congress, Manuscript Division, Washington, D.C.
MIT	Massachusetts Institute of Technology Archives, Cambridge, Massachusetts
NARA	National Archives and Records Administration, College Park, Maryland
NATO	North Atlantic Treaty Organization Archives, Brussels, Belgium
NBA	Niels Bohr Institute Archive, Copenhagen
RFA	Rockefeller Foundation Archives, Sleepy Hollow, New York

American Hegemony and the Postwar Reconstruction of Science in Europe

1

Basic Science and the Coproduction of American Hegemony

If we were a true empire, we would currently preside over a much greater piece of the earth's surface than we do. That's not the way we operate.
Vice President Dick Cheney, Davos, Switzerland, January 24, 2004

"The premise of this essay is that, given the basic inequality of resources [between the United States and Europe] after World War II, it would have been very difficult for any system of economic linkages or military alliance not to have generated an international structure analogous to empire. Hegemony was in the cards, which is not to say that Americans did not enjoy exercising it (once they had resolved to pay for it)."[1] Thus wrote Harvard historian of political economy Charles Maier in the late 1980s. For historians of science and technology his premise is striking, as it reveals the gulf between what diplomatic and economic historians take for granted about the capacity and behavior of the United States to build a world order aligned with its interests and our approach to such an issue (when it occurs to historians of science at all).[2] For there was not simply an imbalance in economic and military strength between the two sides of the Atlantic in 1945; there was also an imbalance in scientific and technological capability. The immense scientific and technological achievements in the United States during the war and the ongoing support for research in the country after 1945 contrasted sharply with the situation in postwar Europe. There, laboratories were ill-equipped, destroyed, pillaged, and (in the case of Germany) strictly controlled; researchers were poor, cold, hungry, and demoralized; and national governments had far more pressing concerns than scientific (and technological) reconstruction. The United States was not simply the mightiest economic and military power in 1945; it was also the mightiest scientific (and technological) power. Given the "basic inequality of resources" for science between the two sides of the Atlantic (and indeed globally), is it not to be expected that any system of U.S.-European scientific and technological linkages established after the war were also part and parcel of an "international structure analogous to empire"? Were those in the United States who wanted to "reconstruct" or "rehabilitate" European science not also engaged in the American hegemonic enterprise? Should historians of science not also take it for granted, as Maier did, that American hegemony structured the rebuilding of scientific capabilities and institutions in Western Europe, just as it did the economic and military spheres? In this book I argue that in science too an enfeebled Europe became enrolled in a hegemonic postwar American project—and tease out "the degree to which the U.S. ascendancy allowed scope for European autonomy."[3]

The place of science in U.S. foreign relations has only recently begun to attract the attention of historians of science.[4] Much work has been done on the multifarious bonds that were established between science, notably physics, and the American state during and after World War II. We have detailed studies of how scientists and their laboratories were enrolled in the apparatus of the national security system as researchers, advisors, policymakers, and intelligence gatherers, making fundamental contributions to the consolidation of U.S. power in the postwar period and during the Cold War. We know a good deal about the role that scientists played in projecting that power abroad in line with aims of U.S. foreign policy, particularly in relationships with the Soviet Union.[5] This history, dominated as it was by superpower rivalry, largely ignores Western Europe, and indeed the rest of the world. Moreover, the intellectual framework that it provides for thinking about the relationship between science and foreign policy ignores the asymmetry of power in which it was embedded. Ronald Doel points out that, particularly after 1945, "international science" was used "as a vehicle to promote American values and interests in the post-war world."[6] Similarly, diplomatic historian Joseph Manzione tells us, "The United States shared science to strengthen the Western alliance against Communism and to preserve technical and scientific preeminence. It shared science to support doctrinal arguments about the superiority of liberal capitalism and democracy over Marxism-Leninism."[7] Doel and Manzione recognize that if internationalism could serve these purposes after the war, if it came to mean something more than simply the circulation of knowledge and ideas within the scientific community itself, it was partly because science had become an affair of state. But they do not emphasize sufficiently that internationalism could only be an effective instrument of foreign policy because of the massive scientific and technological imbalance in favor of the United States vis-à-vis its allies. Combining scientific advantage with economic and political leverage, scientific statesmen, officials in the U.S. administration, and officers in organizations like the Ford and Rockefeller foundations did more than simply "share" science or "promote" American values abroad; they tried to *reconfigure* the European scientific landscape, and to build an Atlantic community with common practices and values under U.S. leadership.

This book is not simply about science and foreign policy, then, but about how science was embedded in, and instrumentalized for, the projection of American power in postwar continental Europe. More specifically, it is about how, in the first decade or two after 1945, the United States attempted to use its scientific and technological leadership, in conjunction with its economic, military, and industrial strength, to shape the research agendas, the institutions, and the allegiances of scientists in Western Europe in line with U.S. scientific, political, and ideological interests in the region.[8]

This chapter has two purposes. First, I introduce the notion of hegemony as used by economic and diplomatic historians to theorize U.S.-European relations in the postwar era.[9] Second, I suggest that basic science, or fundamental research, was the key node articulating American hegemony with the postwar reconstruction of science in Europe. The coupling of science and foreign policy was symptomatic of the new role that science, and basic science in particular, had in the postwar period, and of its presumed significance to economic growth, industrial strength, and national security. In the remainder of the book, I fill out that claim through a series of case studies that follow one another in roughly chronological order and that demonstrate how U.S. scientific statesmen, policymakers, and foundations, in collaboration with elites abroad, tried to rebuild European science to reflect U.S. concerns in the early years of the Cold War.

The Coproduction of Hegemony

The concepts of hegemony and empire as developed by diplomatic and political historians are not bound by notions of territorial acquisition or local rule, hallmarks of the "formal" empires imposed by Europeans on much of the world from the fifteenth to the nineteenth centuries. Tony Judt, for example, deems it "irrelevant" (historically) that the United States "eschews territorial acquisitions." Like the British at the height of their imperial reign, the United States today "prefers to get its way by example, pressure and influence"—even if that does not always suffice.[10] In similar vein, John Lewis Gaddis defines empire as "a situation in which a single state shapes the behavior of others, whether directly or indirectly, partially or completely, by means that can range from the outright use of force through intimidation, dependency, inducements, and even inspiration."[11] Tracing the origins of this strategy back to John Quincy Adams in the nineteenth century, Gaddis remarks that since Adams's day the United States has sought to maintain a preponderance of power (the term is Leffler's)[12] as distinct from a balance of power, then on a continental, now on a global, scale.[13]

The construction of an "informal" American empire in Western Europe after the Second World War was undertaken in collusion and in collaboration with sympathetic elites on the Continent, and with a large measure of mass support. By and large the United States did not use force to impose its methods of industrial production and management, its economic system, its political preferences and models, its military ambitions, or its cultural products on supine European peoples who were too demoralized and disoriented to do anything but accept them (even if Washington sometimes toyed with the idea of armed intervention). Indeed, European leaders who shared the United States' political and ideological objectives asked, at times even begged, the country to remain involved, to be a major economic, political, and military presence. The American empire that emerged was the negotiated outcome of a complex process in which European partners selectively appropriated and adapted features of the U.S. agenda and ambitions for the Continent and made them their own.[14] As Tony Smith points out, "Indirect imperialism of the American sort can only be effective when foreign peoples lend themselves root and branch, and for their own reasons, to the design of the imperial center."[15] To consolidate a liberal, democratic, capitalist regime abroad by resorting as little as possible to the use of force required a transnational elite that linked U.S. policymakers with a "team of partners" in Western Europe whose members "quickly became convinced that their countries' interests, and perhaps their own political fortunes, were best served by alignment in the new field of U.S. strength."[16] It is by virtue of that "alignment" that the United States "perfected the art of controlling foreign countries and their resources without going to the expense of actually owning them or ruling their subjects," as the British and other European powers previously had to do.[17] The specificity of American foreign policy is to be sought in the repertoire of instruments other than territorial expansion and direct subjugation that the United States could use to achieve influence and control after World War II—not in the illusory view that the United States, albeit a great power, "doesn't do empire."[18]

Building an informal (or "quasi") empire by consensus involves a gamble.[19] By eschewing force, and by resorting to threat and blackmail as a last resort, the United States accepted that it could not determine the course of postwar European reconstruction. It could only hope to shape its general trajectory and physiognomy in line with U.S. interests. And therein lay its strength. Europeans' relative freedom of action under the American umbrella, Maier writes, "did not weaken Washington's policies. On the contrary, it allowed the U.S. actions to seem less dominating and less constraining and thus probably made for a more broadly accepted policy. Precisely this possibility for national divergence made American policies more supple and more attractive than they might otherwise have been."[20] The United States, having left centrist European leaders the space to determine their own destiny, aided and abetted by Washington, constructed an "empire by consent,"[21] founded on "consensual hegemony,"[22] that is, a hegemony that was coproduced.[23]

The term *coproduction* is familiar to the science studies community.[24] It is covalent with Maier's consensual hegemony, but goes beyond that term in drawing attention to the creativity of both partners and to the relative plasticity of U.S. policymakers. Coproduction also signals that the United States gave Europeans room to leave their imprint on the hegemonic regime and implies that empire building is a fluid process. As Ann Stoler stresses, imperial formations are states of becoming rather than ready-made, rounded, bounded objects; they are founded on ambitions that trade on fuzziness, ambiguity, and confusion.[25] These nuances are crucial, in my view, not simply because they permit us to grasp better the flow of historical events that I describe in subsequent chapters, but also because they add depth to the brief discussion of the "Americanization" of European science in the final chapter.

The postwar coproduction of an American empire cohered with a Wilsonian view of America's global role in the twentieth century. Its dominant leitmotiv was, in Woodrow Wilson's own words, "to make the world safe for democracy."[26] The view of American exceptionalism—the idea that the United States had a unique role and mission in history and that America's interests were not narrow and parochial but embodied the interests of all—predated Wilson himself. For two or three centuries those who built the New World believed that they were creating a "model, a light shining out to a wretched globe and inspiring it to lift itself up."[27] For them, though, that model would be diffused best by example, not by imposition or proactive promotion. World War II and its aftermath changed that. First German militarism and then the conviction that Soviet Communism was bent on world domination led to the view that the nation's security lay in the expansion of democracy worldwide. Now the United States could not simply watch "failed states" stumble along without leadership. Now economic misery, industrial backwardness, and political instability threatened to create a vacuum that could be filled by forces hostile to democracy and to the global vision that inspired America's leaders. As cooperation with Stalin's Soviet Union gave way to confrontation, and to the Manichean division of the world into two rival political and ideological systems, faith in the United Nations as an instrument for managing the new world order collapsed. Convinced that there was "a clear and present danger to national security,"[28] the United States took it upon itself to make the world safe for democracy. It decided to use "the nation's great power actively and often very aggressively to spread the American model to other nations, at times through relatively benign encouragement, at other times through pressure and coercion, but almost always with a fervent and active intent."[29]

The idealistic fervor that inspired American interventionism is not to be underestimated. Robert Kagan correlates it with the determination to hold totalitarian expansion at bay:

After Munich, after Pearl Harbor, and the onset of the Cold War, Americans increasingly embraced the conviction that their own well-being depended fundamentally on the well-being of others, that American prosperity could not occur in the absence of global prosperity, that American national security was impossible without a broad measure of international security. This was the doctrine of self-interest, but it was the most enlightened kind of self-interest—to the point where it was at times almost indistinguishable from idealism.[30]

The justification for "internationalism" thus lay not simply in the overwhelming military and economic power that the United States had at its disposal in 1945, though that obviously facilitated matters. It was inspired by a definition of America's mission and identity that was deliberately crafted, a definition with moralistic and evangelical overtones that had deep roots in the American psyche, a definition that identified the United States with freedom in a world menaced by totalitarianism. As Henry Luce, the founder and editor/publisher of *Time*, *Fortune*, and *Life* magazines and author of a classic book on the "American century" (1941), explained, "If we had to choose one word out of the whole vocabulary of human experience to associate with America—surely it would not be hard to choose that word. For surely the word is Freedom. . . . Without Freedom, America is untranslatable."[31] The American empire was built to defend national security by promoting democracy and resisting tyranny, and that noble mission implied that it must protect not only narrow U.S. interests but also the interests of all "free men." America shouldered the burden of world leadership not simply, or even predominantly, because it was a major power intent on defending itself from attack and maintaining world superiority. It did so because its global vision embodied the protection at home, and the promotion, or imposition, abroad, of "universal" "freedoms" that were exemplified in U.S. policies and practices and in the daily lives of the American people. As Dean Acheson, Harry Truman's Under Secretary of State, put it, "For the United States to take steps to strengthen countries threatened with

Soviet aggression or communist subversion . . . was to protect the security of the United States—it was to protect freedom itself."[32]

The evangelical idealism that infuses much of American thinking about its role in the world cuts across political party lines. Both Democratic and Republican presidents have seen the United States as unlike all other hegemons, as a nation that wields its power benignly and in the common good. If they differ at all, it is only in hyperbole. Jimmy Carter, suggesting that good governance was based on "ethics, honesty and morality," not Realpolitik, went on to affirm that, accordingly, "there is only one nation in the world which is capable of true leadership among the community of nations and that is the United States of America." Ronald Reagan felt that it was an "undeniable truth that America remains the greatest force for peace anywhere in the world today." In 2002 George W. Bush described America as "the greatest force for good in history."[33]

If the claim that the United States is a benign hegemon making the world safe for democracy is repeated so persistently and with crusading zeal by U.S. leaders, it is also because there is an underlying fear that democracies, free markets, and liberal values would be undermined unless their scope was constantly expanded. As Federico Romero puts it, "A positive confidence in the global reach of modernity is thus intertwined with a keen perception of its fragility, and therefore with a globalist notion of American security, interest and prosperity. . . . As on a bicycle, one either moves forward or collapses."[34] The fragility that stalks the American project arises from a lack of consensus, both at home and abroad, on the universal applicability of the model so avidly promoted by U.S. leaders. The identification of freedom with quintessentially American freedoms, the conviction that the American way is the one best way to organize civil society, and the belief that the United States has a moral obligation to build an empire enshrining its values, these tenets of foreign policy have been repeatedly contested by friend and foe alike. What is more, the soporific effects of the idealism and evangelistic zeal of those who promote them have left the same people bemused and confused when, as has happened so often, U.S. motives were distrusted and the country's claims to be acting in the general interest were dismissed as mere rhetoric. For those who were skeptical of U.S. motives, making the world safe for democracy simply meant protecting social divisions and hierarchies at home and promoting narrow American interests abroad. "As a combination of high-noon sheriff and proselytizing missionary," writes Stanley Hoffmann, "the United States expects gratitude and affection. It was bound to be disappointed."[35]

The coproduction of hegemony has to be understood in this context. To speak of consensual hegemony is not to imply that the construction of a postwar order in Europe that the United States found acceptable was uncontested. America's was after all a hegemonic project that aimed to implant a particular set of practices and values among people who sometimes refused them. Consensus had to be won, and sometimes it was brokered against a backdrop of deep local and national divisions. In other words, consensual hegemony implies only that an influential fraction of a local elite, supplemented by U.S. overt and covert support, and operating in a particular local constellation of the balance of forces, was able to impose its vision of what kind of society should be built in continental Europe after the war. The consensus was circumscribed, and it needed to be consolidated if it was to endure. The intellectual fascination of the notion of coproduced hegemony, then, does not lie in using it as a blunt instrument to label the way in which the United States projected its power abroad in collaboration with local elites who shared its values. It lies rather in identifying the general parameters within which the United States tried to steer Western Europe, in exploring the relative plasticity in U.S. policymaking that empowered Europeans to adopt and reinterpret the United States' aims and ambitions, adjusting them to suit local conditions, and in unraveling the complex mechanisms by which a European life-world was built that was increasingly permeated with U.S. influence.

Hegemony and Basic Science

Hegemony is not a force that is deployed and that determines or dictates outcomes. The American empire, Maier reminds us, generally implied "power to" rather "power over."[36] Hegemony is a capacity, a state of being, a preponderance of power. It permits one, when one wishes, to intervene from a position of strength and to try to influence the course of events along lines of one's choosing. It requires instruments to achieve its objectives and pressure points where they can be applied. Political support and scientific legitimation, supplemented by money for grants, fellowships, and training programs, were the main instruments used by the United States to reconfigure European science after the war. Basic research was the main pressure point to which they were applied. Of course, the distinction between basic and applied science was blurred, notably in a postwar techno-scientific world, in which so much basic research was funded by, and enrolled into, "applied," or "mission-oriented," research agendas. All the same, the distinction was imperative for key scientists and policymakers engaged in the American hegemonic enterprise on the Continent. Those who believed that it was important for the United States to help "rehabilitate" European science did not want to contribute directly to applied research, notably research that was evidently coupled to military rearmament, and above all to the acquisition of a nuclear capability. This was as true regarding a defeated country like Germany as for an ally like France: the fear of a resurgent nationalism and militarism in one was matched by the fear of a newly legitimated Communism, and then of a militantly independent Gaullism, in the other. To support science in Europe it was therefore essential to operationally distinguish between basic and applied research even though everyone knew that the one was inextricably interwoven with the other.

Supporting basic science through grants, fellowships, and education and training programs had several advantages. I mention just a few of the most important at this stage; others will emerge as the narrative unfolds. First, one could capitalize on the tradition of scientific internationalism to enroll national scientific elites on both sides of the Atlantic in the project of postwar European reconstruction. "Internationalism" was intrinsic to the scientific ethos, and international scientific exchange and collaboration was a well-established mode of communication between scientists in different countries. It was an already existing tissue of social relations that could be mobilized as an instrument of foreign policy since it encouraged the circulation of people and ideas in nonclassified areas of research. The Europeans welcomed international scientific exchange, for it helped them close the gap, not to say chasm, in fundamental scientific knowledge that had opened up between the two sides of the Atlantic after the war. The Americans welcomed it because it enabled them to lend a helping hand to a Europe that was down-and-out scientifically, and for whose scientific tradition they had an immense professional respect.

Basic science could also contribute to European economic growth, social well-being, and, eventually, military strength. This broad social role for science was canonized by Vannevar Bush in his famous *Science— The Endless Frontier*, first released in July 1945. New scientific knowledge, he said in his letter of transmittal to President Roosevelt, was "one essential key to our security as a nation, to our better health, to more jobs, to a higher standard of living, and to our cultural progress." Bush's argument was a contribution to the debate in the United States immediately after the war over the proper relationship between science and the political order.[37] It was enthusiastically embraced both by an American "scientific elite who sought to achieve the permanent support for science that they had been trying to garner since the middle of the nineteenth century" and by U.S. industry, which increasingly established R&D programs in the postwar period.[38] It was also a popular theme among those concerned with U.S. foreign policy. Strengthening basic science in Europe was essential to the long-term economic prosperity of the Continent, they insisted, and was the only sure antidote to Communism that flourished wherever poverty and social unrest prevailed.

Strengthening basic, unclassified research in Europe could also make an important contribution to the scientific capital, the stockpile of knowledge, of the U.S. scientists. With the reconfiguration of the relationship between the civilian and the military in the United States after the war, it became increasingly important for U.S. scientists, if only in the interest of efficiency, to collaborate with colleagues in friendly nations whose classificatory regimes were less restrictive.[39] The advantage to the United States of access to science produced abroad cannot be overestimated. Security was not an imposition on U.S. scientific laboratories but a constitutive component of the postwar technical order. As Michael Dennis has put it, "Looking at the civilian in postwar America is much like looking at a map of an archipelago composed of discrete islands of civilian life connected by a larger, largely invisible military framework."[40] Peter Galison has attempted to quantify the extent of that hidden infrastructure. He estimates that the classified universe is now five to ten times larger than the open literature that gets into U.S. libraries. In his equally graphic terms, "The closed world is not a small strongbox in the corner of our collective house of codified and stored knowledge. It is we in the open world—we who study the world lodged in our libraries . . . *we* who are living in a modest information booth facing outwards, our unseeing backs to a vast and classified empire we barely know."[41] Access to basic unclassified research in Europe compensated for the restrictions on publication prevailing in the United States.

International scientific exchange gained further urgency in a postwar world that was dedicated to the elimination of racism, nationalism, and xenophobia in Europe. Science was seen as having a key cultural role to play as a bearer of liberal democratic values. Its epistemological hostility to authority, its putative celebration of organized skepticism, and its critical approach made it an ally in the struggle to de-legitimate and to eliminate authoritarian systems of government.[42] Science and scientists could play a constructive role in combating the twin evils of nationalism and totalitarianism by forging bonds of support and solidarity that cut across political and ideological boundaries. Basic, nonclassified research was an invaluable platform for building a transatlantic scientific community that put the shared pursuit of truth ahead of ideology. It was an apolitical instrument with major political effects: it catalyzed mutual understanding and respect among scientific elites in countries that in some cases had only recently been at war, and so contributed to peace.

For the United States to strengthen science in Europe was not without risks: the more successful the policy, the more Europe threatened to emerge as an independent pole capable of challenging U.S. leadership. To meet that challenge, it was essential to maintain that leadership even as partners abroad gained in strength. In the immediate postwar period, at least, senior officials were reassured that their position was secure by virtue of the immense technological, managerial, and industrial head start that the United States had over Western Europe. Even if major scientific discoveries of economic or military importance were made there, America would be far more capable of taking advantage of them. For example, Lieutenant General Lucius Clay, deputy military governor for the American zone in Germany in 1946, was satisfied that there was no great risk in making information on "German trade processes and advanced scientific thought" available to U.S. allies. For, as he pointed out, "While we are making this information available to all, our own industrial advancement makes it of greater value to us than to the others."[43] Similarly, U.S. Atomic Energy Commissioners who favored the foreign distribution of radioisotopes for medicinal and research purposes in 1947 reassured a security-conscious Lewis Strauss that they could exploit any scientific findings abroad far more rapidly than the foreign recipients of the isotopes. "With its superior technological potential," they pointed out, "the United States can expect to profit more quickly and more fully than any other nation from the exploitation of published findings" in the field.[44] Put differently, many U.S. policymakers in the postwar period were well aware that there was a middle term in the "linear model" between basic research and applied technological development. They realized that none of the new technologies and products developed during the war "could have emerged without the enormous engineering and manufacturing know-how and capabilities of [U.S.] corporations"[45] Therein lay the country's enduring capacity to maintain its hegemonic regime without resort to force. While basic research would benefit European economic growth and stability, it would also enhance U.S. leadership.

It must be stressed again that European scientists and their governments supported the steps taken in the United States to strengthen basic science on the Continent. They needed to close the gap that separated the two sides of the Atlantic and to become once again able and respected members of the international, that is, U.S.-led, scientific community. Desperately starved of resources, lagging seriously behind the research frontier, and aware of the enormous technological and industrial, economic, and military potential of some fields of science, both scientists and their governments welcomed U.S. support with open arms. If they did not catch up quickly, they would trail behind indefinitely. American hegemony, we have stressed, was thus coproduced. Europeans willingly cooperated in the reconstruction of their scientific capacity: they had little choice, and they were given enough latitude to adapt the American model to local circumstances, or even to reject it altogether.

My argument does not deny that U.S. scientists, science administrators, and foundation officers genuinely wanted to rebuild European research and were sincere in their wish to reconstruct science in Europe. But it also recognizes that, while the situation that they confronted was distressing, it also provided them with an opportunity. The reconstruction of Europe was a project that they could share in and shape in line with the values they held dear. And they had the means to do it, thanks to the structural asymmetry in power between the United States and the rest of the world. This lead was in the making before the war, accelerated during it, and was consolidated after 1945 by the mobilization of increasingly important financial resources for scientific research and technological innovation in academic, government, and industrial laboratories. The preconditions for American hegemony were there in 1945, and it was exercised in science and technology as it was in the economic, political, industrial, and military spheres.

This book is not "anti-American." My argument is permeated not with hostility to the United States but with a sense of Realpolitik and its meaning in the Cold War.[46] I reject as morally arrogant and self-deceptive that view of American exceptionalism that holds that whereas "other states had interests, the United States had responsibilities": all great powers have both.[47] Accordingly, I refuse to reduce the motives of key American actors who were promoting science in Europe to personal sentiment or to accept that they behaved simply out of "decency" and in the interests of their friends and allies abroad. I argue instead that the United States, through its formal or informal representatives, used its immense power after World War II to pursue not only its political and economic but also its scientific and technological interests in the European theater, working closely with European elites who shared its overall objectives. This is a study in the coproduction of hegemony in the scientific realm in the interests of rebuilding Europe, but also of maintaining U.S. leadership, of promoting "freedom," and of "making the world safe for democracy."

2

Science and the Marshall Plan

In June 1947 President Truman's secretary of state, George C. Marshall, announced a major foreign policy initiative at the commencement ceremony at Harvard University. Referring to the desperate economic situation and political instability in Europe, he said that the United States would be willing to help European governments draw up a recovery program and to support it financially. With that, the so-called Marshall Plan—officially the European Recovery Program (ERP)—was born. It was another significant step in the growing entanglement of the Truman administration in the future of a continent that was struggling to get back on its feet after a second grueling war. After pumping billions of dollars into Great Britain and continental Europe after 1945, Truman and his advisers came to believe that short-term aid was pointless and that a systematic program of economic restructuring was imperative. Without such a program, Communist parties and the Soviet Union would exploit poverty and despair to advance social strife, destroy democracy, and threaten the security of the United States.

Over the next four or five years, the U.S. Congress authorized about $13 billion in assistance for a variety of measures proposed by representatives of Western European states and implemented in consultation with U.S. administrators. The overall aim of the plan was to stimulate Western European economic recovery and, ultimately, political unity by creating an (economic) "United States of Europe." Increased productivity achieved by adopting American mass production and management techniques and stimulated by mass consumption would bring about general prosperity; class strife and the danger of authoritarian regimes would give way to social harmony and stable liberal democracies.

The Marshall Plan was at heart an economic and political effort to remodel (Western) Europe in the image of the United States. Its implementation required the active collaboration of transnational elites who together worked out how best to build a social order under American leadership that could meet the Communist threat on the Continent. The coproduction of this hegemonic regime has been studied in depth by diplomatic, political, and cultural historians; historians of science have not contributed to the debate. This is not surprising. Short-term economic reconstruction and political stability, not scientific research and its putative long-term benefits, provided the major rationale for the Marshall Plan. But this is not to say that science was forgotten altogether. In fact, within six months of the announcement of the plan, Vannevar Bush and others in the national military establishment suggested that the "rehabilitation of European Science" should be considered part of the program.[1] They argued that sustained economic growth and national security were not possible without a strong underlying scientific capability.

In this chapter I describe the steps taken during the late 1940s to reconstruct Europe economically and politically and to put in a place a military shield to defend Western Europe against what was perceived as Soviet expansionism. This political and diplomatic history is crucial to a rich, contextualized understanding of the intersection between science and U.S. foreign policy in Western Europe in the first decade or so after the war. This political dimension lays the groundwork for my subsequent discussion of attempts by Bush and later Compton to have science included officially as part of Marshall aid.

The Truman Doctrine, the Marshall Plan, and NATO

In the late 1940s the Truman administration took steps that drove the supporting pillars of American hegemony deep into the shifting sands of Western Europe and that threw into sharp relief the geopolitical and ideological boundaries of the emerging Cold War.[2] U.S. policy initiatives were triggered by fears that national Communist parties would capitalize on local misery and unrest, gain power democratically, and impose regimes sympathetic, or even completely subservient, to Moscow's demands. This prospect was all the more alarming since Truman's advisors came to believe that Germany had to play a major role in the economic recovery of Europe yet feared that it might seize the opportunity provided by industrial reconstruction to again impose its will on its neighbors; it might even be tempted to ally itself with Soviet Union to achieve its political objectives. The United States made aid to Britain, France, and Italy (which their governments desperately sought) conditional on their acceptance of its policies for Germany, creating immense anxiety in France in particular. To still French fears of being overrun again and to fill the gap created by Britain's decision to decrease the size of its armed forces, Washington eventually agreed to maintain an important military presence in Western Europe within the framework of the North Atlantic Treaty Organization (NATO). NATO was formed, in words attributed to its first secretary general, Lord Ismay, "to keep the Russians out, the Germans down, and the Americans in."[3]

The Truman Doctrine

Although economic recovery was under way eighteen months after the end of the war, the bitterly cold winter of 1946–47 slowed its advance and took a heavy toll on Europe's exhausted peoples. Credits intended for reconstruction ($4.4.billion to Britain, $1.9 billion to France, $330 million to Italy) were being depleted to provide short-term relief rather than long-term economic revival.[4] In Britain, coal was in desperately short supply, energy consumption was strictly controlled, and mass unemployment temporarily reemerged. Economic output declined sharply in France, and the bread ration was cut to 250 grams, leading to bread riots. One American commentator traveling through the region around this time remarked on the absence of the most basic necessities, everyone being preoccupied by "one state which is universal—complete absorption in the problem of how to live. Every minute is dedicated to scrounging enough food, clothing and fuel to carry through the next 24 hours."[5] In the words of historian David Ellwood, "Old Europe's political passions had gone: now a vote could be captured by a bucket of coal, an ounce of bread, a packet of cigarettes."[6]

Faced with increasing financial difficulties, the British government informed Washington in February 1947 that it could no longer provide economic and military aid to Greece and Turkey, and that it was withdrawing its troops from Greece. (London had previously deemed their presence vital to national security and had supported conservative elements in both countries.) Senior officials in the State Department saw the United Kingdom's retrenchment as just the first indication of Britain's slide from a major to a medium-sized world power (at the time the country still had an armed force of 1.25 million troops and was spending $3 billion annually to fulfill its worldwide commitments—a situation impossible to sustain in the winter crisis). Washington judged that if the United States did not intervene, the Greek Communists would capitalize on the ensuing unrest, seize power, and align the country with the Soviet Union. Democratic forces throughout Europe would be demoralized, domestic Communist parties in Italy and France would gain the initiative, and liberal regimes for which American blood had just been spilt would become absorbed into an expanding Soviet empire without Moscow so much as firing a shot. To avert this impending catastrophe, Truman went before a joint session of Congress on 12 March 1947 and asked for $400 million of aid for Greece and Turkey. This was, he said, a "fateful hour" in which nations had to choose between "alternate ways of life." The United States was determined henceforth "to help free peoples to maintain their free institutions and their national integrity against aggressive movements that seek to impose upon them totalitarian regimes . . . by direct or indirect aggression." Such regimes, he said, "undermine the foundations of international peace and hence the security of the United States."[7] Making the world safe for democracy now meant ensuring that Europe was not (again) dominated by a totalitarian system hostile to the United States and to freedom.

The enunciation of the "Truman Doctrine," and the president's appeal to Congress to provide money to implement it thousands of miles away, was made in somewhat inauspicious domestic circumstances. The Republicans had just gained control of both houses. Ambitious newcomers, like Wisconsin Senator Joseph McCarthy, joined sitting isolationists and nationalists from the Midwest and the mountain states, who were deeply hostile to foreign entanglements and financial sacrifices. Russia itself had begun to show a conciliatory attitude and was not a military threat. Indeed, U.S. military supremacy, secured by the atomic bomb, long-range bombers, and overseas bases in strategic areas, was guaranteed for the immediate future. British Field Marshall Montgomery estimated, late in 1946, that the Russians were "very, very, tired. Devastation in Russia is appalling and the country is in no fit state to go to war."[8] The Pentagon agreed. Faced by a skeptical Congress, Truman exaggerated the danger beyond the particular circumstances that had triggered it. He implied that the United States was embarked on a crusade, not simply to check the possible expansion of Soviet power in the European theater, but to oppose Communism wherever it appeared. In fact, the United States did not intend to, nor could it, police the world against Communism; the administration was quite willing to work with Communist regimes, like Tito's Yugoslavia, whose leaders were not tools of the Kremlin.[9] Truman's resort to universalistic rhetoric was intended to win support for his program among conservative isolationists who were committed to America first but also inveterately hostile to Communism. In so doing, he took an irreversible step along the path that demonized Communism of whatever variety wherever it appeared and laid the foundations for a foreign policy in the 1950s and 1960s that tended to see the world order "as an undifferentiated whole, and to regard Communist threats to that order anywhere as endangering the structure of peace everywhere."[10]

To enroll unwilling Republicans into his plans, Truman's spokesmen also had to reassure senators that aid for Greece and Turkey would not necessarily lead to further requests for funds and escalating foreign entanglements. This was disingenuous. As he said on 12 March, the risk to democracy came from economic distress and social disorder in countries with strong Communist parties. "The seeds of totalitarian regimes," the president told the Congress, "are nurtured by misery and want. They spread and grow in the evil soil of poverty and strife. They reach their full growth when the hope for a better life has died. We must keep that hope alive."[11] Given the dramatic economic situation in Western Europe in the winter of 1946–47 and the growing strength of Communist parties in some countries, it was inevitable that the president would soon extend the scope of his aid program. It was a logical consequence of his universalization of a specific threat and of his analysis of the socioeconomic roots of Communism. Indeed, in a famous memorandum of 27 May, less than three weeks after Truman's package had been finally agreed by the House, an alarmed Undersecretary of State William Clayton wrote to Dean Acheson that an aid package of $6–8 billion annually for three years was needed to save a Europe that was "steadily deteriorating. . . . Millions of people in the cities are slowly starving." Unless the United States offered further and substantial aid, Europe would disintegrate economically, socially, and politically. This would have "awful implications . . . for the future peace and security of the world" and "disastrous" effects on the United States' domestic economy, which would lose an important export market.[12]

The Marshall Plan

Truman's secretary of state, George C. Marshall, acted quickly. In his address at Harvard University on 5 June 1947 he launched the famous program that subsequently bore his name, though at this stage his was an invitation rather than a plan. Describing the desperate situation in Europe, Marshall identified the main need as being to "restore the confidence of the European people in the economic future of their own countries and of Europe as a whole." To this end, he called upon European nations to take the initiative and to agree among themselves on what they needed. The United States would be attentive to their requests, "providing friendly aid in the drafting of a European program and . . . later support of such a program so far as it may be practical for us to do so." European leaders were relieved and enthralled: "We expected them to jump two inches and they've jumped six feet" one American commentator remarked.[13]

Marshall did not limit his offer to Western Europe; it was also open to countries from Eastern Europe and the Soviet Union. "Our policy," he said, "is directed not against any country or doctrine but against hunger, poverty, desperation and chaos. Any government that is willing to assist in the task of recovery will find full cooperation . . . on the part of the United States government."[14] This was seen as one possible way of luring countries like Hungary and Czechoslovakia out of the Soviet orbit. It was also intended to defuse criticism of the scheme by the Communist parties in France and Italy, who were extremely hostile to the United States.

Earlier in the year, moderate leaders in France and Italy had been told that, in return for arranging emergency loans for them, Washington expected them to expel Communists from the governing coalitions. This policy was not driven simply, or even primarily, by ideological concerns. U.S. policymakers were convinced that if Communist parties held the upper hand, American aid, instead of being used to develop a free-market economy, would be directed to quite other objectives. The new French socialist prime minister, Paul Ramadier, accordingly revoked the portfolios of the Communist members of his cabinet on 4 May 1947.[15] In that same month, Italy's Christian Democratic leader, Alcide de Gasperi, under pressure from the United States, but also the Catholic Church and his country's pro-American south, formed a coalition that excluded the Communists from power. In this context, any overt move to limit Marshall Plan aid to Western Europe would only inflame an already tense situation and expose fragile governments to the charge that they were willing to alienate the Soviet Union and divide Europe to get U.S. dollars. In the event, negotiations with the Soviets over the implementation of the plan soon collapsed due to intransigence and suspicion on both sides. Domestic Communist parties in the West were instructed to launch a vitriolic campaign against it. And the Kremlin forbade East European countries from participating in the scheme—as Jan Masaryk, the Czech foreign minister, put it bitterly, after being summoned to be told of the ban: "I went to Moscow as the Foreign Minister of an independent sovereign state. I returned as a lackey of the Soviet government."[16]

Marshall aid was not only conditional on coalition governments marginalizing the influence of Communist parties but was also intended to push Europeans along the path toward closer economic integration. The secretary of state implied that Europeans were to view the problem of recovery as a whole, and see what contribution each could make to the mutual benefit of all. The program, he said, "should be a joint one"; rather than come to the table with individual shopping lists, European governments should think how best to coordinate their requests. Indeed, leading figures in the State Department insisted in private that the only "politically feasible basis" on which the United States could deal with the plan required institutionalized economic cooperation, "perhaps an economic federation to be worked out over three or four years," or even "some form of regional political association of Western European states."[17] In short, the Marshall planners took the American structure of a federation of states with a single market as a "model" for the direction in which Europe should move—not by force, but by "persuasion."

The "German question"

The success of this holistic strategy for recovery involved addressing the extremely thorny "German question." Immediately after the war, Allied policy toward the vanquished state, which was divided into American, British, French, and Soviet zones,[18] was guided by two potentially contradictory considerations. First, there was the need to enforce the demilitarization and the dismantlement of the German war industry, from which the recovered machinery and plant was seized by the occupying powers as war reparations. Second, there was a need for the denazification and democratization of the population to ensure that no demagogue could ever again drag his people into armed conflict. The challenge was to build an economy that allowed for a decent standard of living for the population without encouraging nostalgia for German economic and military domination. If industrial demilitarization went too far, notably in dual-use industries like steel, it could cripple the peacetime German economy, and (American) taxpayers' money would be needed to stop the country's sliding into economic misery and political chaos. If the dismantling of plant and equipment did not go far enough, Germany could once again become a major industrial power, leading to a resurgence of militarism and to a call for rearmament.

By early 1947 American planners were convinced that a radical shift in German policy was required. Levels of industry agreed on between the Allies in March 1946 were crippling the economic recovery of the country, whose overall industrial production had only reached 24 percent of its 1938 level. Coal production was far too small to satisfy Germany's needs, let alone the recovery needs of other countries, notably France. Steel quotas were intolerably low. Germany could not export enough soft goods to pay for its required imports. And the people were hungry: rations in the bizone were down to 1,200 calories a day. It became clear to Marshall and his colleagues that European recovery could not occur without an increased availability of German coal, steel, fertilizer, and chemicals and without the integration of the German market into a West European regional economy. German coal mines and steel furnaces had to be returned to German management, the currency had to be reformed, the three Western zones had to be fused into one, and the country had to be given some measure of self-government. In short, for Europe to get back on its feet (and cease being a continual drain on the U.S. taxpayer), Germany had once again to assume its role as a major economic power on the Continent with a measure of sovereignty and with some authority to manage its own affairs.

This package of measures was meant both to give expression to German aspirations and to contain them within structures that could channel them in line with Western interests. Everybody combined awe for Germany's scientific, technological, and economic potential with the fear that, if unleashed, that potential might once again be mobilized to achieve German supremacy. Once recovery gets under way, warned John Foster Dulles, the Germans "will almost certainly be dominated by a spirit of revenge and ambition to recover a great powers status."[19] The people, said senior policymaker and chief Kremlinologist George Kennan, were "sullen, bitter, unregenerate, and pathologically attached to the old chimera of German unity."[20] What would happen if, once back on its feet, the country maneuvered between East and West to achieve unity, even being willing to align itself to that end with the Soviet Union (which ardently supported German unification)? If the country "went Communist," the entire global balance of power would be upset and Western security gravely threatened by "a Germany controlled by the Soviet Union with German military potential used in alliance with the Soviet."[21] Hence the urgent need for a policy that would both enable German recovery and lock it institutionally, economically, politically, and ideologically into the Western bloc. German unification became secondary to modernization and to the economic recovery of Britain, France, and Italy, along with "West Germany," under non-Communist regimes.

The implementation of the European Recovery Program required getting Britain and France to acquiesce in America's plans for Germany. Britain's foreign minister, Ernest Bevin, was easily swayed and demanded only a few relatively minor concessions: for him, throttling the expansion of Soviet power was now the top priority. The French, by contrast, were utterly dismayed by the scheme. This was partly because it severely jeopardized their Monnet Plan for economic recovery, agreed by the French cabinet in January 1947. Monnet's program called for France to replace Germany as the major steel producer on the Continent, using coal imported from the United States as well as coal extracted from Germany. Marshall's idea that German coal mines be returned to their German owners and that German steel production be increased to 10.7 million tons (from a figure of 5.8 million tons agreed in March 1946) thus horrified Prime Minister Ramadier and his foreign minister, Georges Bidault. They were accused by the Communist left and the Gaullist right of sacrificing French recovery, at the behest of the Americans, to reconstruct the economic and military potential of the former enemy and occupying power. The French were also deeply suspicious of German militarism and feared that "if Germany is resuscitated, her resources and technical proficiency would be used by the U.S.S.R. against France," the country would be overrun by "Russian hordes" and France would once again be occupied.[22]

France's security concerns and growing fears that Europe would fall under Soviet domination led the Truman administration to adopt new measures to impede Communist advances in Greece, France, and Italy. The Soviet Union, opposed to German self-government, its participation in the ERP, and its integration into the Western bloc, tightened its grip on its satellites. Moscow also encouraged domestic Communist parties to do all they could through strikes and social unrest to sabotage the Marshall Plan. This further undermined economic recovery and increased fears in the United States that Europe would degenerate into chaos before the plan was put in effect. Congress authorized $600 million for food and fuel for France, Italy, and Austria in December 1947.

In February 1948 the Communist delegates to the Czech Parliament took power by a maneuver defined by the French ambassador in Prague as a coup—"a model of its genre, a masterpiece of Communist strategy"— engineered in Moscow.[23] It was seen as indicative of Stalin's expansionist intentions and his determination to extend an iron grip over satellites in his sphere of influence. The suspicious death soon thereafter of Jan Masaryk—he reputedly fell from his bathroom window—confirmed for many that Moscow would stop at nothing to achieve its objectives.

The Prague coup ignited fears that democracy was also under siege in Italy, where elections were scheduled for April 1948.[24] The United States was determined not to let the Czech scenario repeat itself there. Every imaginable form of persuasion and pressure was brought to bear on the electorate to ensure that it did not vote Communist.[25] The CIA was authorized to carry out covert operations against the Italian Communists (who were also being financed and secretly armed by Moscow), and the administration "allocated $10–20 million to pay for local election campaigns, anti-Communist propaganda and bribes."[26] The National Security Council agreed that military power might be needed to prevent Italy falling under Soviet domination.[27] Italian-Americans were encouraged to write home begging their friends and relatives not to vote for the Communists; "Freedom Flights" cosponsored by the U.S. Post Office and the airline company TWA carried their ten million letters and cables abroad.[28] The Church got involved. As one historian has put it, the issues before the electorate were reduced to "a series of simple choices: democracy or totalitarianism, Christianity or atheism, America or the Soviet Union, abundance or starvation."[29] In the event, the Christian Democrats won 48.5 percent of the vote (while the Communists and Socialists together had 31 percent) and the political centrists and right wing secured an absolute majority in the Chamber of Deputies.

Developments in Czechoslovakia and in Italy galvanized American public opinion. The suggestion that democratic forces on the Continent were inherently weak and impotent in the face of Communist determination won over the isolationists and protectionists in the House. On 17 March 1948 Truman went before Congress and called for passage of the ERP, which was signed into law on 2 April, for an augmentation of U.S. military personnel in Europe, and for increased defense spending. He indicated that the United States would contribute to European security and that American troops would remain in Germany until peace was assured. In June the French National Assembly, by a tiny majority, accepted the establishment of a new West German government—de Gaulle's supporters were unhappy about the concessions to Germany, while the French Communist Party decried subservience to an Anglo-American agenda. It was an admission that France was no longer a major power, and that to influence policy in the future it had to make unpalatable decisions in the present. "In order to impress upon our Allies the real and legitimate concerns of France," said Foreign Minister Bidault, "it is all the more important that our country does not separate itself either politically or economically from the camp of liberty."[30] The next day, on 18 June 1948, the U.S. military governor in Germany, General Lucius Clay, announced the implementation of currency reform and began to circulate the new deutsche mark in the Western zones, including Berlin.

The Soviet Union watched these developments with growing concern. Unwilling to intervene militarily, it had to find some way to retaliate against the growing division of Germany into two blocs and the strengthening of the Western zones. Sporadic interruptions of the flow of traffic into Berlin that had begun in March were escalated into a full-scale blockade. Electricity and coal supplies were cut off in the city and access to it by road and water corridors through the Soviet zone of the country was progressively impeded and then stopped altogether. Washington and London refused to be intimidated: it was deemed essential to divide Germany, notwithstanding Soviet anger and French anxieties. The "Berlin airlift" got under way. Beginning on 26 June 1948 and lasting until May the following year, American and British pilots supplied two-and-a-half million people in the Western zone of the city with an average of about 5,500 tons of basic necessities every day. The Soviets could have easily interfered with the operation, but they did not. They did not want to go to war over Berlin and sought rather to "harass the Allies out and to convince the starving population, by fair means and foul, that salvation could only be found on their side."[31] The success of the Berlin airlift proved just the contrary and, on 23 May 1949, two weeks after Stalin lifted the blockade, the first steps were taken to establish the Federal Republic of Germany (which formally came into being in September 1949).

The U.S. Congress eventually authorized, by 1951, $13 billion of Marshall aid for European recovery. Each country deposited an equivalent amount to what it received in "counterpart fund" accounts, thus investing their currencies to build national infrastructure. There is a lively debate in the literature over the economic impact of this support on postwar European expansion.[32] Certainly though, the point of the plan, as we have seen, was not merely, or perhaps even predominantly, economic; it was psychological, political, and ideological. Put simply, Washington policymakers aimed to remake "the Old World in the Image of the New" by encouraging an "economic United States of Europe" that, through a single market backed by mass production, would bring increasing prosperity to all, stimulate mass consumption, undercut the lure of Communism, and dissolve class tensions.[33]

The hegemonic regime that resulted was, as I have stressed, coproduced. The European elites who implemented the plan accepted, sometimes with reluctance, its broader geopolitical and ideological objectives and in return maintained a measure of control over how it was put in place and adapted it to local circumstances. Notwithstanding sometimes violent criticism from the left, they retained legitimacy at home while negotiating the implementation of the plan with the United States and its emissaries. Their task was facilitated by two things. First, as Charles Maier has recently stressed, the Soviet Union's refusal to accept Marshall aid and their own adherence to the Moscow line "consigned the Western Communist parties to a political ghetto."[34] They were forced to oppose a popular program that, in the event, led to a palpable improvement in the standard of living in Western Europe by 1950. The Marshall Plan alone surely could not take credit for this, but for many it was the prime mover.

Indeed, as cultural historians have stressed, it was the lure of the American lifestyle that anchored the promises of the Marshall Plan among the majority of the population and legitimated, in their eyes, their governments' willingness to promote Washington's objectives in the European theater.[35] Wagnleitner, who personally lived the effects of economic recovery in postwar Austria, writes:

In 1945, more than ever before, the United States signified the codes of modernity and promised the pursuit of happiness in its most-updated version, as the pursuit of consumption. Whenever real consumption climbed into the ring, chances were high that real socialism had to be counted out. . . . The great majority of European women and men, the defeated as well as the victors, wanted nothing more than to go on with their lives . . . and—after all those years of killing, running for shelter, fearing for loved ones, and scraping along—to have some fun again.[36]

The promise of the Marshall Plan that "You too can be like us" was music to their ears, for that was the way they wanted to be.[37] They wanted to share in the American dream.

The Brussels Pact and the North Atlantic Treaty

Within days of the Prague coup, Britain and France, along with the three Benelux countries (Belgium, the Netherlands, and Luxembourg) signed a defense agreement in Brussels. The Brussels Pact establishing the Western Union was intended to reassure Washington that the Europeans saw economic reconstruction and security as the two side of the same coin.[38] Signed on 17 March 1948, it was in fact a "Treaty of Economic, Social and Cultural Collaboration and Collective Self-Defense."[39] The five contracting parties declared themselves convinced of "the necessity of uniting in order to promote the economic recovery of Europe," coordinating economies, improving standards of living, and enhancing understanding between their peoples. The crucial Article IV stated that if any of them were "the object of an armed attack in Europe" the others would "afford the party so attacked all the military and other aid and assistance in their power." The only specific threat to peace that was identified was Germany; no mention was made of the Soviet Union.

The Brussels Pact was also aimed to entice the Americans to play an active role in the defense of Western Europe. Notwithstanding the increasing tension in Europe in 1948, the United States was not interested in becoming a member of the union, however. Economic aid to Europe was one thing; military engagements, another. The Europeans sought immediate and inextricable entanglement in the event of war, and the United States baulked. Article IV was particularly unsatisfactory since it seemed to require that the United States would be "automatically at war as a result of an event occurring outside its borders or by vote of other countries without its concurrence."[40] This was totally unacceptable. The United States was prepared to extend political and military aid to Europe, but only conditionally.

First, if France wanted American military guarantees "as far east as possible," it had to accept American and British plans for German self-government.[41] As noted, the National Assembly did so by a hairsbreadth in June 1948. Second, Congress had to be assured that Europe was not just expecting military handouts: there had to be signs of burden sharing. On the model of the ERP, the administration insisted that the Europeans come forward with their own coordinated defense plan using the resources currently available to them and identify how best their "collective military potential can be increased by coordinated production and supply, including standardization of equipment."[42] The United States would study the estimates of supplementary assistance needed and react accordingly. Third, Washington felt that the scope of the Western Union was too narrow: it had to be more "Atlanticist" and should include the "stepping-stone countries" of Norway (because of Spitzbergen), Denmark (for its colony Greenland), Portugal (for the Azores and notwithstanding the fact that it was a dictatorship), Iceland (a geographic bridge between the two continents), and perhaps Ireland.[43]

Two main considerations informed these demands for the enlargement of the Western Union. One was military and logistical. If the United States was to intervene successfully in Europe it needed forward bases for air and sea transport.[44] The other was political. The Truman administration, fearing the weight of isolationists in Congress, wanted to decouple a treaty intended essentially for the defense of Western Europe from its regional objectives and show that it was pertinent also to the defense of the North American continent. It was a difficult task, and not only in Congress. The signatories of the Brussels Pact and the Atlantic nations themselves were all most reluctant to accept the enlarged military alliance sought by Washington.[45] Only Canada did not require "cajoling, even intimidation."[46] On 4 April 1949 the North Atlantic Treaty was signed by the signatories of the Brussels Pact (excepting Luxembourg), by all the stepping-stone countries bar Ireland, and by Canada, Denmark, Italy, and, of course, the United States.[47]

The signing of the treaty did not quell disputes between the Europeans and the United States. The first concrete defense plans devised in Washington had the United States effectively abandon Europe at the onset of war. The revised plan placed the defense line at the Rhine: Europeans wanted it at the Elbe. The Korean War, which broke out in June 1950, put an end to the haggling. Was a Communist assault in a divided Korea not simply a prelude to an invasion in a divided Germany? "The specter of 60,000 East German paramilitary troops, backed by twenty-seven Soviet divisions in the Eastern zone," led to a massive increase in aid, the creation of the posts of Supreme Allied Commanders in Europe (based in Paris—General Eisenhower was the first) and in the Atlantic, and, by winter 1951, the decision by a reluctant Congress to send four divisions

to Europe.[48] What had been merely a treaty now became on organization. Thanks to the Korean War, the O was put in NATO and fears of
German rearmament were pushed aside by the need to strengthen Western Europe's defensive capabilities and to consolidate the Atlantic
alliance. Attempts to integrate Germany completely into a European
framework persisted, driven by the deep-seated fear that it was an unre-
liable ally. As the American diplomat Charles Bohlen put it, "The distrust of Germany is reflected in the belief that as soon as Germany
recaptures her freedom of manoeuver she will inevitably begin to play
the West off against the East with the very real danger of coming to rest
on the side of the Soviet Union."[49] On 5 May 1955 the Federal Republic
of Germany was formally recognized as a sovereign state by the Western
occupying powers; on 9 May, precisely ten years after V-E day, it became
a member of NATO.[50]

Science and the Marshall Plan

Truman's public pronouncements identifying European economic recovery as the key to defusing the Communist threat and to building a viable
Western alliance made no direct reference to science. However, by the
time the machinery for implementing Marshall aid was put in place in
April 1948 with the establishment of its official bureaucracy, the Economic Cooperation Administration (ECA) in Washington, support for
science was one of the tasks it was asked to consider. The suggestion
came from the Research and Development Board (RDB), newly embedded in the National Military Establishment (NME).

The NME was one of the bodies created by the National Security Act
of July 1947. This act responded to Truman's changing perceptions of
the Soviet Union and of the threat Communism posed to European stability and democracy. It created inter alia the National Security Council
and the Central Intelligence Agency (CIA), along with the NME, where
the (formerly Joint) Research and Development Board was lodged.[51] The
prime task of the board was to coordinate military research and development that was already being done in each of the armed services. The
wartime head of the Office of Scientific Research and Development, Vannevar Bush was appointed chairman of the Joint RDB in June 1946, and

when it was renamed in 1947, he agreed to stay on for one year. In fall 1948 Karl Compton, president of the Massachusetts Institute of Technology, assumed the chairmanship. His qualifications included his experience as a member of the U.S. scientific intelligence mission to Japan at the end of the war.

The RDB was primarily concerned with weapons development. Its suggestion that the promotion of scientific research should also be among the activities encouraged by the Marshall planners was coherent with Bush's views on the social role of science in the postwar order. It is not surprising, then, that at one of its first meetings in December 1947 the RDB "resolved, that in the implementation of the European Recovery Plan, careful consideration be given to the rehabilitation of European Science."[52] For a man like Bush, without a strong scientific base there could be no sustainable industrial strength, economic prosperity, social peace, or military preparedness. He transmitted this resolution to Secretary of Defense James Forrestal, who reassured Bush that, in his view, "appropriate measures in the scientific field could be included in the implementation of the European Recovery Plan as presently conceived."[53] Forrestal passed on the resolution to Secretary of State George Marshall, asking that it be given "appropriate consideration."[54]

The RDB was, in Michael Dennis's words, "an incredibly successful domestic intelligence gathering apparatus that compiled an inventory of US research and development."[55] It sought to repeat abroad what it did at home, establishing the current capabilities and future scientific needs of individual Western European countries. The Office of Naval Research (ONR), the primary federal agency supporting basic science at the time, was called upon to gather the information.[56] The ONR turned to its Office of the Assistant Naval Attaché for Research in London for help.

In August 1948 the London staff submitted its report on the "Rehabilitation of Science in Europe." The report is twenty-eight pages long and deals with the rehabilitation of science in the United Kingdom, France, Germany and Italy, Belgium and the Netherlands, and the Scandinavian countries.[57] It was supplemented by almost two dozen appendices describing the organization and state of science in various countries—memoranda written by the London office, reports by consultants, trip reports by leading American scientists, articles from professional journals, and some official documents.[58] This supplementary material was heavily biased in favor of physics, as we might expect from a military establishment: chemistry and the life sciences, for example, were seldom, if ever mentioned. In his covering letter to the chairman of the RDB, Chief of Naval Research Rear Admiral Solberg again "emphasized that due regard should be given to the fact that a healthy industrial recovery and recovery of adequate standards of education, health and general welfare may not be secured and maintained without sound and stable conditions with respect to science, both in research and education."

The introductory and concluding sections to the ONR report made some general recommendations regarding the needs of science in Europe as a whole. They stressed that "the destruction of laboratories and libraries and the more or less natural deterioration during the war years have had a profound effect." Scientific training had been disrupted, scientific equipment and materials were sadly lacking, and money was in short supply. While "scientific thought in governmental circles" had been changed by the war—presumably meaning there was greater willingness to invest in science—the ONR also expressed concern about the presence of scientists in "high positions in government" who were "identified with extreme left wing groups in their countries" (an obvious reference to the situation in France). The ONR was also extremely disturbed by "the emigration of young scientists from their native countries to other countries where the situation from a scientific and economic point of view, is much better." In some countries—the situation in Italy was particularly serious—this was "draining the lifeblood of science from the area."[59]

The ONR office stressed the importance of foreign aid to remedy this situation: "One of the primary problems in all the countries concerned is the lack of dollar exchange with which to purchase apparatus, books and periodicals and with which to pay for trips outside the country, either to international conferences or for exchange fellowships." Action of this kind would also relieve "one of the most serious problems in the rehabilitation of science in Europe ... the exodus of scientists both young and old" from their native countries in search of better working and living conditions. War surplus could also be used to help alleviate the equipment gap. National communities should be encouraged to revitalize scientific societies and organizations that could take the lead in scientific reconstruction, and administer any foreign aid program that was put in place.[60]

A top secret appendix to a CIA report published on 27 October 1948 was circulated along with the ONR material. This CIA report is still classified.[61] We know from other sources that it was "primarily concerned with the short term gains which would accrue to the Soviets if they were able to capture the highly trained technological manpower of Western Europe," or with what another source called "Security risks involved in such a program [of foreign aid] if Europe were overrun."[62] The best way to minimize these risks was to concentrate support on "pure research of the academic type" and only fund applied research in very specific cases. The obvious advantage was that basic research was a long-term investment with no immediate or obvious interest to the Soviets, who needed "engineers and productive experts in fields which would receive little stimulus from the ONR program" and who would in time probably train enough scientists and engineers within their own borders to meet their needs.[63]

The focus on basic research had many advantages. As two navy members of the RDB put it, it had "great possibilities for international cooperation, friendship and the ultimate economic independence of Western Europe."[64] All were not only beneficial to Europe but to the United States as well. International cooperation, Compton wrote, would help "increase . . . our basic scientific and technical knowledge," and also "replenish our stockpile of ideas," which had been depleted during the war. He identified the fields that merited financial support and assistance in the form of texts, technical periodicals, and instruments: "Utilization of human resources,—food, medical sciences, chemistry, physics, metallurgy, synthetic materials, geology, meteorology, geography, and soil mechanics." Helping European scientific recovery would also entail "a reduction in foreign requirements on the U.S. economy" and "an increase in technical and scientific intelligence data of great potential value to national security."[65] Isidor I. Rabi, Nobel Prize winner, scientific statesman, and chairman of the Department of Physics at Columbia University from 1945 to 1949, concurred: "We should put a certain amount of pressure on the Marshall Plan countries to support their sciences more soundly," he wrote Compton in February 1949, "if they are to recover and be less of a burden to us and a prospective good ally from the military point of view."[66]

Notwithstanding his institutional position, Compton was opposed to military sponsorship of any scientific recovery program. In his annual report to the president at the end of 1948, Secretary of Defense James Forrestal said that he considered "military aid to Western Europe to be of the highest priority." He nominated U.S. Army Major General L. L. Lemnitzer as his representative to the new interdepartmental Steering Committee on Foreign Assistance Programs intended to develop and coordinate the administration's proposals.[67] In his submission to Lemnitzer, Compton suggested that the rehabilitation of foreign science should not be supervised by the military.[68] The reasons he gave were "fear of resurgence of the German military under such aid, military domination of scientific thinking, and military backing of foreign science might provoke USSR accusations before the Security Council of the UN on the grounds of unwarranted military interference." The military, Compton went on, "should have access to foreign classified technical information and intelligence data" through the soon-to-be-signed North Atlantic Alliance, but it should "not be involved in the administration or supervision of programs or assistance." A program for European scientific rehabilitation, while of "great interest" to the defense establishment, Compton wrote, "should be accepted entirely on its merits in promoting economic recovery" and administered in such a way as to "have no direct military implications." Instead, it should be handled by the ECA "with the advice of a non-military scientific group, such as the National Research Council." The RDB endorsed Compton's recommendation and authorized him to try to get resources appropriated for European scientific recovery under ECA legislation.[69]

The ECA quashed any suggestion that it should take the initiative in supporting science,[70] apparently because, to quote Compton, "the original concept of aid to Europe was of necessity based on the more material requirements to sustain life."[71] The long-term benefits to be derived from basic science were not priorities for those responsible for implementing the Marshall Plan. There was some talk that a European program could be sponsored by the State Department under the president's "Point Four" plan announced in his inaugural address on 20 January 1949—a commitment "to help the free peoples of the world, through their own efforts, to produce more food, more clothing, more materials for housing, and more mechanical power to lighten their burdens."[72] This too came to naught: Point Four aid was for "the backward areas of the world, i.e., Ethiopia, Iran, Iraq, Arabia, etc," not for Europe.[73] Another possible bureaucratic home for the program inside the State Department was the Interdepartmental Committee on Scientific and Cultural Cooperation. Though it lacked resources, the military, it was suggested, might use it as a front to funnel funds for European scientific reconstruction to an agency like the National Bureau of Standards and to the National Research Council.[74] This proposal was also short-lived and it was reaffirmed that the military should not be called on to sponsor any basic science in Europe, even indirectly and under cover of a civilian administration.[75]

These efforts to support the rehabilitation of European science under the auspices of the Marshall Plan had two important consequences. First, they led the State Department to accept that science had become a significant component of foreign policy and that it needed to provide a rationale, formulate guidelines, and establish procedures for supporting science abroad. In July 1949 Lloyd Berkner, a scientific statesman already intimately familiar with the functioning of the RDB, was asked to produce a "Survey of State Department Responsibilities in the Field of Science." This led to the production of a major policy document, *Science and Foreign Relations,* released in May 1950.[76]

Second, the burden of taking the initiative in seeking aid for science was shifted squarely onto the shoulders of the European countries themselves. As Compton put it, "Until or unless requests for such assistance are made on ECA *by the countries themselves,*" there was nothing much that he could do.[77] An official in the State Department suggested a possible procedure. One could try to "persuade the European countries in question to up their dollar expenditure for rehabilitation of their scientific research activities and to increase proportionately their request for ECA funds."[78] Of course the problem was that the administrators of the ECA were not convinced of the need to support science. Compton was determined to change their perception. "Personally, I intend to do all possible to encourage this matter being brought to the attention of ECA," he wrote to Rabi, "in the hope that the vital role of science in economic and social recovery will be recognized eventually as being just as important as food, farming equipment, etc."[79]

Did the European countries act on this, and did the ECA respond positively? Our evidence is fragmentary, but we know that in at least two cases they did, with quite different results. First, Walter Hirsch, who was responsible for the distribution of (counterpart?) Marshall Plan Funds in the Federal Ministry of Economics convinced the American authorities that 41 million deutsche marks (DM) should be used to support industry-oriented research projects. Both industrial and scientific clients, including the Fraunhofer-Gesellschaft and the Deutscher Forschungsrat (DFR [German Research Council]), officially constituted in March 1949, successfully applied for these resources. The DFR received DM 4.1 million from this source in 1950, almost half of its budget of DM 8.3 million for that year, and more than all the German *Land* cultural ministers together could provide for research. ERP support also accounted for 22 percent of the DFR's budget in 1951 (whereupon this source dried up). The strings attached to the money reflected the aims of the Marshall Plan: it had to be used to stimulate the German economy. In this case funds were to be used to buy scientific instruments produced in the country: no foreign equipment could be acquired with it, nor could it be used to pay salaries, to buy books or building material, or to pay for perishable goods.[80] The German authorities thus devised an ingenious way to satisfy both the crying need for scientific equipment and the economic rationale of the ECA.

The Italians were less successful. Gilberto Bernardini, an outstanding physicist in Rome, spent the academic year 1948–49 with Rabi at Columbia University. While there, Bernardini, with active support from Rabi, wrote to Compton asking him whether U.S. authorities involved in the ERP could persuade the Italian government of the need to support science. By now it was clear to Compton that the Italians would have to take the initiative themselves, and he suggested to Bernardini that "there is a possibility that some measure of aid in the form of technical publications, laboratory equipment and funds might be obtained through ECA if these items were included in country requests." He also asked Rabi to encourage Bernardini to ask his authorities at home to apply for such aid.[81] Apparently nothing came of this. Hopes were raised the next year, in September 1950, when Samuel Goudsmit told the leader of the Rome group, Edoardo Amaldi, that the U.S. Atomic Energy Commission was sympathetic to the idea that Italy construct an accelerator using the country's Marshall Plan funds. Amaldi immediately applied officially for $580,000 to buy the necessary equipment to build a Kerst-type 300 MeV betatron at the university's physics institute. It was rejected without ado by the ECA authorities on 17 November 1950. Amaldi was told that the ECA's general policy "disallowed the use of ECA funds to finance research in the field of nuclear physics."[82]

We have no detailed information about the situation in France, but it seems unlikely that administrators there sought Marshall Plan funds to support science. Early in January 1946 French President Charles de Gaulle endorsed Jean Monnet's ambitious "Plan de modernisation et d'équipement économique," intended to mobilize all sectors of the economy in a vast planned program of national reconstruction and modernization.[83] Frédéric Joliot-Curie, nuclear physicist, Nobel Prize winner, and newly appointed high commissioner for atomic energy, was one of the consulting experts. He immediately objected, to no avail, that neither atomic energy nor the refurbishing of France's scientific infrastructure were included in the plan. Joliot-Curie drafted a manifesto entitled "On an Omission from the Monnet Plan," which was widely disseminated among, and supported by, the scientific community.[84]

Joliot-Curie's manifesto made three main points. First, it stressed that every rich and industrialized modern country had a strong scientific and technological research base; expenditure on research should be proportional to industrial power unless firms were to be reduced to developing everything under license from abroad, impoverishing the nation, and reducing it to a "colony of a foreign power." Second, the country's needs in terms of scientists and engineers must be quantified like any other infrastructural resource—raw materials, energy, or machinery—and these people must be educated, paid, and equipped appropriately. Third, the plan must recognize that both pure and applied science were crucial to social recovery and it must allocate resources between them as economic circumstances permitted. Monnet was apparently not convinced. Echoing the argument in Washington, he said that supporting science now was too expensive, that to make a significant impact it required the training of many qualified people, and that it only brought tangible results in the long term. It was therefore not a priority in France in

1947.[85] In this diffident climate, French administrators refused to consider using their funds for scientific recovery, and it was only with the fourth, and above all the fifth, plan (in the early 1960s) that science was taken seriously as a major national resource among others.[86]

We do not know if Britain used Marshall aid for scientific reconstruction. Certainly the reports from the London office of the ONR confirm that the state of science in England was far better than in any of the major Continental countries.[87] Laboratories and libraries had not been looted of scientific equipment, books, and periodicals, as had happened in the occupied countries. Funds and materials were available for building heavy equipment like "cyclotrons, betatrons, synchrotrons, linear accelerators." Scientific instrument manufacturers were being encouraged by the government to produce for export. Policies were being put in place to increase the numbers, and to make better use, of scientists and engineers, and the government had substantially increased its support for research in universities. Expenditure on research and development was expected to climb to £76.5 million, 86 percent of it for defense and industrial research. The London office of the ONR concluded that "the government interest in the financial support of research in Great Britain is such that any external financial support may be redundant. The principal need," it went on, was for the "expansion of laboratory facilities and of living accommodations for students and faculties," and ERP dollars might be helpful for that purpose.[88]

Britain was an exception in another way too. Compton and the RDB were emphatic that there should be no overt military interest or association in Washington's support for European scientific reconstruction. This condition was waived in the case of the British. The board recommended that there be closer cooperation with the British military "through visits of key personnel," an "exchange of technical information," and an "exchange of critical materials and knowledge of substitute materials." By contrast it was deemed "doubtful whether any form of military research and development should be carried out by *continental research institutions . . .* , principally due to intelligence difficulties." In the case of France, for example, with its strong left-wing scientific community, "only basic research should be fostered."[89] In short, the scientific and technical relationships between the United States and Britain in the framework of the Marshall Plan were to be shaped by the far greater strength of British science compared to that on the Continent, by the government's commitment to an expanding scientific and technical base, by the absence of a Communist threat, as in France, or of fears of a resurgence of militarism, as in Germany, and by the "special relationship" that existed between the two Anglo-Saxon nations, notably that embodied in the exchange of nuclear information. Britain's exceptionalism demands an analysis of the exercise of American power there that reflects these realities: it falls beyond the scope of this book, which concentrates on continental Europe.

A Closer Look at Two Countries: Italy and Germany

To complement the broad picture described above, it is useful to look more closely at the situations in two countries that did seek Marshall aid for science, Italy and Germany. Both lived with the historical legacies of fascist regimes and of having been defeated in the war. Scientific reconstruction faced enormous material and political difficulties in both. In one, external support was spurred by the fear that without rapid economic growth, material prosperity, and better funding for science there would be a massive exodus of skilled manpower. In the other, it was retarded by the fear that rapid scientific reconstruction might feed militaristic ambitions and legitimate a scientific community whose democratic credentials were still in question.

This account is not intended to provide a comprehensive survey of the state of science in early postwar Italy or Germany. Rather, its parameters are set by the questions asked and reports written at the behest of the RDB by the Office of the Assistant Naval Attaché for Research in London (and these concentrated almost exclusively on physics). In other words, what follows is a summary of the information that was fed back to policymakers in the RDB about the state of affairs in these two countries and how they reacted to what they learned. My aim is to emphasize that the American project to rehabilitate science in Europe was not only about providing material resources, but also about building structures and changing attitudes and values among scientists in line with democratic values. It is a dimension of hegemony that I discuss in greater detail in the next chapter.

Italy

The RDB was provided four separate reports on Italy in 1948. One was a copy of an article by Edoardo Amaldi, published in the first number of *Physics Today*.[90] E. J. Schremp, a scientific director in the ONR's London office, wrote a twenty-nine-page report on the physical sciences after visiting universities and technical institutes in Turin, Genoa, Rome, Naples, Padua, and Milan around the time of the hotly contested elections in April 1948.[91] External scientific consultant Hans Lewy wrote two brief reports, one explaining the organization of Italian science and the other identifying needs in both Italy and France and the possible form U.S. aid could take under the ERP.[92] Finally, we have a report written in May 1949 by a scientific liaison officer, T. Coor, that describes the state of university physics in Rome, Milan, and Padua at that time.[93]

The difficulties faced by Italian physics after the war, as reported to the ONR, are typical of the situation in that country with respect to support for science as a whole. They dramatically reveal the chasm that had opened up between the United States and Italy between 1935 and the late 1940s. In a little more than a decade the brilliant group in Rome led by Enrico Fermi went from being world leaders in the study of nuclear disintegration to being dispersed, demoralized, and hopelessly underfunded. Indeed, the community might have disappeared completely but for their skill at improvisation, their outstanding ability as physicists, and their decision not to leave their native land (which was reinforced by the security constraints on nuclear research in the United States after the war, experienced firsthand by their new leader, Edoardo Amaldi).

In 1935 it was already clear to the Fermi group that to advance their studies on neutron absorption they needed fluxes far greater than those provided by naturally radioactive sources.[94] Franco Rasetti visited Robert Millikan's laboratory in Pasadena and Ernest Lawrence's Radiation Laboratory in Berkeley to learn what he could about accelerator technology and performance.[95] The next year Emilio Segrè was impressed by the "tremendous potential" of Lawrence's 27-inch cyclotron and promptly used it to discover a new element, technetium.[96] Italy would need this kind of heavy equipment if it was to maintain a leading position in the field, though Segrè suspected that it would be impossible to raise funds for it.

The first thing Fermi's group did was to join forces with the Institute of Public Health in Rome; together they successfully obtained money for a 1 MeV Cockroft-Walton machine, to be housed at the institute, with the nice benefit that the Fermi group had the resources to build a 200 KeV prototype for their *Istituto di fisica* at the university.[97] Then in January 1937 Fermi submitted a proposal to the director of the Consiglio Nazionale delle Ricerche (CNR) for establishing a National Institute for Radioactivity. It would include an "economical" cyclotron and would cost some 600,000 lire to build, plus 230,000 lire annually to operate. It was killed in June 1938. Due to lack of funds, Fermi was told, it would be impossible to support his scheme for the moment; 150,000 lire was set aside for him in the CNR's 1938–39 budget, enough to run a university physics laboratory, but far too little to build even an inexpensive cyclotron. In the months that followed, anti-Semitic laws were promulgated by Mussolini's fascist government, leading to the expulsion of Jews from all public offices, including the universities and the academies. Fermi and his wife, who was Jewish, had had enough. Starved of resources to do frontline physics and disgusted and threatened by the racist regime, Fermi picked up his Nobel Prize in Stockholm in December 1938 and continued on directly to the United States, never to return. He was not the only leading physicist to emigrate: so too did Rasetti, Segrè, and Bruno Rossi (to the United States) while the war drove Giuseppe Occhialini to South America.

The war brought with it the destruction and looting by both Allied and German forces of many physics laboratories. By 1948 these had mostly been rebuilt, but they were sadly lacking in equipment. To make matters worse, the technological gap between the two sides of the Atlantic was growing wider by the day. Immediately after the war the Italians were planning to establish a nuclear physics center furnished with two machines: a Betratron providing 20 MeV electrons, and a 30 MeV cyclotron that had been designed for the Science Hall at a Universal Exhibition scheduled for 1942 but never held. While the Rome group was recycling old designs, Fermi wrote Amaldi that *his* "biggest problem was to imagine enough things on which to spend [his] money." In February 1946 Gilberto Bernardini remarked in despair that any idea the Italians may have had of keeping up with the United States in the nuclear research realm was simply an "illusion." They "were losing ground at kilometres per second, and perhaps without any hope of ever catching up. It is not excluded that we will change direction," the Rome physicist added.[98]

This impression was reinforced when Amaldi visited the United States in 1946. People in Berkeley were putting the finishing touches on their 184-inch synchro-cyclotron and were speaking of reaching 1 BeV in the near future. Crippled by lack of funds (and of industrial capacity) and hopelessly outstripped by the Americans, the Rome group decided temporarily to abandon plans for an accelerator and to concentrate their efforts on cosmic ray physics research, which was far cheaper and still of considerable scientific interest. The small, prefabricated Testa Grigia laboratory in the Alps, funded by the Minister of Industry and Commerce and by private industry, was established in 1948 for this purpose.[99]

In addition to lacking money for even the most rudimentary modern scientific equipment, "the economic conditions of the teaching profession [were] deplorable." As Amaldi put it, even though salaries were about seventeen times higher than they were before the war, the cost of living was about sixty times higher, so that university staff had to resort to other activities to supplement their income. A full professor's monthly salary in summer 1948 was $60 to $80; an assistant professor earned $30 to $45 per month. By comparison, professors in an American university at the time could expect to earn $500 to $1,500 per month.[100] Gerard Pomerat touring Italy for the Rockefeller Foundation in 1948 confirmed the "small salaries" and "real financial sacrifice" of the professors, but he was also struck by their ability to have both paid and voluntary assistants, and to get married even though none of them could afford to do so: "it remains a mystery how they can manage things," he added.[101]

The CNR's budget for physics research in the country was also abysmal: according to Amaldi, for 1947 it was "theoretically" 200 million lire, though in practice it was half of that, about $174,000. In 1949 an ONR officer stated that for that fiscal year the CNR had allotted about $300,000 for physics. Bernardini, visiting Isidor I. Rabi at Columbia University in February 1949, wrote that "the total [annual?] budget of the Rome Institute, including the salaries of the research personnel, is

$20,000."[102] By comparison the *monthly* payroll for staff at Lawrence's Berkeley Laboratory in February 1946 was $194,000; the semiannual budget of the Radlab at that time, still funded by the Manhattan Engineering District, was $1.37 million, and the building was packed with surplus war material, including 750 radio generators.

Nor did there seem much hope of the situation in Italy improving soon. "The government," wrote one ONR consultant, "seems to consider a strong science a luxury to be afforded by richer nations than Italy, and is willing to support its scientific institutions to the same extent that it supports other relics of a glorious past." Another consultant agreed. According to the physicists he spoke to, the development of their field was hampered by the bad economic situation in the country, the government's lack of interest in research, and a general animosity to the CNR from the minister of education and other state departments. The ill-feeling was mutual. Gustavo Colonetti, who chaired the CNR, reputedly remarked that it was just as well that the minister had no mandate to support basic research because he had no idea how to do so anyway, and if he meddled in the CNR he would ruin Italian science.[103]

Popular wisdom at the time was Malthusian and held that "overpopulation" was the cause of Italy's woes. It aggravated poverty and provided a fertile recruiting ground for the Italian Communist Party, which aligned itself with the Catholic Church in opposing family planning. Physicists and the government had different solutions to the problem. Bernardini saw salvation in science. Italy could "save itself only by increasing in every direction the intellectual level and technological performance of its population."[104] Prime Minister Alcide de Gasperi saw salvation in emigration. He allegedly said that he would gladly give up all Marshall aid if he could "export" 400,000 people a year—the number of babies born annually. As far as he was concerned, population reduction would far more effectively meet the Communist threat than would the economic restructuring demanded by the ERP.[105] In this inauspicious official climate, "the greatest and most unavoidable danger [was] the continuing departure of physicists to other countries, especially to the United States, where they may have better research equipment and living conditions."[106]

The desperate economic and political situation in Italy, the prospect of a massive scientific exodus from the universities and research centers, and the increasing influence of Communism dismayed the ONR. Emigration would cut off "one of the most promising methods of putting the Italian economy on its feet—through the stimulus to production and employment which comes from developments in science and engineering."[107] The lack of intellectual leadership in academia was leading to "an alarming increase in the effect of Communism upon the thinking of students as well as those in responsible positions in the universities."[108] The Communist threat in Italy (and in France), said the renowned Hungarian émigré and aeronautical engineer Theodore von Kármán, made it urgent to help these countries "create conditions favorable for scientific life and creative work within their own boundaries." It was "definitely in the best interest of the United States," von Kármán said, to "stop or at least to throttle the exodus of all European talent to the United States or the USSR, which is drying up Western Europe and Italy."[109] Another consultant linked Communist inroads into Italian academia with the general failure of U.S. policy to stem the Red tide in the country, quoting Senator Henry Cabot Lodge to the effect that "we have spent more than eighteen hundred million dollars in Italy since the war and during that same period the membership of the Communist Party there went from 60,000 to 2,500,000."[110]

All these efforts to persuade U.S. authorities to support Italian physics failed. Even the weight of the U.S. Atomic Energy Commission (USAEC) did nothing to move the Economic Cooperation Administration to change its policy in 1950 and support nuclear physics. Rebuffed, from September 1950 onward Amaldi devoted all his energies to securing a leading role for Italy in the European physics laboratory project (later CERN), which was just beginning to take shape (see chapter 3).

The Italian situation reveals the limits of the rhetoric that surrounded Marshall aid and of the debates inside the U.S. administration over the need to use the plan to rebuild European science. Prima facie, Italian physics was an ideal candidate for support. The Rome community aggressively pursued every opportunity for financing that presented itself and exploited their considerable international network for political and technical support. Some of the best Italian physicists, including Enrico Fermi and Emilio Segrè, had already left. Some of the best still in the country, notably Edoardo Amaldi, were sometimes tempted to leave.

Even Prime Minister de Gasperi suggested people should go. A vacuum was being created in academia that the Communists seemed to be filling. Yet the Rome group's appeal to the ECA was turned down. The argument that a strong science base was necessary for economic reconstruction was apparently trumped by the security concerns surrounding the field of nuclear science and the fears that left-wing university scientists could not be trusted.

Germany

The rehabilitation of German science was subject to the same changing policies as was the economic and political reconstruction of the country.[111] Before the war was over, in 1944, U.S. Treasury Secretary Henry Morgenthau famously suggested that to prevent "a new upsurge of the Teutonic fury" the German people should be reduced to a nation of farmers tilling the soil. However, the production of even the barest essentials for the survivors of the war called for a measure of domestic industrial recovery; Germany would otherwise be an endless drain on the U.S. Treasury. Retribution had to be balanced with reconstruction. Too little could drive Germans into the arms of the Soviet Union, which was making a major effort to attract educated elites. Too much could lead to a resurgence of German nationalism and militarism. Thus the pillage and the carrying away of equipment, patents, and people as war booty that characterized the immediate postwar period gave way to a more controlled management of German industrial and scientific recovery by the occupying powers. The decision to allocate millions of deutsche marks of Marshall aid to German science must be situated in the context of the granting of limited sovereignty to Germany that the implementation of the plan demanded and the increasing relaxation of limitations on research.

The initial draconian program for controlling German science was inspired by Morgenthau's determination to cripple Germany irreversibly. As embodied in the famous document issued by the Joint Chiefs of Staff in May 1945 (JCS 1067/8), the program required Eisenhower and his deputy military governor in the American zone, Lieutenant General Lucius Clay, to control German research in three phases. The first step was to close at once "all laboratories, research institutions and similar technical organizations," except those necessary for securing public health. Thereafter all "laboratories and related institutions whose work has been connected with the building of the German war machine" were to be "abolish[ed]," except for those of technological interest to the United States. Their personnel were to be detained, and their equipment removed or destroyed. In a third phase, scientific research could resume, but under tight regulations and only if the authorities were sure that it would not contribute to enhancing Germany's war potential. The specific kinds of research that were permissible had to be defined. No one who had previously held a key position in "German war research" would be allowed to resume a scientific life. There would be frequent inspections of every laboratory, and all research results had to be freely disclosed. If these requirements were violated the control authority had the right to "impose severe penalties, including permanent closing of the offending institution."[112]

A less punitive regime was suggested by a blue-ribbon committee chaired by Roger Adams, the chair of the Department of Chemistry at the University of Illinois. The committee was set up by the National Academy of Sciences in response to a request from Vannevar Bush, then chairman of the OSRD, himself responding to pressure within the State Department to thwart the Morgenthau Plan. Adams's committee was made up of a number of senior business executives from research-based industries like Bell Telephone, Carbide and Chemicals Corporation, and General Electric, as well as several scientists, including Isidor Rabi and Hugh Dryden, the chief physicist at the National Bureau of Standards. Its report landed on Bush's desk on 6 July 1945. The committee's basic approach was the antithesis of the Morgenthau Plan. It argued that science and German scientists had a key role to play in the economic reconstruction of Germany. Unduly restrictive measures, rather than securing permanent peace, would breed resentment, deceit (an obsessive fear of the Americans at the time), and the desire for vengeance. Military research had to be barred, of course. However, basic and applied research, notably in chemistry and physics, as well as industrial-prototype design and pilot-plant construction should be encouraged and be visible and open, thus facilitating control. The closer the research and development was to processes and products that could be diverted from economic to military ends, the tighter the controls would have to be.

The Adams committee's proposals deviated from the policy called for by JCS 1067/8 in two important ways. First, JCS 1067/8 insisted that nonmilitary research be permitted only in specified areas. Adams's committee was willing to allow any research that was nonmilitary, using on-the-spot inspections, licensing, and budgetary control to monitor what was being done and to ensure that it was not being turned to bellicose ends.

Second, JCS 1067/8 wanted to exclude a priori people who had been involved in "German war research" and to ban them indefinitely from scientific life. Adams's committee, desiring to rebuild rather than to restrict, was more indulgent. On the whole, it said, the scientists who had stayed in Germany during the war were "as little influenced by Nazi doctrine and teaching as any group in the population." They "comprised an island of nonconformity in the Nazified body politic" and were pre-dominantly preoccupied with maintaining their professional status. They contributed little to politics or to victory, withdrawing rather into "the traditional ivory tower which offered the only possibility of security."[113] In short, while the JCS looked on German science and scientists as a dangerous political and ideological threat to security that needed to be kept under strict control, Adams's committee saw them as apolitical professionals who were essential to securing a permanent peace and who needed to be encouraged and unobtrusively monitored.

In November 1945 Roger Adams himself arrived in Berlin as an "expert consultant" to the War Department and as an official representative of the National Academy of Sciences. He stayed for about four months (leaving soon thereafter to do the same job in Japan) and laid the foundations for what became Allied Control Law 25, on the control of scientific research, promulgated on 29 April 1946. This law combined some of the key ideas developed earlier in the Adams committee with the requirements of JCS 1067/8 and cohered with the gradual shift in U.S. policy away from destruction and tight control to reconstruction and limited autonomy.

Consistent with the spirit of the Adams committee (and with the approach strongly promoted by the British), Law 25 did not specify what one could do, but what one could not.[114] Its main aim was to prohibit military research, or research that contributed to the nation's military capacity, and to bar people who had been dedicated to the Nazi cause or who had played a major role in the Nazi war machine, from going back to their laboratories. It distinguished fundamental from applied research in terms of goals. Fundamental research was dedicated to the discovery of new knowledge, laws of nature, or substances. Applied research used fundamental research for industrial purposes, to scale-up from the laboratory bench to a pilot plant, or to improve a production process. Fundamental research was allowed as long as it did not call for massive or specialized items of equipment that could be turned to military ends. Applied research was allowed, except for areas listed in an appendix to the law that defined what constituted military research, including applied nuclear physics, applied aerodynamics and hydrodynamics, anything of direct value to the navy and the air force (for example, rocketry, radar, sonar), encryption methodologies, war chemistry (for example, explosives, toxic gases, rocket fuels), and potential biological weapons. The appendix also listed some areas of applied research that were deemed dangerous but that could be authorized (for example, synthetic rubber, radio activity for nonmedical purposes, and industrial explosives).

Any laboratory or institution that wanted to engage in any research activity at all had to ask permission to do so. On pain of being punished, even executed in serious cases of infringement, researchers had to describe the work they wished to undertake and write detailed reports every four months (as well as an annual nontechnical report). There would be regular inspections of their premises. No senior officials in the previous regime, members of the Nazi Party, or of other pro-Nazi action groups could be employed. On the other hand, the fact that a scientist had done weapons research during the war did not automatically exclude him from pursuing a scientific career.

It proved impossible to implement Law 25 rigorously. It was essentially an American policy statement. The other Allies were inclined to be more pragmatic in their zones, and many of the best scientists simply moved to the British zone to avoid its too-rigid interpretation. The meaning of denazification slithered between prohibition and exclusion, as required by law, and reeducation and reinsertion, necessary for scientific reconstruction.[115] The demand that all laboratories, including those in important industries, be open to inspection was seen as perpetuating the large-scale appropriation and pillage of trade secrets that had characterized the early postwar period and discouraged many firms from engaging in new research.[116] The welcome given by the United States to the brightest and the best émigrés who had worked for the Nazi war machine was deeply resented by scientists and engineers who were far less compromised and who had stayed in Germany only to be mistrusted and monitored.[117] As a result, two years after the war, when the Marshall Plan was announced, the apparatus of control on German science was crumbling. Researchers were becoming more strident in their demands for autonomy and more outspoken in their contempt for the occupation, especially in the American zone.

Scientists and religious groups in the United States, it should be said, were equally disturbed by the government's efforts to bring over tainted German researchers and possibly offer them citizenship. (The French too were considerably reluctant to employ German scientists, engineers, and technicians in their factories and laboratories.)[118] The Federation of American Scientists and the *Bulletin of Atomic Scientists* (*BAS*) were particularly active in opposing these measures. At a protest meeting at Cornell University, in January 1947, the director of the graduate school of aeronautical engineering violently opposed the relocation of "Nazi scientists" to American universities and industries. He described them as mostly mediocre researchers, ruthless collaborators with the Hitler regime, and opportunists who had used "intrigue and political machinations" to rise to positions of importance in research and development.[119] He was not alone in universally condemning those who had worked under the Third Reich, regardless of whether or not they might settle in the United States. In a provocative review of Samuel Goudsmit's book on the German nuclear program that was published in the *BAS* in December 1947, Phillip Morrison, also at Cornell, wrote that what bothered him was not that his German colleagues had worked for the military—everyone did that—but that "they worked for the cause of Himmler and Auschwitz, for the burners of books and the takers of hostages." "The community of science will be long delayed in welcoming the armorers of the Nazis," Morrison went on, "even if they were unsuccessful."[120] Max von Laue, a leading German physicist and director of a Max Planck Institute, was incensed, all the more so since he was internationally recognized as someone who had courageously refused to collaborate with the regime. He replied at once and scathingly attacked Morrison for the "monstrous suggestion" that German scientists were responsible for the holocaust.[121]

In the strained circumstances of the time no one was capable of a balanced assessment of the relationship between German science and Nazism. The Adams committee painted far too rosy a picture of a politically detached community that had retreated to its ivory tower; Morrison erred in the opposite direction. Contemporary scholarship suggests that German scientists and engineers, like their counterparts elsewhere, ably adapted themselves to the political and ideological context in which they found themselves. They promoted their scientific, professional, and institutional agendas within the polycratic structure of the Nazi state, relying on a variety of patrons for the financial support that their research needed, and benefiting notably from it when their findings could contribute directly to the regime's war aims. Physical scientists were particularly compromised, as their research bore immediate technological fruit. Indeed, Mitchell Ash has suggested that internationally recognized research was primarily supported by the Third Reich for its "actual, projected, or imagined relevance to Nazism's core projects—world conquest and the creation of a new German race."[122] This was the price of patronage, whether or not the scientists shared the aims of the regime.

Richard Courant and Dr. N. Artin were deeply disturbed by what they saw as the unrepentant collusion of some physical scientists with the Nazi regime. They were asked by the London Office of the ONR to go to Germany in 1947 to evaluate the situation for Vannevar Bush's RDB. Courant, who was Jewish, was forced to leave Göttingen University shortly after Hitler came to power.[123] He arrived at New York University in 1934, where he rapidly established himself as an outstanding applied mathematician, later heading his own Institute of Mathematical Sciences. His companion was probably Natalie Artin, the (Jewish) wife of another fine German mathematician, Emil Artin, who had decided to leave Hamburg for the United States when the Nazis came to power and had moved to Princeton from Bloomington in 1946. For five weeks in June and July 1947, Courant and Artin visited a number of institutions and spoke to scientists and officials in the British and American zones in Germany, also stopping by Austria. They shared their findings with friends in England and in Denmark, in particular with Niels and Harald Bohr.[124]

Courant and Artin wrote two reports for the ONR.[125] One dealt exclusively with the state of science in Göttingen and Hamburg. Its aim was not simply to assess the quality of the work done there, but also to establish whose research was of "interest from the point of view of work sponsored by the Navy" and whose knowledge "deserves more exploitation." Part of their mission was thus to determine who could be useful to the U.S. military. The second report placed greater emphasis on the attitudes and values of German scientists, "top-ranking research men as well as students, assistants, and technicians" with a view to exploring "possibilities of scientific rehabilitation and cooperation."

The two émigrés were particularly impressed by the situation in Göttingen. The university and the Kaiser Wilhelm/Max Planck Institute (the name was being changed to the latter) had been relatively unscathed by the war, and the British occupation authorities were doing all they could to reestablish the institute to its previous outstanding international standard. Many leading German scientists, including Otto Hahn, Werner Heisenberg, Max von Laue, and Carl F. von Weizsäcker had settled there. Artin and Courant noted the interest "for our navy contract" of the work that Heisenberg and von Weizsäcker were doing on fluid dynamics, specifically turbulence. Ludwig Prandtl's work on gas dynamics had been impeded by military government restrictions, but he was devoting his efforts to dynamic meteorology. Among the mathematicians they selected the work by W. Magnus on electromagnetic waves and wave guides as of especial interest to the navy.[126]

The situation in Hamburg was rather less promising. The city had been badly bombed, and the university and academic institutes had suffered considerable damage. One engineer they met, Günther Diedrich, had worked on the German V-1 rocket, and they suggested that his help could be "advantageously used in the Project Squid under the ONR."[127]

In their scientific report Artin and Courant only made periodic references to the political attitudes of the people they mentioned (for example, in Göttingen they "found a relatively large number, maybe even a majority, of people whose political and human attitude seemed all right"). Their second report was far less charitable. While pointing out that some academics had an admirable record of intellectual autonomy and moral courage (notably Max von Laue), Artin and Courant also insisted that these were the exception, not the rule. There were few Nazis among the scientists; however, as many as 90 percent of them had "willingly cooperated and gladly benefited" from the Nazi regime and had made "unnecessary concessions to Hitlerism." These "aggressive nationalists" and "opportunists" were dangerous in the long run—"ready to build a new German nationalism" and militarism. "Along with the majority of Germans, they do not really understand their position in the world," wrote Courant and Artin. "They have no clear conception of the misery inflicted by Nazi Germany on her victims. Self-centred, they indulge in criticism of the Allies and are unwilling to see the present plight of Germany as a consequence of Hitlerism rather than as of Allied mistakes," they added.[128]

Artin and Courant were shocked by these attitudes, which were in fact common among the scientific community in postwar Germany.[129] They were particularly disturbed since they were not only held by "men of medium calibre but also by well-known people of authority." They were repeatedly told, Artin and Courant said, that the Allies were deliberately starving the German population; food shipments from the United States were dismissed as propaganda. They quoted people as saying things like "We disliked Hitler, but the British and Americans are worse. They have replaced Hitler with Stalin." One of the "most renowned scientists" flippantly remarked that "even the Nazis fed the inmates of concentration camps."[130]

Appalled as they were, Artin and Courant stressed that some modest steps should be taken at once "with a short range view to saving the life of German science in a critical moment when waiting longer may lead to a state of disintegration beyond repair." They realized that a more liberal, less repressive approach was controversial given the state of mind of many German scientists. But they believed that material support as well as travel and exchange opportunities should be given to senior people with respectable political credentials and to carefully selected students. There were several reasons for this. Germany had the strongest concentration of scientists in Europe, and this reservoir of skill had an essential role to play in reconstruction. Second, considerations of "elementary justice as well as self-interest" demanded that "honest and courageous German scientists" be encouraged. The study of science might also help clear the "intellectual morass left by Nazi indoctrination" of a cohort of young people among whom "one finds an appalling lack of sense for real values." Finally, there was the Soviet threat. If they were not supported, German scientists, even though now leaning toward the West, could be driven into the "Russian camp" by "an American policy unsympathetic or cool to German scientific rehabilitation." Housing, heating, and the official food rations in the Western zones were hopelessly inadequate; books, periodicals, and outlets for scientific publications were almost nonexistent; and paper for printing or even writing was in short supply. The Soviets were consistently trying to lure scientists away by offering better living conditions and opportunities for scientific work (just as they were doing with artists and writers, it should be said).[131] The more opportunistic Germans might be tempted to switch sides by virtue of "disillusion, hunger, cold and further deterioration of conditions for scholarly work."[132]

Officials in the ONR London office made a number of suggestions for rehabilitating science in Germany on the basis of Artin and Courant's reports: rebuild universities as quickly as possible using prefabricated structures if need be, increase the official allocation of paper, provide foreign exchange, and encourage scientific exchange with the United States. They also suggested that the military or the foundations should directly support research. Taking up a recommendation by their rapporteurs, they suggested that "financial support of certain research projects in the American zone should be provided, possibly through contracts placed by the Office of Naval Research or failing that, by support from such organizations as the Rockefeller Foundation, Guggenheim Foundation, and so on."[133] These recommendations, vague as they were, were significant in that they called for U.S. financial support for research in Germany, including research of military significance. The specter of German rearmament was receding, replaced by a growing perception of the country as an essential ally in the Cold War conflict with the Soviet Union. Indeed Hans Bethe from Cornell University, who visited Germany at the behest of the ONR in August 1948, a year after Artin and Courant, found that the "attitude towards English and American occupation was extremely friendly," at least in Göttingen. There was a reason for this: the risk of a possible conflict with Russia weighed "heavily on all minds." Everybody believed firmly that "the country would not survive another war and

especially the prospective occupation by the Russians"; most were prepared to sacrifice Berlin in return for a slice of the Russian zone, like
Thuringia. All but one were horrified at the thought of the withdrawal of
occupation troops from their zones. Most were also willing, if reluctantly,
to give up the idea of German unity in return for having West Germany
integrated into a "Western European Union."[134] In short, German scientists, along with their government, were now seeking closer ties with the
Western, non-Communist, bloc in line with the objectives of the Marshall
Plan: the Soviet Union was a dangerous threat, not a desirable alternative.

The recommendations emanating from London were read through the
prism of the RDB's policies as defined in Washington. As pointed out
above, Compton and the board decided that the military should definitely not sponsor any kind of research within the framework of the
Marshall Plan. That source was thus excluded. But what about the foundations? And the national authorities who, Compton told Rabi, were the
only ones in his view who could demand that Marshall Plan funds be
released for science?

As I point out in chapter four, the Rockefeller Foundation rapidly
moved to support the reconstruction of the physical sciences in the
Netherlands and in France after the war. The officers were also under
pressure to help in Germany. Germans researchers, complaining increasingly of isolation, economic deprivation, and the "ghettoization" of their
scientific community, expected the foundation to step in and help; foundation officers drew back, appalled by the "unrepentant opportunism"
of some of those who applied for grants notwithstanding their war
records.[135] As the chorus of complaints mounted, two trustees, John D.
Rockefeller III and William Myers, from Cornell University, toured Germany and Austria in August 1946. They reported that the U.S. military
government and the State Department were keen for the foundation "to
get back into Germany as soon as possible."

A conference of foundation officers was held early in April 1947.
While several of them felt that something had to be done for Germany,
those directly involved in funding scientific research were inclined to be
prudent. Alan Gregg, the director of medical sciences, simply refused to
set foot in Germany after the war, visiting it for the first time in 1949; he
came home disgusted at the unrepentant attitudes he found among med-

ical researchers.[136] Speaking at the officers' conference, Warren Weaver, the director of natural sciences, asserted that "any support that we give is potentially support to military strength." He too was concerned above all about the mindset of the research community: "it is not science that makes war," he said, "but a variety of circumstances that make people use science for war."[137] His distrust in German scientists was reinforced by a reading of the report from Courant and Artin prepared for the ONR, which may have been forwarded to him by Compton or even by Courant himself (they were close acquaintances).[138] The report, an internal memo said, spelled out "the reasons why WW thinks it is too early for us to step in and help Germany."[139]

Notwithstanding their doubts, the officers of the medical sciences division sharply increased their grants to Germany between 1947 (6 worth $62,500) and 1948 (45 grants worth $455,311). In 1949 they gave fewer awards, but the average size of each was double that of the two years before (18 grants worth almost $355,000).[140] The main catalyst for the program was said to be "the omnipresence of Russia." It was suggested that the Soviet Union would be welcoming and generous and that the ideologically fickle younger generation of medical researchers, whom the foundation supported in the hope of promoting a democratic Germany, might be tempted to turn to Communism.

As for the national authorities, remember that it was Hirsch in the Economics Ministry who managed to have about DM 40 million of (counterpart?) ERP funds made available to stimulate the German scientific instruments industry. Again it may have been the Communist threat that loosened the purse strings. In October 1949 Compton wrote to Roger Adams expressing concern that "Russians are treating scientists better than Americans." In his view, even a small gesture—a subscription to a scientific journal or assistance with laboratory apparatus—might have "frequently spelled the difference between a Communist or an American Collaborator among people in important scientific positions, but whose ambitions for their science may exceed their ideological loyalty."[141] The use of ERP funds to stimulate the scientific instrument industry would have allayed Compton's fears. It also satisfied the thinking of his RDB and the Marshall planners at several levels. It could be taken as confirmation of the view that economic development required a sound scientific infrastructure. It stimulated a traditionally important sector of German industry through bodies like the DFR. And it would undermine Soviet attempts to buy allegiance with equipment.

The DFR replaced the German Scientific Advisory Council set up by the British Control Commission in Göttingen in 1946.[142] The advisory body was modeled on those in the United Kingdom, like the agricultural and medical research councils. In 1948 Werner Heisenberg proposed that the council be transformed into a body to plan science in the bizone and to "represent the requirements of scientific research for the administration of West Germany."[143] This suggestion came up against the traditional prerogative of the cultural ministers of the German states, who promptly resurrected the Weimar-era Notgemeinschaft der Deutschen Wissenschaft (NDW [Emergency Association of German Scholarship]), which was (re)founded in January 1949. Not to be outdone, Heisenberg garnered support from powerful allies in Göttingen and the Max Planck Society to press ahead with an organization to advise the federal and state governments. The DFR was officially established in March 1949 and was composed essentially of elite scientists. It claimed to be the sole representative of German science both at home and abroad, representing the country in foreign and international bodies. It also aimed to participate in the promotion of scientific research at home, notably by soliciting and distributing public funds for research. It matters little for our purposes here that the DFR and the NDW were fused in 1951.[144] ERP support for the DFR was significant—DM 4.1 million, or 47 percent of its budget, in 1950, and 22 percent of its budget in 1951—and some Marshall Plan funds were distributed to the Notgemeinschaft by Heisenberg's Forschungsrat. This gesture legitimated the pretensions of the DFR to represent German science as a whole, it reflected the close ties that had been built between Heisenberg and the first West German Chancellor, Konrad Adenauer, and it was a symbolic expression of the birth of the Federal Republic. It was also a strong affirmation of the need to rebuild the country's strength in basic science, an affirmation that was reinforced a few months later when Isidor I. Rabi proposed that this new and still only partly sovereign state be integrated into a multinational European physics laboratory—a story to which we now turn.

3

The Place of CERN in U.S. Science and Foreign Policy

The story of the origins of CERN, the European Organization for Nuclear Research (now the European Laboratory for Particle Physics), has been recounted so often that readers can be excused for wondering why it needs retelling.[1] However, most of the work done to date by myself and others, including some of the physicists engaged in launching the project, has concentrated almost exclusively on developments in Europe from the late 1940s onward. Far less attention has been paid to the United States' role in the process and to how the political configuration of the intergovernmental laboratory dovetailed with the United States' scientific and foreign policy in the region, notably the overall goals of the Marshall Plan. This chapter aims to fill that gap in our understanding and to show how CERN was not simply a boon to European science, but also a coproduced instrument of European and American political interests in the early Cold War.[2]

Isidor I. Rabi is the main actor in the story. In consultation with the State Department, the eminent Columbia University physicist shepherded an important resolution through a meeting of the United Nations Educational, Scientific, and Cultural Organization (UNESCO) in Florence on 7 June 1950.[3] The resolution called on the organization's director-general "to assist and encourage the formation and organization of regional research centers and laboratories . . . in fields where the effort of any one country in the region is insufficient for the task."[4] In the debate on the motion Rabi identified physics as a particularly suitable field for research, while the British delegate, Julian Huxley, suggested that such a center be located in Western Europe and that the Economic Cooperation Administration (ECA), which administered Marshall Plan funds, be approached for financial support. In a press conference after the meeting, Rabi identified accelerator-based physics in particular as one possible focus for the center and added that Western Europe could establish a facility "as good as the best in my country" if a number of countries, including France, Italy, and West Germany pooled their resources to build it. A center like this he added, would help "to preserve the international fellowship of Science, to keep the light of Science burning brightly in Western Europe" and was "one of the best ways of saving western civilization."[5]

Rabi was already deeply engaged in European affairs by the time he arrived in Italy as an official member of the U.S. delegation to UNESCO's fifth general meeting. As a young physicist, he had spent a couple of years on the Continent in the late 1920s and had come to know the giants in the field—Niels Bohr, Werner Heisenberg, Wolfgang Pauli, Erwin Schrödinger. Rabi performed his first ever molecular-beam experiment in Otto Stern's laboratory in Hamburg and there laid the foundations for his future Nobel Prize–winning research program.[6] After the war he was immediately involved in defining American policy for science in Germany. He served on Roger Adams's committee, appointed by the National Academy of Sciences in 1945 to advise the State Department on how to treat the defeated nation. As noted in chapter 2, this committee insisted that, rather than seek retribution and denazification, the United States should promote reconstruction taking advantage of German scientists, who, the committee claimed, had mostly retreated to their ivory towers during the war. Rabi also personally supported Bernardini's efforts to get Marshall Plan funds for Italian science in 1949 through Compton and the Research Development Board (RDB) when the Rome physicist was his guest in New York. The recovery and strengthening of European science was then an abiding concern for Rabi. As he put it years later, surveying the Continent in 1950, he found the lack of facilities for advanced research "really pitiful" and Europe "down and out, not just in property, but the whole scientific level." Hence he wanted "very much to help remove a sense of frustration which, very understandably, [was] growing among scientists who do not have the material means that we have in the United States."[7]

The belief that America should help Europe to stand on its own feet was central to Rabi's thinking about policy in the area. He was aware of

the possibility of American hegemony, if only as an unintended consequence of the balance of forces in the region. Rabi was the only scientific
member of a group set up by the highly influential think tank, the Council on Foreign Relations, to study the long-term implications of the Marshall Plan. The group met for the first time in January 1949 with the
Columbia University president, himself a European veteran, in the
chair—General Dwight D. Eisenhower. Its overriding question was,
What kind of Europe does the United States want in the long run? Its
overriding concern was to respect European autonomy while protecting
U.S. interests in the region.[8] Should Western Europe be wholly independent—friendly but not under American control—or should it be partly
dependent, enabling the United States to "maintain a kind of American
hegemony"?[9] Since Europe so desperately needed aid, perhaps dependence was inevitable—either it would fall under Soviet control, if the
United States did not provide enough assistance, or it would precipitate
American hegemony by failing to solve its own problems, becoming
increasingly reliant on the United States.

We do not know precisely what Rabi made of these debates. However,
his general line of approach as regards science is clear: he wanted the
United States to assist European recovery without sapping European initiative. He did this because he cared personally for Europe and hoped, as
he put it in a 1973 interview, "that Europe may build up strongly enough
so that we won't have to bother with them—they'll be independent and
we'll have somebody to talk to . . . we'd have civilized people to talk
to."[10] Rabi sought a privileged relationship with European science, one in
which Europe and the United States would complement each other without the latter using its superior resources to dominate its partner. This is
one reason why he was glad that Marshall Plan (counterpart) funds were
not used as seed money for a "regional center," as suggested by Huxley.[11]
On leaving Italy, he passed by the office of Milton Katz, legal counsel for
the ECA in Paris to seek his support. Katz was out of his office and Rabi
dropped the idea. It was never resuscitated.

Whether or not Rabi himself desired it, "hegemony was in the cards."
For Europeans would not embark on a collaborative project in the nuclear
field without a favorable signal from the United States. There were several
discussions already under way on the Continent for a multinational

"atomic" venture, but nothing could be done without a green light from Washington. Scientists needed technical advice and support from their American colleagues. Governments that sought privileged bilateral relationships with Washington needed reassurances that their scientists' schemes were coherent with U.S. aims.[12] Dependency left Europeans vulnerable and gave Rabi and the State Department the opportunity to promote a project that was both compatible with U.S. interests and palatable to European governments and scientists (for if this was hegemony it was not to be imposed but coproduced).

In this chapter I describe the dynamic surrounding the definition of the scientific and political configuration of what became CERN. Rabi gave an enormous boost to European science and to European political integration by winning support for his idea that the laboratory should be restricted to doing unclassified research using particle accelerators, that its collaborating partners should be restricted to Western Europe, and that West Germany should be among the member states. He also consolidated U.S. scientific and foreign policy interests in the region and helped integrate European physics into the emerging Western alliance.

The French Multinational Project: A European "Brookhaven"

Rabi's proposals in Florence were not uniquely his, though he insisted to his dying day that he alone deserved credit for the precise form that they took. European physicists and science administrators in several countries had similar ideas of their own on what to do about the gulf that separated them from the United States.[13] The best-documented and most sophisticated was the scheme being discussed at the time in France.

From the time of its inception in 1946, senior scientists in the French Commissariat à l'Énergie Atomique (CEA [Atomic Energy Commission]) sought to build an international collaborative effort in the domain of atomic energy, including a reactor. On 15 December 1948 the CEA's first experimental pile, Zoé, went critical.[14] With this achievement behind him, Lew Kowarski, a member of the CEA's Scientific Services, approached J. Robert Oppenheimer to discuss what might be done next. At the time, Oppenheimer was the chairman of the General Advisory Committee to the U.S. Atomic Energy Commission (USAEC); Rabi was

one of its distinguished members. Kowarski was received sympatheti-
cally. "Oppie" thought that it might be possible to establish a European
research center for nuclear energy. He suggested that the United States
might help in building an accelerator, might provide uranium metal for a
reactor, and could probably also offer some technical and financial assistance.[15] There were two problems, though. One was the fear that the
reactor would necessarily become integrated into a nuclear weapons program. Hence Oppenheimer insisted that the reactor not produce plutonium. The other concern was that the high commissioner of the CEA,
Frédéric Joliot-Curie, could not be relied upon to be loyal to the Western
alliance. His communist affiliations were well known, as was his admiration for the Soviet Union (see chapter 4).

Joliot-Curie's outspoken pro-Soviet stance was a blessing in disguise
for the State Department. They made it clear that as long as he was head
of the CEA the United States would not provide France with any help in
the atomic energy field and that only the French were to blame for this.[16]
Truman's announcement in September 1949 of the explosion of the
Soviet A-bomb suggested that this argument might lose weight: the "pretence that there was an important 'secret' in pre-1948 atomic science and
technology" could no longer be sustained, Kowarski suggested, nor
could the fears that key information would leak to the Soviets.[17] In fact,
the main U.S. argument for not aiding France collapsed completely in
April 1950, when the French government relieved Joliot-Curie of his
office. Capitalizing on the opportunity created by the departure of their
scientific director, in "April/May 1950" Kowarski and some of his colleagues at the CEA drafted the "Note on the Creation of a Cooperative
Organization for Atomic Research in Western Europe."[18] It was intended
to stimulate the French authorities, in consultation with other governments, to take the lead in setting up a nuclear research establishment
"within a structure of common interests of the signatories of the Atlantic
pact."[19]

Kowarski's "Note" argued that to have a meaningful atomic energy
research program, individual European countries must combine their
efforts, either through a coordinated construction program (France would
build a powerful research reactor, Switzerland a large synchrotron, the
Netherlands a linear accelerator, and so on) or by grouping several pieces

of big equipment in one inter-European center. Kowarski believed the latter approach to be more effective. The center would be provided initially with a 500 Mev–1 Bev accelerator for meson physics and a 10–20,000 kW uranium and graphite or beryllium reactor. This would be used to study chain reactions, to produce radioisotopes, and to run tests on heat and power generation. No applications would be explored, and all military work would be scrupulously excluded. Finally, the center would have a radiochemistry laboratory with all the necessary precautions for dealing with highly radioactive materials (including plutonium). Kowarski went on to spell out a possible budget, to suggest a site, and to identify what special contributions national industries in France, Belgium, the Netherlands, Sweden, and Switzerland could make. As for Anglo-Saxon reactions, he was confident that scientists in Britain and the United States would welcome the move, and he suspected that the U.S. administration would take a neutral rather than a hostile approach to the venture.

Kowarski's scheme was scientifically and politically ambitious. Indeed, in terms of equipment it was not unlike, and was surely inspired by, Britain's Atomic Energy Research Establishment at Harwell and the United States' Brookhaven National Laboratory, then under construction on Long Island, New York. The latter, cofunded by a consortium of nine East Coast universities, was being equipped with the giant 3 GeV Cosmotron, being assembled even as Kowarski wrote his "Note" in spring 1950, and a 10 MW graphite research reactor, which went critical a few months later, in August 1950.[20] When Rabi was in Florence, then, there was already a project being promoted by French physicists, with official support, to establish a collaborative European "atomic" research center similar to those being constructed at the time in both the Britain and the United States. Was Rabi aware of this scheme? If so, what did he make of it?

Rabi's "Brookhaven" for Europe

It is hard to believe that Rabi was not aware of the moves being made by scientists in the CEA to equip continental Europe with a "Harwell" or a "Brookhaven." He was extremely close to Oppenheimer. They worked together on the General Advisory Committee to the USAEC, and "Oppie" may well have told him of the discussions he had had in 1948 with Kowarski and other French scientists (and diplomats) over the possibilities of building a nuclear research center in Europe.[21] Rabi was one of the distinguished guests at the Rockefeller-funded CNRS conference in Paris in March 1950, dedicated to the physics of fundamental particles, where he rubbed shoulders with Blackett, Bohr, and Fermi, not to speak of a galaxy of other leading American physicists (see chapter 4). It is more than likely that Kowarski or another senior European nuclear physicist discussed the project with him on that occasion. Finally, it is probable that Pierre Auger, French cosmic ray physicist and director of UNESCO's Department of Exact and Natural Sciences, mentioned it to Rabi when they met in Florence. Auger had been engaged in the earlier conversations with Oppenheimer. He may well have been in touch with the ideas being discussed by the CEA scientists.[22] Before Rabi took the floor, he consulted with several European physicists, including Auger, about what to suggest, and in the debate on his resolution he actually identified Auger as the ideal person to carry the project forward. Whether he knew about the French plans or not, Rabi's UNESCO intervention had one immediate effect: it more or less stopped Kowarski's plan dead in its tracks.

On the face of it this may seem strange. Rabi often suggested that Brookhaven had served as a model for him.[23] Did this mean that when he rose to his feet in Florence he was supporting the French scheme? Emphatically not. He was categorically against a reactor, he said afterward, since it would be impossible for such a collaborative platform to survive the centrifugal pull of national interests. As he put it to Kowarski, "nuclear energy smelled very much of military on one side and commercial on the other, and it would have just been full of rivalry."[24] But there was another reason too: Rabi wanted West Germany to be included in the new regional center. In 1950, however, West German scientists were still not permitted to do "applied nuclear research," including reactor physics. So for Rabi, unlike Kowarski, Brookhaven was not the model as far as equipment was concerned, and he was not in favor of the Europeans building an accelerator *and* a pile.

In fact it was the institutional structure of Brookhaven that inspired Rabi, the idea that the Europeans could build together a laboratory "with governments replacing universities." Rabi placed increasing emphasis on this *political* significance of the laboratory as he grew older.

Speaking to Amaldi at Columbia University in 1983, he said that in Florence he "decided, on the basis of the Brookhaven experience, to do the proposal of the creation of an European laboratory [sic] mainly as a step towards the unity of Europe." He added, as Amaldi put it, that "the main reasons [sic] in favour of the European laboratory was the Unity of Europe, that, in his mind, had to replace Great Britain in the role of World Power."[25] What Rabi did in Florence, then, was to send a strong signal that the United States would favor the construction of a European laboratory where scientists could collaborate on frontline accelerator physics while governments worked together to build a united Europe that included the new state of West Germany.

Kowarski apparently did not resent the demise of his project, which was stripped scientifically by the removal of a pile and expanded politically by the inclusion of West Germany (and Italy).[26] In a characteristically nuanced account of the origins of CERN, he wrote, "The domain of common action [had to] be chosen so as not to infringe directly on the taboos of uranium fission, but close enough to it, so as to allow any successes gained internationally in the permitted field to exert a beneficial influence on the national pursuits. For this role of close but distinct neighbour, meson (or high-energy) physics was the *most obvious and attractive candidate*" for governments and scientists alike.[27] As Kowarski knew only too well, the potential research agenda was not that clear-cut, at least not until *after* Rabi had taken the floor in Florence. There, the kind of project that would be acceptable to Washington began to take shape, a project that Kowarski *subsequently* defined as the "most obvious and attractive candidate"—as indeed it was, if one wanted the support of Rabi and the State Department.[28]

Italy, West Germany, and a European Accelerator Laboratory

Rabi specifically identified six countries in the UNESCO press release that might together construct a regional research center in Western Europe. They were France, Italy, the Netherlands, Belgium, Switzerland, and West Germany. Interestingly, four of these had a central role in Kowarski's proposal for an "atomic" research laboratory (Italy and Germany did not).

Rabi included Italy because he was acutely aware of the desperate plight of physics in that country and the risks associated with a mass exodus of gifted scientists as well as of the immense influence of the Italian Communist Party both in and out of academia.[29] As shown in chapter 2, Gilberto Bernardini had apprised him personally of the situation while he was at Columbia University in 1948–49. Rabi had supported his plea for Marshall aid to Compton and the RDB. We do not know what transpired between Amaldi and Rabi while the latter was in Florence (except that they subsequently squabbled over who first had the idea for a European high-energy physics laboratory).[30] However, soon after the UNESCO meeting, Amaldi became actively engaged in remedying the Italian situation by ensuring that his country and his physics community had a key role in the new supranational project. As soon as he learned that the ECA would not financially support a major instrument in Italy, as Goudsmit had hoped, Amaldi threw himself with complete dedication into the preparatory phase of the European project. He withdrew only in 1954, when the CERN convention had been ratified by the member state governments, including his own.

Rabi's promotion of West Germany was consistent with his conviction that German science should be reconstructed and reintegrated into the "international" community and that the country had a key role to play in rebuilding a Western Europe that was aligned with the United States in the Cold War.[31] It was a political affirmation of the legitimacy and (limited) sovereignty of the new state and fully in line with the policy of the State Department at the time. On the boat to Italy, Rabi discovered that his colleagues in the U.S. delegation had been instructed to make a number of proposals for projects in West Germany. He found all of them somewhat trivial and demeaning for a great nation.[32] He called Washington to discuss his proposal and was given the green light to add it to the list of official initiatives to be included on the agenda. When Rabi stood up at UNESCO's fifth general meeting, he knew that he would be backed up by his administration. He also knew that he had to identify accelerators as a possible item of equipment and not mention a pile, for if West Germany was in, piles were out.

The suggestion that German scientists and engineers participate in the construction and use of a major European accelerator amounted to a radical relaxation of the formal constraints imposed on German physics.[33] Allied Control Authority Law 25, promulgated in April 1946, had severely limited the kinds of research that could be done (see chapter 2). It specifically prohibited fundamental research with instruments that could be used to produce war materials, thus nominally excluding cyclotrons that could be used for uranium enrichment. In March 1950, just three months before Rabi arrived in Florence, Law 22 (on the monitoring of materials, equipment, and fittings in the field of atomic energy) confirmed a ban on the operation of accelerators with energies exceeding 1 MeV. These laws were interpreted differently within and among the American, British, and French zones, leading to considerable confusion and resentment. The only functioning cyclotron at the end of the war was in Walter Bothe's Institute for Medical Research in Heidelberg, and the U.S. authorities restricted him to producing short-lived radioactive elements for medical research— although in October 1947 he hoped shortly to start up a 30 MeV Van de Graaff.[34] When Hans Bethe visited Göttingen in the British zone in July 1948 at the request of the Office of Naval Research, he found that a very fine 6 MeV betatron was in operation there. The group desperately wanted Siemens to build them a 30 MeV device, a project that Bethe strongly supported but that was being blocked by the American military government in Germany.[35] In the fall of 1949 a 1.4 MeV cascade generator began operating at the University of Mainz, in the French zone. It was built from parts salvaged from the Kaiser Wilhelm Institute for Chemistry in Berlin, which had been severely damaged in spring 1944, and from material shipped from the Muller Company in Hamburg, in the British zone.[36] Rabi may well have been aware of the confused situation in Germany, of the frustration caused by American reluctance to allow German physicists to do respectable accelerator research (Heisenberg had decided to do cosmic ray research instead in Göttingen), and of the harm that American inflexibility was doing to the United States' image in the German scientific community.[37] His intervention in Florence, if interpreted as a commitment by a high-ranking American official, amounted to a radical reassessment of formal U.S. policy on permissible levels of accelerator research by German physicists and engineers.

One consideration facilitating this policy shift was that German physics would be "contained," at two levels, in the supranational environment of

a European laboratory. First, it would be impossible for German physicists to reorient their work to military objectives: they would be included
in international teams doing unclassified research. Second, the collegial,
international environment itself would detach them from past allegiances,
instill in them democratic values, and enroll them in the political and ideological framework of the West. This at least is how some of the scientists
in Göttingen saw the future redemption of Germany when Hans Bethe
visited them in July 1948. As he put it in his report to the ONR: "Most of
these people would welcome a Western European Union of which Germany would form a part. In fact several of them told me that this seemed
to them the only way that a real democracy could be introduced into Germany. Only by forming part of a larger unit which is overwhelmingly
democratic, would they themselves learn what democracy is about."[38]

In promoting the scientific and political and reintegration of West Germany into the European physics laboratory, Rabi was being fully consistent with the aims of the Marshall Plan. In a world increasingly divided
into two camps, CERN was also unambiguously aligned with the West.
Moreover, it did not simply confirm existing alliances. By including Tito's
Yugoslavia among the original twelve member states, the founders deliberately tried to drive a wedge into the Soviet bloc and lure other satellites
to follow the example of the rebellious Yugoslav leader. In 1951 a group
of leading academics in the Boston area prepared a special (classified)
report on "political warfare" for the State Department.[39] They recommended that Yugoslavia "be given not only emergency aid but every possible support in developing an economic and political life independent of
Russia." The satellites on the other side of the iron curtain, they went on,
"struggling to exist under the pitiless demands of the USSR, will observe
the growing progress of Yugoslavia. No argument against the Russians
could be more persuasive."[40] CERN was not simply an instrument to
promote the aims of the Marshall plan in Europe. It was also a platform
on which to build a Western alliance under American leadership.

U.S. Physics and a European Accelerator Laboratory

Rabi's support for the construction of a major European laboratory
equipped with a powerful accelerator was widely appreciated by scientists

on the Continent. A machine of this kind was also of interest to the American physics community. Indeed, if American scientists were going to have "somebody to talk to" in Europe, as Rabi put it, the communities had to have a shared framework for discussion. This meant, first of all, working in an unclassified area of physics. Everybody agreed that the new laboratory should be a "maison de verre" (glass house) from a security point of view and that research conducted there should have no (direct) military implications. This was also a sine qua non for its being sited in Geneva in (politically neutral) Switzerland. However, having something to talk about also meant doing cutting-edge research using instrumental platforms and practices that were standardized on both sides of the Atlantic.[41] In the late 1940s the research frontier in meson physics was being defined by U.S. cyclotrons, notably the 184-inch cyclotron at Ernest Lawrence's laboratory in Berkeley, commissioned in 1946.[42] Speaking of the achievements of U.S. accelerators at a workshop in Pocono Manor in spring 1948, Richard Feynman remarked, "Just as we were apparently closing one door, that of the physics of electrons and photons, another was being opened wide by the experimenters, that of high-energy physics."[43] If Europeans were to cross this threshold, if their conversations with their American colleagues were not to degenerate into little more than idle chatter, if they were to gain scientific credibility and cognitive authority, they too had to enter the domain of high-energy, or meson, physics using controlled sources of energy. The exotic, individualistic fieldwork of cosmic ray studies, at which they excelled, had to give way to the disciplined, bureaucratized laboratory work of accelerator physics, about which they knew little and which many of them disliked.[44]

A meson physics laboratory in Europe had a number of advantages from an American point of view. Most important, it could provide a valuable stimulus to work in the United States, adding to its "stockpile of scientific knowledge." This point was stressed repeatedly by representatives of the U.S. scientific community who in the early 1950s had to deal with a domestic political climate that was hostile to international scientific exchange and deeply suspicious of foreign scientists. The report prepared for the State Department by Lloyd Berkner and a distinguished panel that included the Rockefeller Foundation's Warren Weaver and Rabi himself made the point clearly. Partly released in May 1950, just before Rabi left for Florence, "Science and Foreign Relations" affirmed that it would be fatal to underestimate the importance of foreign scientific developments to the progress of American science itself. Such science was invaluable because, the panel wrote, "it enables the American scientist to exploit directly the discoveries and lines of thought delineated by other scientists and to build upon their accomplishments . . . [and] provides constant stimulation to the American scientist with respect to his own thinking."[45] Alan Waterman, the director of the new National Science Foundation, made the same point to a Commission on Immigration and Naturalization in December 1952: "We must not imagine that America does not need information and inspiration, and cooperation, from outstanding scientists in friendly foreign countries," he said. "We do not have any monopoly on scientific talent or the emergence of new discoveries in science."[46]

There was more to it, though. For Europe had a long tradition of doing outstanding fundamental research that could stimulate U.S. science (and vice versa). Many people felt at the time that the Europeans were, in fact, better at making new discoveries than their American colleagues, whose strength lay in development and organization. Europeans had the creative ideas that "Yankee ingenuity" and pragmatism applied to useful ends.[47] The Berkner report was explicit about this. It drew on the distinction between basic and applied science that was reified by Bush in his *Science—The Endless Frontier* and mapped it onto the Europe/U.S. divide.[48] "American pre-eminence as demonstrated thus far is in the application of scientific discovery," the report claimed, and "it is to our practical advantage to promote the fullest scientific intercourse."[49]

This global division of scientific labor also had important implications for the security of the nation. As Waterman reminded the congressional commission at the height of the Korean War, "The development of some of the most vital weapons in our armament stems from open, unclassified fundamental scientific research abroad. Radar, the atomic bomb, jet aircraft, and penicillin were perfected in the United States on the basis of discoveries and research in foreign countries to which we were given ready access."[50] America's capacity to wage war with advanced technologies would be enhanced by Europe's capacity to do outstanding unclassified basic research.

The need for unclassified research in meson physics in Europe was all the more pressing in 1950. By a secret order issued on 10 March, President Truman committed the United States to the development of the hydrogen bomb. He had yielded to incessant pressure from Edward Teller, in particular, amplified by the explosion of the first Soviet fission weapon in August 1949 and the exposure of Klaus Fuchs's treachery in February 1950. Teller, who insisted that the world was in a situation at least as dangerous as the one it faced in 1939, told American scientists to go back to the weapons laboratories: "Our scientific community has been out on a honeymoon with mesons. The holiday is over. Hydrogen bombs will not produce themselves. Neither will rockets nor radar."[51] Rabi was also pleased when he learned that Lawrence had decided to mobilize his Berkeley laboratory for weapons work. "It is certainly good to see the first team back in," he said, adding that "You fellows have been playing with your cyclotrons and nuclei for four years and it is certainly time you got back to work!"[52] While many U.S. physicists narrowed their focus to work again on the development of gadgets, the Europeans could add to the stockpile of unclassified knowledge of nuclear properties and behavior. This was all the more important because of the conviction in some quarters that weapons work was "applied" and added little to basic understanding. As Oppenheimer put it after the war, "In the scientific studies which we had to carry out at Los Alamos, in the practical arts there developed, there was little of fundamental discovery, there was no great new insight into the nature of the physical world."[53]

An "open" European laboratory could benefit national security in another way—by serving as a site for scientific intelligence gathering. This was addressed in a classified appendix to the Berkner report.[54] It stressed that scientific capability had increasingly become a measure of the power of states and that every effort had to be made, from scouring publications to capitalizing on personal contacts at meetings of "UNESCO, the international scientific unions, and international scientific congresses and conventions" to find out what scientists in other countries were doing.[55] The "prime target" for intelligence gathering was, of course, the Soviet Union, but "other areas are also of major importance, first, because research and development results in those countries may contribute to our own scientific and technological advancement, and second, because such discoveries may become known to the Soviet Union and so be of potential use against this country."[56] Berkner and his panel stressed that it was highly desirable that qualified American scientists be enrolled in intelligence gathering and that they do so informally and without raising suspicions. "The emphasis should be on the free and open discussion of the content, procedures and mechanisms of the science involved," the panel wrote.[57] CERN was of course not mentioned in the report; but it was clear that an international laboratory doing unclassified work in high-energy physics, and eventually gathering together the most outstanding scientists and accelerator engineers in Europe, would be an ideal venue for capitalizing on free-wheeling technical discussions between peers to evaluate the strength of science in other countries.[58]

Coproduced Hegemony (and the Limits of Consensus)

The leitmotif of this chapter has been that Rabi and the State Department used the opportunity provided by the UNESCO meeting in Florence to promote a U.S. scientific and foreign policy agenda in Western Europe. That policy was articulated around the construction of a major accelerator in a regional West European organization that included West Germany and Yugoslavia among its member states.

From a political point of view, the timing was perfect. Early in May 1950, just a month before Rabi's statement in Florence, French Foreign Minister Robert Schuman announced that his government proposed to place the whole of Franco-German coal and steel production under an international authority.[59] Rabi's plan and the Schuman Plan were similarly structured. Both required that national governments cede some of their sovereignty in return for the benefits of pooled resources, both used a key asset with links to energy production as a platform around which to forge union, and both were limited to states on the Continent. (Rabi never mentioned Britain as a potential partner in the physics laboratory, just as Britain was not among the "Europe of the Six" that emerged from the coal and steel community, later to form the European Common Market.) This convergence must have appealed to governments that were at that time promoting European unity. Rabi's scheme for a supranational physics laboratory that included West Germany dovetailed perfectly with the first halting steps toward economic supranationality in Europe, just as the subsequent inclusion of Yugoslavia can be read as a symbolic affirmation of the significance of the Atlantic alliance recently institutionalized in NATO.

The venture was also scientifically appealing. Accelerator-based high-energy physics was one of the most promising unclassified research domains at the time, and Europe was falling badly behind the United States in terms of equipment. Of course French scientists, in particular, saw more interest in trying to provide Europe with its own Harwell or Brookhaven, endowed with an accelerator and a pile. But if the United States was against that kind of structure, so be it. In summer 1950 there were distinct political advantages in France to including West Germany in a supranational laboratory, even if this had to be done at the expense of a pile.

This does not mean that the European laboratory project met with universal political approval. There was a hotly fought referendum in the Canton of Geneva, promoted by the Communist Party, which claimed that having a nuclear laboratory there was a danger to the population and would impugn Swiss neutrality.[60] The Communist Parties in France and in Italy were also strongly against the participation of their governments in the center. In a number of highly critical articles in Communist magazines and newspapers in Paris in summer 1953 it was noted that two leading French scientists and Nobel Prize winners, Frédéric and Irène Joliot-Curie had been excluded from the international and internal discussions over the shape of the project. Extracts from the Berkner report were used to suggest that CERN was part of a general U.S. attempt to establish in Europe a "Vast Network of Scientific Espionage under the Cover of Cultural Exchanges." The influence of "Nazi professor Heisenberg" was remarked upon. And the limitation of membership to the countries of "Atlantic" Europe, along with Tito's Yugoslavia, did not go unnoticed.[61]

Several elite European physicists—Niels Bohr, James Chadwick, Hendrik Kramers—were also most unenthusiastic about the laboratory at first.[62] Bohr and Kramers argued that it would drain much-needed funds away from their national scientific efforts, which were desperately in need of support. It is also quite likely that they balked at the scale of the laboratory, the engineering ethos that would suffuse its halls, and the loss of individual creativity in the organized, bureaucratized teamwork that doing physics with such massive equipment would entail.[63]

Notwithstanding the doubts and the opposition, the basic agreement on CERN was provisionally adopted by the majority of its member states in February 1952. The physiognomy of the laboratory was exactly what Rabi had wanted. A cable to Columbia University signed by almost two dozen leading European physicists, including Amaldi, Auger, Bohr, and Heisenberg, read: "We have just signed the Agreement which constitutes the official birth of the project you fathered in Florence. Mother and Child are doing well, and the Doctors send you their greetings."[64] It was an explicit recognition of the role that Rabi had played in shaping the European laboratory project in line with his conception of what it should look like and of the success that a handful of European scientists, engineers, science administrators, and politicians had had in building support for that conception on the Continent. Rabi was so touched that he kept the telegram framed on a wall in his office at home.[65]

4

The Rockefeller Foundation in Postwar France: The Grant to the CNRS

The previous two chapters have described the steps taken by high-ranking scientific statesmen with close links to the U.S. administration to promote the reconstruction of science in Europe using political incentives to link it to the aims of the Marshall Plan. A striking feature of the initiatives by Bush and Compton in the Research and Development Board and by Rabi, in consultation with the State Department, was the reluctance to provide financial aid even though Europe was "down and out" scientifically. Political support and legitimation was forthcoming, but European governments had to find money for science themselves.

The foundations, however, operated in a different, if overlapping, space. Economic support, in the form of grants and fellowships, was their traditional mode of encouragement and their main means of pressure. These were sources of both material and symbolic capital for their beneficiaries, but they had no explicit political weight. On the contrary, foundations like the Rockefeller and, perhaps to a lesser extent the Ford, had an image of independence vis-à-vis the administration, which was particularly important in the early years of the Cold War. Raymond Fosdick, the president of the Rockefeller Foundation, stressed in May 1948 that the principle on which the foundation had always acted was "that our assistance is given without regard for race, creed, color or political opinion."[1] This gave the foundation a legitimacy abroad that Washington simply did not have. The administration's room to maneuver politically was restricted both by European suspicions that the U.S. government was interfering in the domestic affairs of sovereign states and, as the Cold War got into stride, by assaults from nationalists and isolationists in Congress who were instinctively hostile to "foreign entanglements." During the McCarthy era the

State Department in particular was extremely prudent about taking any initiatives that might be construed as supporting Communism; the foundations could afford to be more ambitious and could increase their credibility by affirming their "independence" of the prevailing xenophobic climate. Senator John F. Kennedy told Ford Foundation officer Shepard Stone that the Polish refugee program that he had launched in 1957 "was showing the way to the government." Given the prevailing mood in the country, Kennedy was convinced that it was "easier for a private foundation than for the government to take up relations with Eastern Europe, and he thought it a very statesmanlike and very important political development."[2]

It must be stressed that this autonomy of action was implemented in a framework of assumptions about where America's interests abroad lay—assumptions that the foundations shared with liberal elements in the State Department and even the CIA. The mutual embrace of these branches of the U.S. administration and big philanthropy can be traced back to the period just before World War II. In May 1938, in response to the growing influence of the Soviet Union and National Socialism, the State Department announced the creation of a special division dedicated to international cultural relations.[3] The State Department invited the major foundations to coordinate their activities with the needs of U.S. foreign policy. After the war, the growing consensus in the United States that the future of democracy and national security lay in integrating West Germany into a larger European community and in unyielding opposition to Communism and the Soviet Union forged an alliance between likeminded senior foundation officers and government administrators. By virtue of this ideological consensus, the goals of one became intermingled with the goals of the other: the philanthropies, as a senior Ford Foundation official put it, "allowed themselves to be used as private instruments of public policy."[4] The next three chapters will give substance to what he meant by exploring several initiatives taken by the Rockefeller and Ford foundations on the Continent in the first decade after the war.

A Few Words about the Rockefeller Foundation

The Rockefeller Foundation was established in 1913 with the aim of "promoting the well-being of mankind throughout the world." This

international dimension differentiated it at the time from the other Rockefeller philanthropies, which had restricted their activities to the United
States. For about a decade the foundation defined its mission in terms of
promoting general education and public health. From 1923 to 1929 it
placed increasing emphasis on medical and scientific education. A major
reorganization of the Rockefeller philanthropies in 1928 formalized this
evolution, and five divisions were created within the foundation: Natural
Sciences, Medical Sciences, Social Sciences, Humanities, and International Health. To avoid being accused of "dictating the course of scientific research," foundation executives preferred to support already
prestigious colleges and universities with capital grants intended to
"make the peaks higher." The awarding of fellowships to individual
researchers was turned over to the National Research Council.[5]

In 1932 Warren Weaver was recruited by foundation President Max
Mason to be the director of the Natural Sciences Division. Mason and
Weaver had known each other for many years and had collaborated on a
book about the electromagnetic field. Weaver had also published occasional papers in mathematics, mostly on probability and statistics. Somewhat reluctantly he decided to leave his post as chairman of the
Department of Mathematics at the University of Wisconsin to build a
natural science program coherent with the foundation's goals. It took
him some time to work out with the trustees the most productive way to
spend the resources he had—about $1.5–2.5 million annually from 1935
to 1938. The program he settled on privileged research that demanded
the application of the new techniques of molecular physics and chemistry
(radioactive isotopes, the ultracentrifuge, electron diffraction, X-ray
crystallography, and even particle accelerators) to the problems of
human biology. In this way he fused his own scientific competence with
the general concern of the Rockefeller philanthropies to develop the "Science of Man," and the life sciences in particular.[6]

The type of grant also changed, originally in response to the economic
downturn in 1933–1934. The most important innovation was the project
grant. Averaging about $20,000 for three years, it was problem-oriented
and represented the foundation's commitment to an individual who was
working in a new field of research that often cut across disciplinary
boundaries. "Group grants" were also introduced—awards to leading

researchers from different fields in a single university who were willing to collaborate on projects that satisfied Weaver's criterion of relevance.

Weaver's program was temporarily suspended during the war, when he was appointed by Vannevar Bush to be the head of the Applied Mathematics Panel of the Office of Scientific Research and Development. In this role, Weaver made major contributions to what was later called operations research, notably in the area of antiaircraft fire control and bomber effectiveness (see chapter 8).[7] He returned to the foundation shortly after the war. In the early 1950s it was decided to broaden the scope of the Natural Sciences Division to include agriculture, notably in the third world, while phasing out activities which governments in the industrialized countries, alerted to the importance of science and technology as strategic resources, were increasingly willing to fund. When Weaver eventually left the Rockefeller Foundation in 1959, he estimated that he had watched over the distribution of about $90 million for research in the field of experimental biology.[8]

Weaver's approach to grant giving differed radically from that prevailing in the foundation in the 1920s. The director of the Natural Sciences Division was deliberately interventionist and tried actively to shape the trajectory of research in biology, albeit "from the warped perspective of the upper rungs of the Comtean ladder of the sciences."[9] Weaver himself coined the term "molecular biology" in 1938, and his programs undoubtedly played a role in the emergence of the new discipline, though his one-sided emphasis on the importance of technique also led him to ignore promising and eventually more productive lines of research.[10]

In making an award, Weaver and his colleagues, trained scientists themselves, not only had to draw on their scientific competence; they had also to build into their decisions judgments of character, of integrity, and of the motivations of applicants. As Weaver put it, "All that really counts is the intellectual competence of the individual—his imagination, his character, his dedication to science—and this is to be found in his record and in him, and it is not to be found in his [grant] application."[11] Such judgments required a deep understanding of human nature and a wide knowledge of the best people in the field. To achieve its ends, Weaver said, the foundation had to have "intellectually capable and socially acceptable officers each one of whom spends a great deal of time traveling, who spends a great deal of his time talking to scientists all over the world."[12] During these travels, foundation officers meticulously recorded opinions and assessments night after night in diaries that were carefully indexed and made available to all the officers in the organization. By this means, Weaver and his colleagues built up a shared and documented who's who of the field and an appreciation of potential applicants' intellectual and personal strengths and weaknesses.

Without trust and openness between officers and applicants, this system of evaluation could never have functioned successfully, if at all. Indeed the intimacy with which the foundation's officers communicated with fellows meant that their relationships often went beyond mutual respect to genuine affection. It would be inaccurate to say that this was a system of personal patronage. But personal considerations were at the core of the assessments—necessarily so in the eyes of the foundation's officers. In this respect, Weaver's mode of patronage and his activist system of management were quite different from the more impersonal and putatively "objective" procedures put in place after World War II, notably the system of "blind" peer review.[13]

The Significance of France for Weaver

The pitiful state of science in continental Europe created a window of opportunity for the Rockefeller Foundation. Ostensibly there were two main candidates for support on the Continent in 1945–1946, namely, (occupied) Germany and France. As noted in chapter 2, Weaver and the other officers in the foundation were extremely reluctant to intervene in Germany, notwithstanding pressure from the trustees to do so. They preferred to work in France, where there was a needy, gifted, and politically courageous scientific community and where liberation had released energies and a determination to place scientific research on a new footing.

The foundation had a long history of activity in France. It had a European "field office" in rue de la Baume in Paris from which it managed almost $50 million worth of grants for health and medical projects in eighteen countries before World War II. Its earliest project on the Continent was an antituberculosis campaign in France during World War I, and it was actively, if not always successfully, involved until 1941 in promoting medicine and public health in the country.[14] French physicists and biologists had also benefited from foundation support. In the 1930s Weaver agreed to contribute to the costs of an accelerator for Frédéric Joliot-Curie, notwithstanding his doubts about the grantee's sincerity in stating that the equipment would be used for biological research. He also gave fellowships to scientists at the Rothschild Foundation's Institute for Physico-Chemical Biology in Paris, including Boris Ephrussi, Emmanuel Fauré-Fremiet, Louis Rapkine, and René Wurmser. Here, physicists, chemists, and biologists could work together—an arrangement that particularly appealed to Weaver.[15]

This chapter and the next will study in detail two grants made by Weaver in France in the first few years after the war. The first was a major grant for equipment and international scientific colloquia, negotiated by Louis Rapkine and made to the Centre National de la Recherche Scientfique (CNRS) early in 1946. The second was a significant award to Boris Ephrussi, intended to help him set up a new genetics research laboratory at Gif-sur-Yvette, just outside Paris.

Weaver liked the novelty, in the French context, of the programs being promoted by the beneficiaries of these grants. The CNRS was reestablished after liberation by President Charles de Gaulle, with Frédéric Joliot-Curie and Georges Teissier as its new director and assistant director, respectively. Joliot-Curie was not only an outstanding scientist; he was also highly respected for his courageous work in the Resistance during the German occupation. Furthermore, he was determined to reconfigure French science, through the CNRS if possible, by linking basic research more tightly with university teaching, on the one hand, and with applied research and industry, on the other. Ephrussi's project was historically unusual in that, before the war, genetics was a marginal research endeavor in France, held back by a combination of academic conservatism and an allegiance to Lamarckism. Indeed, it was only in 1946 that the first university chair in genetics was created (at the Sorbonne), and it was Ephrussi who filled it. There was a new atmosphere in French science after the liberation, then, a determination to break with the rigidity and inertia of the past, and Weaver saw both of these grants as catalyzing that process.

Historians of science have previously studied the Rockefeller Foundation's awards to the CNRS in 1946 and to Boris Ephrussi in 1948.[16] While helpful, this work fails to situate the awards within the foundation's broader social and political agenda in the country, nor does it consider carefully the effects of anti-Communism in the United States on the kind of people the foundation would likely fund. Weaver and the foundation were not simply interested in supporting good science and new directions in France. They wanted to use their financial leverage to steer French scientists along quite definite lines. Weaver in particular believed that the French were parochial and inward-looking. He wanted to transform them into outward-looking, "international" researchers, using techniques and tackling questions that were current above all in the United States. It was a vision inspired by the conviction that, without a radical remodeling of the French scientific community on American lines and the determined marginalization of Communist scientists in the field of biology, the country could never hope to play again a major role in the advancement of science.

Rapkine's Plea to Help Rebuild a Divided Community

On 4 August 1945 Louis Rapkine wrote two long and rather similar letters to Warren Weaver and his vice-director, Harry ("Dusty") Miller.[17] He had recently corresponded with both of them, and with P. O'Brien in the Paris office, who had "vaguely suggested that I should make a rapid journey to New York." Senior colleagues and officials in France had agreed. The letters to Miller and Weaver were intended to sound them out on the advisability and usefulness of the journey. Its main purpose would be to discuss how best the Rockefeller Foundation might help rebuild postwar French science.

Louis Rapkine was born in Russia in 1904, moved with his family to France for a couple of years, and then settled with them in Montreal in 1913.[18] He attended McGill University as a medical student, became disenchanted, and decided to move to Paris in 1924 to concentrate on biological research. His life was hard until he reluctantly accepted a fellowship from the Rockefeller Foundation of $90 a month. Rapkine never forgot how humiliated he had felt to accept a grant, nor how important the eighteen months of support had been to his material well-being and his research project on the cell. His ensuing intimate, almost obsequious, relationship with the foundation's officers was crucial to his subsequent successful dealings with Weaver. Rapkine remained an obedient servant of the Rockefeller Foundation until his premature death from lung cancer in 1949.

Rapkine first became engaged in supporting science in 1936, when he and French physicist Edmond Bauer set up a committee to provide relief for penniless and desperate savants fleeing Nazi Germany. After the collapse of France in June 1940, Rapkine, now a French citizen, devoted his energies to assisting French scientists escape the country. He set himself up at the New School of Social Research in New York, where once again the Rockefeller Foundation provided some of the funds he needed for the émigrés. In December 1941 General de Gaulle nominated Rapkine to be head of the scientific office of the Free French movement in the American city.[19] Rapkine's aim, which proved more difficult to achieve than he had thought, was to insert fleeing French scientists into war research in the United States. His efforts were met with suspicion, and he decided that he had a better chance of helping the allies if he moved his operation to England, where there were fewer concerns about security. In August 1944 Rapkine moved to London with seventeen other French scientists who gave up positions in the United States and Canada. There he set up and headed the French scientific mission in Great Britain. With Paris liberated and with the war dragging to a close, Rapkine redefined his task as consolidating links between the scientists who had remained in France and those who had left the country to work abroad. His efforts on behalf of French science during the war had earned him respect and credibility with a national scientific community that now faced the challenge of rebuilding its infrastructure and dealing with its internal divisions. His status as the head of the French missions in New York and London legitimated him as a spokesman for French science in the eyes of the foundation.

In his letters of 4 August 1945, Rapkine left Weaver and Miller in no doubt as to the significance their intervention would have. The Rockefeller Foundation, he said, was awaited by many "practically like the Messiah." This was not simply because it had financial resources to disburse. Rapkine had little time for those who saw the foundation as something of a "gigantic Holy Cow" to be milked for whatever one could extract from it. No, it had a larger role. It was a "unique Institution," and people in Europe "expected it to assume certain responsibilities which it had never had to assume in the past (or at least on a very small and localized scale)." At the core of those responsibilities, as Rapkine saw it, was the reconstruction of a fractured French scientific community—an aim which has, surprisingly, been more or less ignored in previous studies.[20] Rapkine's letters, in fact, hardly mentioned the material needs of individuals and laboratories. That would come later. He first wanted to describe the complex and delicate situation (as he saw it) inside a community divided against itself, one that needed external help if it was to put the immediate past behind it and unite to build a strong French scientific presence with international credibility.

Reconstruction was not going to be easy, Rapkine said, and the foundation would have to proceed cautiously. France was "definitely poor and ill—and proud." This was not the pride and self-confidence that came with the glow of success or with a sense of achievement. It was a pride born of defiance. The French had "undergone a hell of a lot, much more than one can learn from the Press, much more than one can learn from the few dramatic comptes rendus," Rapkine said. Theirs was a pride that "compensates certain feelings that they would rather hide,— the feeling of having been over-run by an all-powerful enemy; the feeling of having been crushed into silence for five long years." The humiliation of defeat and Nazi occupation and the material and spiritual decimation that ensued left the French vulnerable to seeing external support as a form of charity. And charity would be resented and rebuffed. The task now was to help the members of the French scientific community "regain their self-confidence, help them heal their wounds, help them by creating in them the belief that they are themselves getting back on their feet." Intervening in this situation was a "delicate task" indeed, but one that the foundation, with its outstanding reputation for fairness and having "wisdom and warmth a-plenty," could pull off with success.

Rapkine sketched some of the parameters of the fragmented community in his letters, adding a few elements in subsequent discussions with Weaver and Miller. There were two main divides. First, there was a gulf between those who had left the country and those who had not. Second, among those who had remained, there were differences that reflected the range of attitudes to occupation, from collaborating with the enemy or the Vichy regime to outright opposition in the Resistance.

The gap between those who had gone abroad and those who had stayed at home was most apparent in scientific knowledge and experience; those who had stayed behind lacked the authority bestowed by such currency. French scientists who had been engaged in wartime projects overseas had worked at the research frontier in some of the most important areas of science. They had participated in the wartime fusion of theory and practice, in the dissolution of the basic/applied distinction, in teamwork and in interdisciplinary research, in the transformation of existing fields, like nuclear science, and in the birth of new ones, like operations research. Rapkine himself had learned the last in England and had played a role in introducing it to the French military after the liberation. The scientists who had been in France, by contrast, had worked under difficult conditions with limited resources. In a system where authority and legitimacy rested primarily on doing original, internationally recognized research, the people who had worked in France during the war were, temporarily at least, at a serious disadvantage vis-à-vis those who had worked abroad.

Rapkine was extremely sensitive to this issue. He took especial pains to discuss the cases of two brilliant French scientists who had left the country, Léon Brillouin and André Weil. Brillouin had moved his physics research from the Collège de France to the University of Wisconsin in 1941. Immediately after the war Brillouin had apparently suggested to the Rockefeller Foundation that it help establish a "sort of French-American University in France." The faculty, as Rapkine understood it, would comprise only people who had had extensive experience in America; those who had remained in France would not normally teach there. Rapkine warned the foundation off the venture. Part of the reason was material. More fundamentally, though, a faculty comprised only of people who had had extensive experience abroad, and so who had been out of France during the war years, would not be welcomed in the country. There had to be a "shuffling and mixing" of the returning scientists with the community that had stayed in France if what Rapkine called "psychological" problems were to be avoided.

The bad odor of "dishonor" suffused the Weil case. André Weil, brilliant mathematician and founder member of the Bourbaki group, was a conscientious objector who fled to Finland to avoid military service. He was imprisoned on being sent back to France and tried for treason. Weil secured his release on the condition that he join the army. This was a mere pretext: he escaped to the United States as soon as he could, teaching at Haverford College and Swarthmore College in Pennsylvania as of 1941. After speaking to Rapkine about his case, the Rockefeller Foundation officers recorded that he had "been completely forgiven" by the end of the war—a phrase that is indicative of the anger his departure had provoked. Indeed, French scientists "were eager to have him return to France" from São Paulo, Brazil, where he had gone in 1945. Weil returned to the United States in 1947, joining the faculty at the University of Chicago, and completed his career at the Institute for Advanced Study at Princeton University.[21] Brillouin likewise never returned to his native land. He moved to Harvard in 1946, became an American citizen, worked briefly for IBM, and then joined the faculty of Columbia University in New York.

Rapkine was so concerned about the hazards of reintegrating French scientists who had worked abroad during the war that he engaged in some "shuffling and mixing" himself. As he explained to Miller, "psychologically it was of extreme importance that the French scientists who were outside France should quickly mix with those who were less fortunate or who chose to remain inside France during the occupation of the enemy." Rapkine had established close contacts with eighteen scientists whom he had brought back from the United States with their families, and six others who were in the United Kingdom. He had arranged for every one of them to come for a short visit to Paris. There they not only met up again with their scientific colleagues but were also encouraged to revisit their *quartier* and speak with the locals, "le concierge, le marchand de journaux, le boulanger, etc" (the concierge, the newspaper seller, the baker).[22] Rapkine also arranged for traffic in the opposite direction. As of August 1945, sixty-eight hand-selected scientists from

Paris and the provinces, covering a wide variety of scientific disciplines, had gone to London under the auspices of the French scientific mission. In his estimation, these exchanges had a "marvellous" effect, overcoming inferiority complexes and rebuilding confidence all around.

The fear of being ostracized for being abroad and the comfort that came from being reaccepted into the scientific community after liberation extended to Rapkine himself. He was pleased that a new department of cell chemistry was being created for him at the Pasteur Institute. He listed over a half-dozen others who had been with him abroad and who were similarly fortunate. They included Pierre Auger, who had been among the contingent of French physicists at Chalk River in Canada, and Boris Ephrussi, who had collaborated extensively with George Beadle in T. H. Morgan's "fly" group at Caltech before the war.[23] Auger had become the director of higher education, and Ephrussi had been appointed to a chair in genetics at the Sorbonne. It was "gratifying," Rapkine wrote Miller, to know that people like this had "not only . . . been reinstated, but [that] a good number of them ha[d] obtained important posts . . . good proof that their departure was not frowned upon."

In fact, the scars were deeper than Rapkine realized, and resentment against those who left the country still simmered just beneath the surface for several years after the war. When, in the late 1940s, the foundation did not give an award to Desnuelle, a biochemist, Claude Fromageot, one his closest colleagues, bitterly blamed Rapkine for the setback. The injustice was cruel, Fromageot reputedly said, since while Rapkine "was safely fighting France's battle in England or abroad," Desnuelle was "in the French Army and later a P.O.W."[24] As it happens Rapkine *had* advised the foundation not to fund Desnuelle, casting doubt on his war record: there was, he said, "suspicion over his release as a war prisoner."[25] Whatever the truth, it is clear that the reintegration of French scientists who had been out of France during the war years was a delicate issue inside the community for quite some time after the liberation.

There were obviously also political differences among those who had stayed within the country. Collaborators and members of the Resistance occupied the two extreme poles. Rapkine provided Weaver with a list of seven French scientists who were, as Weaver put it, "in at least temporary disrepute because of their behavior during the war."[26] Six of them were in physics, mathematics, or related disciplines.[27] The seventh, and reputedly the worst, offender on the list was Moricard, whose field was not mentioned, but who was "definitely bad." The foundation "would do well to stay away" from these people "for at least a year or two," Weaver suggested, presumably on Rapkine's advice.

At the other extreme were those scientists who had been "splendid in courage and fortitude." Rapkine singled out André Lwoff of the Pasteur Institute as a "pillar of Resistance" and mentioned that his wife too had "risked more than she talks about." He spoke of Joliot-Curie and his group at the Collège de France as performing "miracles" and mentioned the physicist's key role in the Front National, the Resistance group based in the universities. Others singled out for special mention by Rapkine were Aubel, Courrier, Lacassagne, Monod, and Teissier. The moral authority and social status of these men was undisputed in the early days after the liberation. But there was a downside. By virtue of their prestige, "certain aristocracies ha[d] grown up, based on whether one was more or less active in the resistance movement."[28] This would obviously affect the distribution of power in the postwar reorganization of French science.

Rapkine's description of the various fractures in the scientific community was complemented by a positive assessment of what the future could hold. The French, he stressed, were "a wonderful people, intelligent, sensitive and enthusiastic." Among the scientists there were a "remarkable lot of men of good will; of men who were willing to learn; of men who have remained faithful to high ideals; of men who appreciate the international aspect of science." And this was true both in the capital city and in the provinces, in places like Nancy, Strasbourg, and Lyon. Indeed, he managed to convince the foundation's officers that old wounds were healing rapidly. The reintegration of men like Auger, Ephrussi, and Rapkine himself, and the positions of responsibility that were given to them, were indicative of this.

Rapkine used three arguments to promote his credentials as a privileged interlocutor between Paris and New York. First, he had firsthand knowledge of the "complex and delicate" situation (as Rapkine saw it) inside the French scientific community, or at least that part of it that was closely connected to the universities and the newly reborn CNRS.

Through his position in the French scientific mission, he had come to know the structural fractures and personal resentments that scientists had inherited from the war. He was also attuned to French sensitivities, particularly about receiving help from abroad (after all, had he himself not originally refused Rockefeller Foundation support "at 21, young, proud and slightly foolish!"). His personal experience and insights were wonderful assets that, he suggested, the foundation could use to guide it through the politico-intellectual minefield of postwar French science.

Second, his views were representative of the opinions of many French scientists, including senior men like Auger and Joliot-Curie. Together with them and others he had discussed the current difficulties faced by science in France and had suggestions for improving the situation. Joliot-Curie himself had encouraged Rapkine to make the trip to New York (also hoping that this would also pave the way for an invitation from the foundation to him and his wife, Irène).[29] Henri Laugier, who had worked closely with Rapkine in the French scientific mission in New York in the early 1940s, also wanted him to go. Laugier had been tipped to be director of the new CNRS. De Gaulle's choice of Joliot-Curie led to his being appointed head of the cultural relations department of the French Foreign Office instead. It was in this capacity that he offered to finance Rapkine's trip to America.

Finally, Rapkine made much of his personal ties to high-ranking foundation officers—Miller and Weaver, in particular. He noted with gratification the affection that senior men in the Natural Sciences Division had for him. He thanked Miller for his "touching signs of friendship" and concluded his letter to his "old buddy" (*vieux frère*) with a "handshake in all affection and friendship." To Weaver he almost groveled. Referring to earlier correspondence, he remarked that "the way you addressed your letter to me in the terms of 'dear Louis,' and the way you signed it 'Warren' [had not] passed unnoticed," and he rejoiced in the fact that he had rightly sensed Weaver's closeness, or "rapprochement," to him for some time. Rapkine commiserated with Weaver, who was at home at the time recuperating from a severe attack of Ménière's disease.[30] When he eventually visited the foundation's headquarters, he had several long meetings with Weaver at his home, three-quarters of an hour by train from New York.[31] Rapkine expressed great interest in Weaver's son, and met his daughter. Weaver's wife sent the Rapkine family a parcel in Paris around Christmas 1946. These mutual gestures of friendship consolidated the intimacy that was established between Rapkine and the two most senior officers in the Natural Sciences Division and surely aided his cause.[32]

Rapkine's Visit to New York, November 1945

The Rockefeller Foundation's officers treated Rapkine's request to see them as a matter of extreme urgency. President Raymond Fosdick had given instructions for Weaver not to be disturbed. Miller decided all the same to pass on both letters from Rapkine to his boss, not even waiting the two days until Fosdick was back in the office to get his permission. His pretext for doing so was that Rapkine was replying to a letter received from Weaver the month before. Miller did not hesitate to add that he was "very fond" of Rapkine and that he knew him well enough to "tu-toi" him.[33] The weight of Rapkine's personal bonds to the foundation had made itself felt already.

Weaver was all in favor of sending Rapkine not just a standard letter saying that he would be welcome to visit New York, but a reply that explicitly and enthusiastically accepted his offer to come to the United States. This positive response was, it seems, not inspired by some grand intellectual or political design: it simply reflected the esteem and affection in which Rapkine was held. Writing by hand from his sickbed, Weaver stressed Rapkine's credibility as a rapporteur on, and representative of, French science. His letters, Weaver remarked, showed that he was "extraordinarily (perhaps uniquely) well informed" about the "physical and human" state of science in France. Many French (and British) scientists and science administrators trusted him. He was not "trying to manipulate himself into some position of prestige or authority"—had he not emphasized that what he wanted to do above all else was to get back to scientific research in his new laboratory? Weaver also remarked that Rapkine, by virtue of his Canadian origins and fluency in English, was a reliable interlocutor notwithstanding his "extreme emotional commitment to France and the French." He had "a certain objectivity and understandability that no Frenchman could ever have."[34] Rapkine was thus ideally placed both to speak for the concerns of the French scientific community with great passion and to detach himself from too close an identification with parochial infighting and narrow interests. What is more, it was not only French interests that he would represent. By virtue of his "intense devotion" to the foundation, going back to the days when he was first awarded a fellowship, he could also be guaranteed to "do absolutely anything in his power to advance the things we stand for," Weaver wrote Fosdick.[35] At the end of August, Dusty Miller invited Rapkine to New York as a guest of the foundation, to fill out the "incomplete" and "fragmentary" picture his earlier letters had given and to share with the officers the "benefit of [his] wealth of information and [his] counsel."[36]

Rapkine arrived late in October in the United States, spent the month of November there, and left around 6 December. He "worked like a slave," discussing the French situation with several departments and "writing report after report" for the officers.[37] In the midst of his visit he confided to his wife that Weaver was "most sympathetic" to his cause, that he had inspired a "great deal of confidence" in the officers of the foundation, and that he had "great hopes that France [would] be treated with priority number one, as far as the countries of Europe that [had] been occupied [were] concerned."[38] He was not to be disappointed. Weaver told Fosdick that, in his view, "of the continental countries, French science presents the strongest combination of appeal (hardest hit and most worth rebuilding)."[39]

Two foci for financial support emerged in these discussions. The first, and the one strongly promoted by Rapkine, was the "rehabilitation and reconstruction" of scientific laboratories. The second, Weaver's brainchild it seems, was support for visitors from the United States and some selected European countries, to participate in brief conferences in France, and to make extended visits of laboratories in the country.[40]

The Equipment Grant

The need to refurbish laboratories by acquiring material from abroad was urgent. The heritage of occupation was pitiful and postwar reconstruction was agonizingly frustrating. Consider Joliot-Curie. When France fell, the support that the Nobelist was receiving from the Rockefeller Foundation (FF 60,000 in stipends for a qualified man and an assistant to help on his cyclotron) was interrupted, on orders from the American government.[41] His laboratory was "occupied," but on terms negotiated by Joliot-Curie that gave him considerable authority and freedom. He "worked alongside" Wolfgang Gentner, who had done research with Joliot-Curie from 1933 to 1935 and who was in charge of five German scientists. Together, and with the help of German experts in radio technology and the Swiss firm Oerlikon, they improved the radio frequency system that was plaguing the French instrument. Around 1942, when Gentner was sent back to Heidelberg (where he promptly built his own machine), it was producing a usable beam of alpha particles (a successful example of "technological collaboration" that could have cost Joliot-Curie dearly had the Nazis chosen to build a bomb during the war).[42] However, struggling with a radio frequency system in a cramped basement in the Collège de France, engaging in periodic acts of sabotage to deflect the Germans away from using the cyclotron for studying the fission of uranium, building transmitters and forging papers, and being active in the Resistance, which Joliot-Curie joined in spring 1941, was hardly conducive to scientific productivity.[43] Being a scientist politically engaged in the fight against fascism in occupied France meant something quite different from being a scientist whose opposition to Nazi power was expressed through work on weapons programs at MIT, Chicago, or Los Alamos. It made quite different demands on one's time, had quite different effects on one's research, and posed quite different challenges to laboratory life.

Even if one was determined not to let the war and occupation cripple one's scientific activity, one always lacked resources. Bacteriologist Pierre Lépine, head of the virus service at the Pasteur Institute as of 1940, was committed to using physical techniques like ultracentrifuges and electron microscopy for his work.[44] A Parisian machine-tool firm built him a simple centrifuge in 1941: it was manufactured without the optical system that was essential for visualizing the sedimentation process. In 1943–44 Lépine embarked on a number of electron microscopy projects in collaboration with a French firm that had an in-house capability in electronics, the Compagnie de Télégraphie Sans Fil (CSF), and with support from the CNRS. Any hope of building a device modeled on the latest prototypes available from Siemens or RCA was abandoned "since lamps, stable high-tension current, and know-how could not be assembled."

Conditions after the war were hardly easier. Dominique Pestre has described the difficulties faced by a researcher in Louis Néel's magnetism laboratory in Grenoble—difficulties that everyone had in the first couple of years after the war:

> Many basic materials were lacking (one needed a ration voucher [*bon d'attribution*] to get a liter of oil for a vacuum pump in May 1946), it was not easy to get hold of most metals (aluminum or copper wire), one had trouble with industry (to forge the magnet yoke, for example), the workshop had to deal with electricity cuts (there was no electric power for two days per week during the winter of 1947), and there was a desperate need for experimental equipment: in 1947 a galvanometer was still something almost impossible to find (and even impossible to build).[45]

The gulf between the two sides of the Atlantic after the war was truly enormous; it was clear that if France was to catch up with the best work being done in other countries, notably the United States, a crash reequipment program was required.

To refurbish these laboratories in 1945–46 one had to look beyond the hexagon. Liberation brought little immediate relief to the material suffering of the French people. Life in Paris in summer 1945 was as "dreadfully difficult" as it had been the previous winter: "There [was] a lack of *everything*."[46] Those fortunate individuals who traveled abroad shopped for their families (Rapkine returned home in December 1945 with "dried milk, candy and a bit of chocolate . . .") and, with equal if not greater enthusiasm, for their labs. The head of the French mission in London, with the permission of the British authorities, took home about a ton of journals, books, and instrument catalogs along with a few chemicals. Rapkine ordered a number of scientific instruments (spectrographs, centrifuges, fluorometers, Ph meters, and so forth) for his new lab at the Pasteur Institute using an "installation grant" of $7,000 from the French authorities.[47] Louis Bugnard, who was sent to the United States by the Ministry for Health and Population at about the same time, likewise shopped for "techniques, materials, instruments and books," noting the addresses of instrument makers and the types of equipment that they could supply.

Shopping (and shoplifting) in occupied Germany was another important source of inexpensive (or free) scientific equipment. Joliot-Curie, in his capacity as the director of the CNRS, sent two scientists into Germany in March 1945 along with the advancing French army. André Lwoff, the brilliant Pasteurian biologist, led the operation and was assisted by André Berthelot, a physicist in Joliot-Curie's group at the Collège de France. Their mission was twofold: to establish the state of German research and the quality of the results achieved there during the war, and "to recover as much scientific equipment, machine tools and raw materials as possible to re-equip our laboratories."[48] A delegation from the department of technical education assisted with the operation, as this sector urgently needed machine tools for its training programs. As the French became entrenched on German soil, a large number of researchers from a variety of scientific disciplines volunteered to go a-hunting for the spoils of war. In July Joliot-Curie expanded the scope of the program to include not just the "capture," but also the purchase of new equipment in German factories, equipment that was intrinsically interesting, "thanks to its low cost and also because the French market could not supply it." Until spring 1946 the French paid only the cost of transporting the material they seized. After that, their authorities made the French scientists pay for their equipment (albeit at greatly reduced prices), and they shopped in their own zone as well as in the other allied zones and Austria. By October 1946 about five hundred machine tools had been obtained for technical education, and much important laboratory apparatus, including items like electron microscopes, had been acquired. The value of the material seized as reparation for the looting and destruction of French laboratories was officially estimated to be above FF 100 million (about $800,000).[49] In addition, about FF 8 million of new material was bought, its market value being perhaps as much as four times that figure (that is, about $270,000).

While Germany was being scoured for equipment, New York was being seduced. Toward the end of November 1945, after three weeks of discussions with Rapkine, Weaver wrote a memo to Fosdick spelling out the needs for France as he saw them.[50] He proposed that at least $100,000 be allocated for the purchase of "research apparatus, chemicals and literature." The kind of thing he had in mind was "modern recording spectrographs, Warburg apparatus, Tiselius apparatus, temperature controlled centrifuges, research microscopes etc: to a lesser extent for chemicals and other minor items of glassware: and least of all for a few critically important items of recent literature." Weaver was careful to stipulate that this money would be held as credits in New York and London for buying equipment that France could not make and that its government would be loath to purchase abroad so as to save precious foreign currency and to encourage local instrument makers and chemical firms.

Within a fortnight the estimate for the equipment budget had doubled, as a careful assessment was made of which laboratories could benefit from support. Just before he left early in December 1945, Rapkine and Weaver drew up a list of people and their needs for equipment "of the sort that they [could] not possibly get in France, nor possibly get French external credits for."[51] Rapkine distributed about $200,000 between researchers in chemistry (mostly physical chemistry—total $36,000), physics (including non–atomic energy work by Joliot-Curie, cosmic rays, X-ray crystallography, and magnetic work—$48,000), biology (exclusively genetics, $30,000), biochemistry (including Rapkine himself—$30,000) and physiology ($25,000). Rapkine also felt strongly that support should be provided for biological stations at Roscoff in the north ($20,000) and Banyuls in the south ($10,000) of France, the second being something of a compensatory gesture for the damage American forces had done to the center during their southern invasion of the country.

On returning to France, Rapkine discussed the equipment program with the senior officers in the CNRS, including Joliot-Curie and Teissier (himself a biologist) as well as Irène Joliot-Curie and Pierre Auger.[52] The basic outlines of the project he proposed to Weaver, and the global budget figure, remained intact. They refined the list of fields that needed support and decided on a working figure of thirty-five laboratories receiving, on average, $6,000 each. The privileged sites were seen as those that would be run by particularly energetic senior researchers who could attract a circle of pupils and train them on the best modern equipment available. Joliot-Curie and Teissier reiterated, in their official request to the foundation for support at the end of February, that if they did not buy chemicals and equipment abroad their scientific situation, already lagging the research frontier, "would really be a hopeless one."

Outside aid, they pointed out, had become all the more important after Rapkine came home from New York. To cope with an economic crisis, the franc had been devalued and even tighter controls had been imposed on the flow of money out of the country.[53] Weaver responded generously. The board of trustees of the Rockefeller Foundation, meeting on 3 April 1946, approved his request for $250,000 to be awarded to the CNRS to buy "items of special equipment" for thirty-five laboratories, receiving an average of $7,000 each, doing research in a variety of fields, including physics, chemistry, biology, and physiology.[54] Just three months before, and soon after Rapkine's visit, Weaver was toying with the idea of an equipment grant of $150,000.[55] He had not only increased it substantially, but had actually asked for $40,000 more than the French had requested. The dramatic situation as sketched by Joliot-Curie and Teissier and Rapkine's impassioned plea (the equipment grant "would simply be a godsend")[56] had powerfully affected him. As Weaver himself wrote to Teissier, "I may say that very few times indeed, in my experience as an officer of The Rockefeller Foundation, has a recommendation been made to our Trustees for a sum larger than that requested!" The exceptional measure was justified, he went on, by the prevailing situation in France and the large number of laboratories that needed help.[57]

The Conference Grant

The need to refurbish French laboratories had been uppermost in Rapkine's mind when he visited New York. During his conversations with Weaver, however, another, complementary, issue emerged. Weaver suggested that the foundation could also support a series of small, informal scientific conferences for two or three years. These conferences would be attended by about fifteen "mature" scientists, including a few non-French participants—say a couple of Americans and two or three non-French Europeans. They would last for at least three or four days, but no more than ten, and would allow for extensive informal interactions between the participants. To take full advantage of the presence of visitors on French soil, the Americans would be encouraged to travel around the country for four to six weeks, the Europeans for at least two weeks, visiting various research sites. The Rockefeller Foundation was willing to pay the travel and living expenses of the non-French visitors, while the French themselves were expected to finance the local infrastructure and the costs of French scientists (and younger observers) who attended the symposia.

Weaver stipulated (in the docket prepared for the trustees) that "something like half of the conferences will be held in provincial universities," hoping thereby to decentralize excellence in a country where intellectual life, he felt, was far too dominated by the capital city.[58] He wanted them to be organized around "some modern problem or group of problems, rather than with respect to classical fields." Finally he wanted the visitors to make a point of telling their French scientific colleagues about the most important avenues of research that were being pursued abroad.

The French were unanimously enthusiastic about the plan. The conferences, wrote Joliot-Curie and Teissier, would be an effective way "to provide ourselves from time to time with foreign experts who will bring us up to date with regard to scientific developments." They would help their country overcome the setbacks suffered "through unusual years under enemy occupation" and give French scientists a "good chance of contributing to that international furtherance of science which does not and should not stop at national frontiers."[59]

The CNRS

As already noted, Weaver preferred to distribute grants to select individuals who were pursuing projects that cohered with his overall techno-scientific agenda. The desperate material situation in France and the "complex and delicate" situation inside the community immediately after the war called for a different approach. It seemed premature to offer fellowships to select individuals at this point. As Weaver put it, it looked as though it would take "one to three years before appropriate candidates (men say 25–40 who have definitely demonstrated a capacity for investigative work and who strongly promise to furnish future leadership) [would] be available in any number."[60] Then of course there were the frictions and divisions inside the community itself that Rapkine had described at such length. Certainly, "The troublesome issues of personal tension—who was a collaborator, who was a hero, who was mildly unwise—[were] disappearing rapidly," Weaver wrote to Fosdick, and "the minor hierarchies in the list that begins 'tortured patriot' and ends 'traitor,' [were] being wiped out and forgotten." All the same, some scientists were still "in at least temporary disrepute." This was an intellectual and political morass into which the foundation would step at its peril. Better then to allocate funds to a local organization to distribute among the most deserving candidates. The "new" CNRS seemed ideal for the purpose.

The CNRS was in a state of transition when the officers of the Rockefeller Foundation first engaged with it at the end of 1945. The body was created by a formal decree on 19 October 1939, six weeks after the outbreak of World War II.[61] Its aim was to improve scientific output by coordinating laboratory work, particularly in basic and applied physics research. To this end, it fused two existing bodies, the National Fund for Scientific Research (funding young academic researchers), and the National Center for Applied Scientific Research, set up in 1938 to mobilize science for the impending war. The CNRS survived defeat and occupation, and its mission was refined under Vichy as bending science to production.[62] In 1944 Joliot-Curie replaced Charles Jacob, a professor of geology at the Sorbonne, who had headed the CNRS under Vichy. The nuclear physicist's outstanding scientific credentials mattered of course. So too did de Gaulle's wish to avoid a confrontation with the French Communist Party (PCF) and the Conseil Nationale de la Résistance (CNR), the umbrella body of several Resistance organizations that was calling for an administrative "cleansing."[63]

Within a week of the liberation of Paris, on 24 August 1944, Joliot-Curie called together leading members of the French scientific community for a series of meetings to define the policy of the new CNRS.[64] He was faced with choosing between two broad conceptions of how France was to move into the future. One, promoted by the CNR, demanded that the bankruptcy of the Third Republic and the shame of Vichy not be forgotten, and that those who had been complicit in either be duly condemned. The other was the vision promoted by de Gaulle, who reduced Vichy to a localized and unrepresentative aberration and called for national reconciliation and the need to discreetly overlook wartime behavior in the pursuit of national unity.[65] Joliot-Curie chose the latter path, notwithstanding his outstanding credentials in the Resistance and his membership of the PCF, which dated from around May 1942.[66] Thus the outline of the structure of the governing board of the CNRS, defined on 7 September 1944, specified its composition as comprising "personalities who were young in spirit and of recognized achievement, [and] who wanted to collaborate in a common enterprise in the best spirit of national unity." Ten days later Joliot-Curie publicly stated that all men of science should work together, and that the backsliding of those who had shown some sympathy for Pétain, like his predecessor Jacob, should be overlooked. For Joliot-Curie, science and technology were to be forces for the reconstruction and independence of France, instruments that would give back France "its grandeur and its liberty."[67] To that end, the scientific community should seek reconciliation, not revenge, stigmatizing only the most abysmal agents of Vichy.

Joliot-Curie had a well-defined intellectual agenda for the CNRS. Like many others at the time, he was determined to overcome the deep schism in interwar France between pure and applied research, between science and industry.[68] He admired the centralized organization of science in the Soviet Union and believed that in France too science should be directed "by a prestigious and competent central body" that would "ensure that important fields would not be neglected and that useless duplication of research would be eliminated."[69] The new CNRS agenda also reflected Joliot-Curie's nationalism and patriotism: he believed that the pure/applied divide was one reason for the ignominious defeat of June 1940.[70] Joliot-Curie actively courted bankers and industrialists during the Vichy regime with a view to establishing close links between science and industry in the postwar period.[71] As soon as he took the reins of office, he stressed again that those responsible for the various fields to be supported by the CNRS should favor links between fundamental and applied work, should engage the universities in preparing students for a research career, and should promote interdisciplinary studies on key problems. He also insisted that it was important for researchers to have well-trained and highly motivated specialized technical support in the laboratory—hence his determination to get as many machine tools as he could from occupied Germany in 1945. Above all, as Girolamo Rammuni puts it, Joliot-Curie in September 1944, "was driven by a sort of urgency to break with the immobility of the past."[72]

A year later, when the Rockefeller Foundation began to think about supporting French science through the CNRS, Joliot-Curie was already on the way out. His vision for the CNRS (which depended on the Ministry of Education) as evolving into a sort of overarching Ministry for Science crumbled as one independent organization after another devoted to basic and applied research in different fields, and financed by different ministries, was established in postwar France.[73] In addition, the predominance of university people on its committees, and an economic conjuncture that seemed to favor applied research, led members of the CNRS to lay increasing stress on the importance of basic science. On 3 January 1946 Joliot-Curie took office as the high commissioner of the newly established Commissariat à l'Énergie Atomique (CEA), an institution in which his desire to fuse science and engineering to promote French grandeur and independence was more likely to be satisfied. In February 1946 Georges Teissier, a leader in the French Resistance, the holder of a chair in zoology in Paris, and a man who was convinced of the need to promote basic science to "prepare the future," succeeded him as director.[74] For Teissier, it should be said, that future was Communist-inspired, a civilization in which research would be genuinely valued by the state and no longer "under the control of [capitalist] monopolies for whom the profit motive alone counts."[75]

The CNRS had two main functions, to carry out its own research and to assist research at universities. Its mission, as Teissier said in 1948, was to support the cutting edge and to provide particularly expensive equipment, "not to pay the Museum's gas bill or buy test-tubes for the Collège de France."[76] Thus the provisional 1946 budget, established in October 1945, planned not simply to extend and reequip some major existing research laboratories, but to develop a new "Center for Pure Research" at Gif-sur-Yvette, south of Paris.[77] This center would feature modern equipment in an environment conducive to scientific work at a location easily accessible from the capital. The laboratories would house researchers in a variety of disciplines—physics, chemistry, biology—and also in genetics, which would be the first to be set up. Weaver was particularly struck by this development. Writing to Fosdick about the plans for Gif, he remarked that there would be an "institute of genetics!" adding that "[t]he exclamation point is justified by the fact that up to now, there has never been so much as a professor of genetics in France."[78] It was symptomatic, in his eyes, of the break with the past that Joliot-Curie and Teissier were promoting.

The Rockefeller Foundation's officers, surveying the CNRS and initiatives like the Gif plan, were impressed with the spirit of renovation, the desire to place the organization of research on a new footing and to break with the rigid and inflexible structures of the past. The body "had assumed a position of real national leadership in science," wrote Weaver, and was representative of all shades of political opinion, from the Communist Joliot-Curie to the "ultra-conservative" Maurice Caullery, biologist and president of the Académie Française.[79] He liked the division into about thirty-five sections that covered a wide variety of fields and that were "not based on old classical divisions of science." The section on genetics, for example, included biologists, chemists, and doctors. He was impressed by the determination of the officers to "build up the strength and prestige of the provincial universities, against the classic tradition that nothing really counts in French intellectual life but Paris." Research groups in Strasbourg, Lyon, Bordeaux, Marseille, and Nancy—all were seen as legitimate candidates for support. Weaver also remarked that many of the leadership were "youngish" and that many of them (in the fields of interest to him) had held foundation fellowships. Even the departure of Joliot-Curie did not bother him particularly. Teissier had given "an excellent impression personally," and "with all the subsidiary committees formed of such excellent men," Weaver felt "fairly certain that the Centre [was] firmly launched" by January 1946.[80] It was an ideal instrument for the reconstruction of the dynamic outward-looking French scientific community the foundation hoped to foster.

Ensuring that the French Scientists Wanted the Grants

The discussion inside the foundation regarding the wisdom of supporting French science was generally positive. Doubts about the weakness of the franc and the movement at the top of the CNRS were easily quelled. Still, one fundamental concern remained for the senior staff. They feared that a proposal that the foundation support the reconstruction of French science would be resented and seen as an insulting act of charity rather than as a gesture of solidarity. Rapkine had made much of French pride, and Weaver was well aware of the danger of stepping on their toes. So was Alan Gunn, head of the Medical Sciences Division, and John Marshall, the associate director of the Humanities Division. To Weaver Marshall wrote that he was "still a bit inclined to expect that, to regain the prestige that they feel they have lost, they may prefer to do such things from their own resources. . . . My only hesitation, then, concerns our moving in only when it is certain that RF aid is wanted."[81] Weaver concurred. In November he had told President Fosdick that it would be made clear to the French that the foundation would only consider supporting projects that "completely correspond to their own enthusiasm [that is, the French's]."[82] He underlined the same point early in a memo in January 1946: "Louis Rapkine understands very clearly that, in talking of these matters [back in France], he is to be most discreet in suggesting these actions as possible one's to discuss, just providing the French officials are, *without any question* tout à fait d'accord. Otherwise, these suggestions are to be dropped, and they will offer alternative ideas."[83]

This concern was exacerbated because the grant was not going to an individual but to an organization that was part of the state apparatus. This was doubly dangerous for the foundation. They were handing over the authority to make awards to another body whose choices could provoke a backlash not just against the CNRS itself, but against the foundation too. In addition, since funds were being disbursed by a source that represented the French government, the grant risked not only provoking the kind of individual feelings of shame at accepting charity that Rapkine had felt, but of inflaming nationalistic sentiments.[84] The Rockefeller Foundation wanted to avoid an escalation of the stakes and a politicization of its award at all costs.

The official reassurances that the foundation sought regarding the projects came in a letter received in New York early in March. It was cosigned by Teissier as director of the CNRS and by Joliot-Curie in his capacity as former director. They explained that they had discussed the results of Rapkine's visit to New York at length with him and had asked him to present the proposals that were in the air to all the section heads of the CNRS. The conclusions reached were thus representative of researchers in France. They had given their "total and unreserved approval" to the proposal to support the purchase of essential modern equipment and to pay the costs of foreign visitors to about thirty conferences and symposia over the next three years. To emphasize the point, Joliot-Curie and Teissier added, "Each of these projects is important, but one could say that taken separately they are insufficient, but taken together they supplement each other admirably." It was perfect ammunition for Weaver's justification for the twinned grant.

In presenting his paper to the trustees, Weaver noted that he was not creating a precedent: in November 1945 they had supported a special emergency grant in aid of $80,000 for medical and natural science laboratories in the Netherlands.[85] Now it was France's turn. Her scientists, who had been isolated from the rest of the world "in an atmosphere of secrecy and fear" and who had returned to ravaged laboratories, desperately needed two things: "They must have something to work with; and they must have their spirits refreshed, and their scientific ideas reoriented through contacts with the rest of the world." An equipment fund would satisfy the material need, a conference fund would provide the spiritual complement. The new CNRS would be the beneficiary of the resources and was likely to do "an outstanding job." The trustees voted $250,000 for the former and $100,000 for the latter, to be used by 30 June 1949. This was a substantial contribution: the entire CNRS budget in 1946 was only about $4–5 million, so the foundation's grant amounted to almost 10 percent of all the funds it had at its disposal.[86]

The grant was received with enthusiasm in France.[87] Teissier wrote both an official letter to the foundation and a personal letter to Weaver to thank them on behalf of the CNRS and of "the whole of French science" for their generosity. The grant, he said, "would help us to overcome current difficulties and to make up as quickly as possible the time lost by long years of oppression."[88] *Le Figaro* was similarly positive. "Such a gift, so disinterested," it wrote, "which—without any rhetoric—serves all humanity, is in no way humiliating." France was so poor on account of the war, *Le Figaro* went on, that it could not support its own scientists adequately: "This was not a poverty to be ashamed of. And it is with no embarrassment, but with true gratitude," the article concluded, "that we thank our friends in America."[89] This was not "charity," but a gift that helped a nation's science shake off the postwar legacy of years of oppression.

Weaver was careful to point out to Teissier that the foundation left the CNRS full discretion as to how the grant was spent. Of course he would like to know what scientists or laboratories were eventually chosen for equipment support but was "content to leave all matters of choice to you." Similarly, as regards the conference grant, "Here again we would wish the officers of the CNRS to have complete freedom and responsibility in choosing the topics for these conferences and the men who are to be invited."[90] This gesture was all the more remarkable in that Weaver knew quite well that some of the money would be used to support research in fields that were not usually priorities for the foundation, notably areas of physics. This laissez-faire approach had three advantages.

First, it shielded the foundation from accusations that its interests, and not those of French science, were behind the grants. *Le Figaro*'s assurance that the American award was made "without any rhetoric" and to serve "all humanity" said it all. The Rockefeller Foundation had no ulterior motives, it was "disinterested," and it certainly was not an arm of the American government or an instrument of American foreign policy. Correlatively, if the foundation had tried to influence the choice of topics and people to whom the money went, it could easily have been accused of selectively promoting certain fields when all needed support, and of steering French science along directions that were coherent with its preferences and not necessarily with the needs of the French themselves.

Second, having been apprised by Rapkine of the "complex and delicate" situation inside the French community, it was obviously wise not to try, just yet, to favor some people over others. Political mistakes could be made. Jealousies could be aroused. And the foundation could jeopardize the image of fairness and objective detachment on which its credibility rested inside Europe. If there were any criticisms to be made of how the funds were disbursed, those criticisms could only be leveled at the French community and would not reflect on the policies or prestige of the foundation itself.

Finally, the policy cohered with the overall aims of the foundation and of Weaver in particular. The grants were not meant to promote a particular line of research, as generally happened with individual grants, but to reorient the French scientific community as such and to integrate it better into the international scientific milieu. The instrument grant, intended "at

least in major part, to purchase certain critical items of modern equipment," would give them the means to tackle questions and to develop
hypotheses using the latest techniques available in the United States and
other European countries. The conference grant would help them build
personal contacts with leading non-French scholars in the field.[91] The two
awards were intimately coupled, in that a fruitful discussion between
equals was only possible if the parties shared material techniques and
practices and addressed themselves to related problems. "Science is
rooted in the experimental method, of course," one foundation report
noted, "but unless the experimenters are able to communicate their
results to others—unless they can meet with their peers in research and
discuss their findings, obtain the benefit of criticism, be put on their met-
tle to defend their interpretations and conclusions—their work is likely to
suffer the defects that usually afflict work done in a vacuum."[92] The
grants were intended to break the "vacuum" in which much of French
science had traditionally been done and to bring down the walls of the
fortress behind which French science had (allegedly) protected itself.

Spending the Money

The officers of the CNRS rapidly allocated the bulk of the equipment
grant made available to them by the Rockefeller Foundation, leaving a
contingency fund of $5,500. Table 4.1 gives the distribution between sec-
tions and group leaders, using data provided by the CNRS and by Rapkine.[93] In sharing out the money, the CNRS said it concentrated on
laboratories that were expected to expand rapidly, those in which there
was already an active nucleus of researchers, and those outside Paris that
seemed particularly promising and worth encouraging. This was in line
with what the foundation had wanted. Support for the medical sciences
was somewhat delicate, since it was normally the responsibility of the
Medical Sciences Division inside the foundation.[94] The CNRS treaded
carefully here: they had to make a gesture to some of the medical
research people who had worked hard on the committee but did not
want to trigger a huge number of requests from others in this area. The
solution they came up with was, apparently, to limit grants in medical
research as such to $6,500 in total.

Rapkine told Weaver that the atmosphere at the meetings at which the distribution was made was good, and that even one though a couple of grants were not based entirely on scientific merit, "one or two doubtful cases out of thirty-seven, [was] not too bad for a human undertaking."[95] For their part, foundation officers Dusty Miller and Gerard Pomerat, a newly recruited assistant director in the Natural Sciences Division with previous experience in the Pasteur Institute, were enthusiastic about the result. About the CNRS directorate, Pomerat wrote, "They have attempted to be just and impartial, they have weighed the general and collateral problems carefully, they have shown themselves willing to arbitrate and to concede whenever it seemed essential or politically desirable, and they have responded almost entirely to every little suggestion which we made." Indeed, Pomerat and Miller were of the view that "these actions come as near to being an ideal solution as we could have hoped to obtain after long and careful investigation."[96]

By and large the French scientific community seemed satisfied with the grants; at least there are few complaints recorded in foundation officers' reports (though it must be said that they strongly discouraged researchers gossiping to them about differences within the community, so even if there was dissatisfaction they would be the last to know about it). The biochemist Claude Fromageot, as mentioned above, felt that Rapkine had far too much influence. Even though a beneficiary himself, he was not keen on the idea of using the CNRS as a conduit for foundation money, fearing the "French love of paper work and the political angle." Oleg Yadoff, an ex-White Russian partly trained in France and then at Columbia University, was also persuaded that political affiliation, not scientific merit had played a big role in the distribution of the awards.[97] A few years later, with the Cold War in full swing and the foundation under pressure from virulent anti-Communists in the U.S. Congress, these accusations sparked some concern, as I discuss in the next chapter. Now Pomerat resented the criticism and ignored the charges.

The Rockefeller Foundation had wanted the equipment grant to be spent on essential modern instruments and special chemicals, rather than on basic infrastructure like laboratory benches or ordinary glassware and consumables that could be obtained in France. Spectrophotometers, stereoscopic microscopes, optical pyrometers, temperature

Table 4.1
Beneficiaries of Equipment Grant in CNRS

	Researcher	Field	U.S.$	Location
I	F. Joliot-Curie	For non–atomic physics work only	6,500	Paris
	I. Joliot-Curie	For non–atomic physics work only	6,500	Paris
	F. Perrin	Theoretical electronics	6,500	Paris
	Cabannes	Spectroscopy and ultrasonics	10,000	Paris
	Auger	Cosmic rays	3,000	Paris
	Cotton	Magnetic reactions	6,500	Bellevue
	Wyart	Crystallography	6,500	Paris
	Néel	Magnetism	6,500	Grenoble
	Kastler	Optical physics	6,500	Paris
II	Lafitte	Kinetic chemistry	6,500	Paris
	Letort	Chemistry	6,500	Nancy
	Prettre	Physical chemistry	6,500	Lyon
	Kirmann	Organic chemistry	6,500	Strasbourg
	Audubert	Electrochemistry	6,000	Paris
	Sadron	Large molecules	6,500	Strasbourg
	Bauer	Physical chemistry	10,000	Paris
	Tréfouël	Director, Pasteur Institute	5,000	Paris
	Champetier	General chemistry, polymers	3,000	Paris
III	Teissier	Marine biology station	18,000	Roscoff
	Ephrussi	Genetics laboratory	18,000	Paris
	Grassé	Evolution	6,500	Paris
	Courrier	Endocrinology	2,000	Paris
	Lacassagne	Radium Institute	3,000	Paris
	Wolf	Embryology	6,500	Strasbourg
	Benoit	Physiology	5,000	Strasbourg
	Mangenot	Biology	6,000	Paris
IV	Wurmser	Physico-chemical biology	6,500	Paris
	Lwoff	Pasteur Institute	6,500	Paris
	Aubel	Biochemistry	6,000	Paris
	Fromageot	Biochemistry	6,500	Lyon
	Roche	Biochemistry	6,500	Marseille
	Rapkine	Pasteur Institute	6,000	Paris
	Fessard	École Pratique des Hautes Études	2,000	Paris
	Schaeffer	Biology	6,500	Paris
	Cahn	Physico-chemical biology	6,000	Paris
	Terroine	Food and nutrition	6,000	Paris
	Grabar	Pasteur Institute	6,000	Paris

baths, potentiometers, distilling apparatus, Warburg apparatus, Waring blenders, vacuum pumps, special chemicals, and so on were all ordered. But so too were tygon and rubber tubing, resistors, capacitors, drawing instruments, and even white ruled cards and binders for a filing system, which were unobtainable due to a paper shortage.[98] The material situation in the country was still so serious at the end of 1946 that Weaver's requirement could not be respected to the letter. He did not object.

Doris Zallen has listed the conference grants, and I shall not duplicate her work here.[99] Weaver's requirement that they be limited to fifteen or so participants was shattered at the first conference on optics in October 1946, which was attended by 360 delegates. So too was the suggestion that participation be restricted to the United States and a few countries of Western Europe. Delegates from twenty-six or twenty-seven nations, including Czechoslovakia, Poland, Romania, and Hungary (though no one was invited from Austria, Germany, or Spain), were at the conference. The scientific credentials of the foreign invitees were outstanding, and several meetings played host to a galaxy of Nobel Prize winners. The list of delegates to the meeting organized by Bauer on the chemical bond (April 1948) reads like a who's who of the world's leading physical chemists at the time: Max Born, Hendrik Casimir, Charles Coulson, Egil Hylleraas, Christopher Ingold, Jan Ketelaar, Nevill Mott, Robert Mulliken, Linus Pauling, Ilya Prigogine, Nevil Sidgwick, and Y. K. Syrkin.[100] A conference in March 1950 on fundamental particles was attended by no less than six Nobel Prize winners in physics: Patrick Blackett, Niels Bohr, and Erwin Schrödinger from Europe; Enrico Fermi, Isidor I. Rabi, and Hideki Yukawa from the U.S.[101] Those invited also included such distinguished scientists as John Wheeler, Robert Marshak, Robert Serber, Edward Teller, Eugene Wigner, John von Neumann, Gregory Breit, and

Table 4.1 (continued)
Source: "Liste des bénéficiaires des subventions pour achat de matériel étranger," unsigned, undated, RFA RG 1.2, series 500D, box 3, folder 31. See also Gerard Pomerat Diary, "19 May 1946 (Paris)," RFA RG1.2, series 500D, box 3, folder 31.
I: Physics, Astrophysics, Geophysics, and Mineralogy ($58,500); II: Chemistry and Physico-chemistry ($56,500); III: Animal and plant biology ($65,000); IV: Physiology, Biophysics, Biochemistry, and Microbiology ($64,500).

Samuel Goudsmit, all from the United States, and Homi Bhabha from India. Later that year, in July 1950, John Van Vleck (Nobel Prize for physics, 1977) and Roman Smoluchowski, from the United States, attended Louis Néel's conference on ferromagnetism and antiferromagnetism in Grenoble. Other guests included three of the most outstanding researchers at Bell Laboratories, including William Shockley, and most of the leading British and Continental researchers in the field.[102] The American economists and (later) Nobel Prize winners Milton Friedman and Paul Samuelson were both invited to a meeting on risk theory in econometrics organized by Darmoy, Roy, and Allay in 1952.[103]

The conference grant was for three years (that is, to cover expenditure up to June 1949). It was eventually extended for another three years, as all of the money had not yet been spent, whereupon it was supplemented by a new grant of $40,000 to be available as of 1 July 1952.[104] The equipment grant, by contrast, was not renewed after its funds were exhausted. The foundation preferred to go back to its traditional method of awarding grants to individual researchers that its officers deemed worthy of support.

The Rockefeller Foundation and the Coproduction of Hegemony

The ravages of war provided the Rockefeller Foundation with an opportunity to intervene in continental Europe in 1945 to rebuild its scientific capability. Within months of the cessation of hostilities the trustees agreed on a "Special Emergency Grant in Aid Fund" to refurbish medical and natural science laboratories in Dutch universities and technical schools.[105] Six months later it was France's turn. "With German science under a shadow for an unpredictable number of years," as Weaver put it discreetly to the trustees in April 1946, "France offer[ed] the one continental European opportunity for a large and important development of the natural sciences."[106]

There were two complementary dimensions to the project, an equipment grant and a grant for international conferences and workshops As for the latter, Weaver was "convinced that nothing else would do as much to stimulate, rejuvenate, and *re-orient* French science."[107] This "reorientation" was partly in terms of scientific approach: Weaver wanted to encourage problem-based rather than disciplinary-based research in France, in line with his general strategy as director of the Natural Sciences Division. Hence the kinds of topics he suggested for the meetings included chemical genetics, protein structure, enzyme chemistry, cellular physiology, recent advances in statistical techniques, magnetic theories, and the structure of metals. However, his aims went far beyond that. Weaver also wanted to take advantage of the fluid situation in France to change the community's attitudes and values, to transform it from being inward-looking and parochial into a full-fledged participant in the international scientific enterprise.

The French scientific community was in flux immediately after the war. Leaving the country after 1940 meant avoiding the humiliation of occupation and the risk of dying at the hands of the enemy and having the chance to work close to the research frontier in several fields. Staying behind had meant personal discomfort, danger or death, and vanishing opportunities for research. Rivalry and resentment were inevitable, and Rapkine for one did all he could to keep their divisive consequences in check (out of love for France, but also to ensure that those who had left the country, like himself, could hope to build a scientific life there after the liberation). For the foundation, the fact that many fine scientists had escaped abroad during the war provided a rare opportunity to reconfigure the community as a whole. Since some of her best scientists had worked in Britain and America, where they had learned to speak English, it "seemed likely that France [thus] catapulted out of her traditional insular attitude by the experience of the war, might welcome a chance to reestablish scientific activities on a somewhat more international basis than had been characteristic of her previous history." Their eye-opening experiences during the war "had taught many of the [French] scientists that cooperation and communication with the rest of the scientific world was a real necessity."[108] War had not been an unmitigated disaster for the French, then; on the contrary, it had also provided them with an opportunity to rebuild their community afresh on a new basis.

When he made his grants in 1946, Weaver had a picture of the French scientific community as being "insular and self-satisfied."[109] It was hidebound by tradition, inward-looking, dominated by a gerontocracy who perpetuated classical lines of study, who were concentrated in the metropolis at the expense of the provinces, and who spoke only their native tongue. It was a picture he hoped to change. If he was so impressed by the new CNRS, it was because he thought it could be the basis for the "development of French science much more modern, international, and balanced than that which existed before the war—distinguished in certain fields, weak in others, and often provincial in spirit."[110]

Weaver's antipathy to the centrality of Paris and his insistence that the awards be distributed geographically with a view to strengthening alternative and rival research poles reflected his conviction that dynamism was fostered by internal competition. His aim to transplant this American conception of how science was best organized into French soil indicates the extent to which his envisaged reform of French science was inspired by, and intended to reproduce, an American model.[111] He was particularly impressed with the younger people in senior positions in the CNRS, many of whom had received foundation support in the 1930s. These people had studied outside France, "had a knowledge of and sympathy for science in other countries," and were eager for outside contacts, "quite unlike the older bearded académicien." In short, Weaver's grants were not only intended to help reequip French laboratories desperately in need of modern instruments and technologies. These dispersed, competitive, newly equipped research sites were to provide the material infrastructure around which the French scientific community would be restructured internally, and reoriented outward.

The conference grants were an integral and essential part of the process of reintegration into the international community. In fact, Weaver hoped that the French, now properly equipped, would align their research programs around the kinds of problems of interest to the foundation and that were being pursued in other countries. In the docket accompanying the grant request to the trustees, Weaver said that it was desirable that visitors from United States, as well as Britain and Scandinavia, be supported. It was they who would go to France at the foundation's expense, attend the conference, and travel around the country for a week or two discussing "the most fruitful lines along which work can now proceed."[112] The "international" community that Weaver wanted the French to join was essentially the Western scientific community, of which the United States had emerged from the war as the undisputed leader.

It is instructive that when Weaver persuaded the trustees to make an $80,000 equipment grant to the Dutch, he did not think it necessary to supplement it with a conference grant. But then the Dutch scientific community did not need to be "reoriented" either. As one Dutch physicist put it in 1948, "The Dutch have always been very internationally-minded, accustomed to traveling all over the world."[113] In France, by contrast, the conference grant tapped into the yearning of "long-isolated Frenchmen hungry for restoration of their citizenship in the world republic of science and eager for knowledge of the frontiers of discovery."[114] According to a grant evaluation made for the trustees in 1952, the conferences organized in France had played "an important role in overcoming the traditional French reluctance to visit abroad (particularly in the United States) or to seek a really strong international rapport."[115] If this enthusiastic report is to be believed, Weaver had achieved some of his objectives.

There is a spirited debate in the literature on how to conceptualize the relationship between foundations and their beneficiaries. The challenge is to grasp both the objective power that the officers and the trustees had over how to spend the foundation's money, and the role of informal, face-to-face negotiations between officers and potential grantees in which together they crafted a proposal that was likely to be supported. Coproduced (or consensual) hegemony does this task admirably, neatly encapsulating the complex bond between boardroom and field that is mediated by the foundation's officers. It recognizes Pnina Abir-Am's demand that the asymmetry of power between officer and grantee be respected, but it also allows for Robert Kohler's emphasis on the "intimate partnership" that characterized an informal relationship that was based on mutual respect and trust.[116]

I am not the first to describe the relationship between foundations and the sciences as hegemonic. In the 1980s this concept was used at the level of "grand theory" to argue that the support of the social sciences by the Carnegie, Ford, and Rockefeller foundations served both "the reproduction and production of [Gramscian] cultural hegemony," and to perpetuate a system of "cultural imperialism."[117] The foundations' aim, it was

argued, was "the preservation and maintenance of the social order," a capitalist social order that embodied the interests of a ruling elite whose
"sophisticated conservatism" was shared by both officers and trustees and
that supported changes "that help[ed] to maintain an international system
of power and privilege" against the interests of minorities, the working
class, and third world peoples.[118] Subsequently Abir-Am (using Foucault)
and Lily Kay (using a "Foucaultian and neo-Gramscian framework") have
related the Rockefeller Foundation's "Science of Man" program in the
1930s and its implementation in Weaver's promotion of a biology "ration-
alized" by the use of physical techniques to this big picture.[119] What these
meta-level theories accentuate is the power that foundations have to shape
the social order along lines congenial to their officers and trustees. What
they miss is the dynamic, face-to-face interactions of officers and their
scientist-beneficiaries, that actor-centered microhistory that Kohler's "intimate partnership" seeks to grasp and that is embodied in the notion of
coproduction.[120]

The relationship between the foundation and the French scientific
community was hegemonic. The desperate need in postwar France for
scientific equipment and for a renewed contact with the research frontier
created an opportunity for the Rockefeller Foundation and Warren
Weaver to actively intervene in the postwar reconstruction of French science. Certainly one would balk at the use of the concept of hegemony if
the grants were *simply* inspired by the wish to provide essential instruments and materials and to help pay for international conferences on
French soil. But Weaver wanted to do more than to revitalize; he also
wanted to use his economic leverage to reorient and to reconfigure
French science. His grants were made against the backdrop of a view
(which he and his officers did not divulge to the French, of course) of
what the French scientific community was and how it ought to change.
Crudely put, that view used a conception of the American scientific community—polycentric, richly endowed with the best equipment, at home
in English, outward looking, and highly competitive—as a benchmark
against which to measure the organization and internal dynamics of
French science. He wanted the French to put what he saw as their
parochialism behind them and to become full and respected members of
an "international" scientific community on the American model.[121]

To achieve his objectives, Weaver needed an interlocutor whom he could trust, someone who was in touch with the situation in France and yet detached from it. Rapkine was the ideal candidate. His intimate personal relationship with Miller and Weaver was the platform on which the coproduction of hegemony turned. As Rapkine carried messages back and forth in 1945 and early 1946, he was reassuring the French that the foundation would do nothing without their agreement (and that Weaver expected to be invited to help them along the lines that he had suggested). He was also reassuring the foundation that the French would welcome them with open arms (and would present the program as if it was entirely of their own making). As everyone knew, this game masked the imbalance in power between donor and recipient and made of the French willing and grateful colluders in Weaver's scheme. They "invited" the hegemon into their fragile community because he had the means to provide them with the basic nourishment they needed and because he was willing to let them disburse his money as they chose. The constraints were few and easy to respect. The Cold War changed that.

5

The Rockefeller Foundation Confronts Communism in Europe and Anti-Communism at Home: The Case of Boris Ephrussi

The specter of Communism haunted American policymakers in Western Europe in the years after World War II. As pointed out in chapter 2, the long-term stability of a liberal democratic order in the region was seen to require the "double containment" of Germany and the Soviet Union and the creation of an economic and political regime that cut the ground from beneath domestic Communist parties. France had a key role to play in this process. It had been of paramount strategic importance in two wars in which aggressive neighbors had threatened U.S. interests and "twice in twenty-five years the United States ha[d] been compelled to pour out its blood on French battlefields."[1] Now, once again, its stability was threatened—this time from within, by a well-organized and militant Communist Party. In elections held on 2 June 1946, the French Communist Party had received five million votes and had been supported by over 25 percent of the electorate. What is more, the party leadership, according to the State Department, was "extensively influenced and inspired, if not dominated" by the Soviet authorities and was extremely suspicious of, if not downright hostile to, an American presence in Europe. France had become a bridgehead for Soviet penetration into the democratic West through a manipulated and obedient Communist Party:

Today, though at peace, France is the scene of an internal political battle, the outcome of which is of the greatest importance to the United States, and whose ramifications will extend to all the countries of Western Europe. The world drama of Russian expansion is being played in miniature on the stage of France.[2]

The United States had to counter this offensive and creatively help build a France that was "friendly and strong, serving as a bulwark of our security, our concept of democracy and our material interests on the continent of Europe."[3]

The officers of the Rockefeller Foundation were aware of the growing divisions inside Europe and of the dangers that Communist power posed to American interests on the Continent. Echoing the sentiments of the State Department, in October 1946 J. H. Willits, the director of the Social Sciences Division, wrote of the situation in France in almost identical terms:

A New France, a new society is rising up from the ruins of the Occupation; the best of its efforts is magnificent, but the problems are staggering. In France the issue of the conflict or the adaptation between communism and western democracy appears in its most acute form. France is its battlefield or laboratory.[4]

A few months before, Warren Weaver had agreed to make a major grant to the CNRS with the hope that it would rejuvenate French science and anchor a community that he regarded as parochial and inward-looking in the international scientific community. The presence of Communists in key positions in France, notably Frédéric Joliot-Curie and Georges Teissier, the two men with whom he had negotiated the grants, did not bother him particularly. But in 1947, as the fault lines of the Cold War began to emerge, the foundation's officers could not remain indifferent to the new political situation, which had repercussions for science both at home and abroad and posed serious dilemmas for them.

The Rockefeller Foundation's officers believed that the most fruitful way to achieve their objectives was to promote good science. Their view of who merited support combined scientific judgment with informally gathered information from peers and face-to-face contact with potential applicants. In this process, the political or ideological beliefs of scientists were irrelevant. Indeed, they were embarrassed and angry when interviewees brought such matters up. The politics of researchers had no bearing on their ability to do good science—at least in principle. Indeed, although Raymond Fosdick insisted, as noted earlier, that the foundation's "assistance is given without regard for race, creed, color or political opinion," he could not but add, "It is a question, however, how far the new conditions of modern life qualify the application of this principle."[5]

As noted in the previous chapter, the principle had already been overridden to suit local contingencies in postwar France. Rapkine gave

Weaver a list of politically dubious French scientists, men who were deemed to have had an unduly close relationship with the occupying enemy, and the officer understood at once that if the foundation was to be respected in France, it had to have nothing to do with these people. Their wartime behavior alone disqualified them from receiving grants. This decision was easy, however, since the Nazi regime was thoroughly discredited in 1945, and those intellectuals who had supported it were widely despised. The Cold War posed dilemmas of quite a different kind.

The basic problem for the foundation was that there very different perceptions on the two sides of the Atlantic regarding the threat that Communism posed to science and to the social order. Immediately after the war the Communist movement was viewed by many on the Continent as having an honorable historical legacy, as socially legitimate, and as being compatible with the functioning of a democratic society. Many Europeans did not share the deep distrust of Communism that was so widespread in the United States, nor did they regard it as necessarily subversive. In France, moreover, many intellectuals were Communist or at least left wing. To retain credibility in Europe, Weaver and the foundation's officers had to respect and work with these features of European scientific life. At the risk of provoking anti-Americanism, they also had to show Europeans, and the French scientists in particular, that they were independent of the American government and that they were not slaves to U.S. foreign policy.[6] At the same time, to maintain credibility in the United States they had to be sensitive to the political balance of forces in their own country and the attitudes current in the Truman administration. As the Cold War picked up momentum and the tide of anti-Communism grew in the United States, the foundation's officers found themselves having to choose between retaining a credible posture abroad, where fears of Communism among scientists were far more muted, and retaining trust and support at home, where concerns about the loyalty and security of scientists grew increasingly strident.

In this chapter I chart the changing attitudes inside the foundation regarding the pertinence of the Communist or left-wing sympathies of researchers to whether or not they should be given support, treating in depth the case of one of Weaver's most favored European fellows, the French geneticist Boris Ephrussi. I show how the domestic context in the

United States, where fears about subversion, security, and loyalty came to be of overwhelming importance, had a major effect on the foundation and, by the early 1950s, played an important role in decisions about whether or not to make awards. The foundation's officers and trustees accepted the constraints imposed by domestic concerns about Communist infiltration, which they shared up to a point, and did not want to alienate the American authorities. Accordingly, they made a deliberate political decision not to support Communists abroad. Fosdick's ideal was refashioned to meet "the new conditions of modern life."

This step was not taken lightly. Indeed some officers anguished over the clash between principle and practice—and solved it by doing some important "boundary-work" (the concept is Gieryn's) of their own. They dissolved their dilemma by mapping the science/nonscience demarcation onto the democracy/totalitarianism distinction. In an intellectual world carved up in this way, which was of course not unusual at the time, a Communist could not be a man of science. Thus he was not even eligible for a grant. Denying him a grant did not mean that political views had interfered with an assessment of his merit; it meant that his political views had destroyed his merit altogether. In foundation officers' eyes, at least, merit was still the measure. They could sleep comfortably at night, noble principles intact and fears of being accused of supporting subversive activities laid to rest.

The First Steps toward the Politicization of Awards

Political and ideological considerations were not uppermost in the minds of the Rockefeller Foundation officers when they made the major grant to the CNRS in April 1946 that I described in the previous chapter. They uncritically accepted the standards of legitimacy adopted by the French scientists in the CNRS. In the eyes of the foundation's officers, the political hue of senior officers in the CNRS was a French affair. What mattered was that the organization distributed its funds on the basis of merit and need, not political affiliation or ideological belief—with the exception of the shamed collaborators.

Joliot-Curie's Communism did not bother Weaver as long as he stayed in France; when it came to visiting the United States, however, where

Communism was already being demonized, he had to be more prudent. This issue arose at once in the early negotiations with Rapkine. In one of his very first interviews with Weaver, in November 1945, Rapkine reported that Joliot-Curie and his wife, Irène, the daughter of Marie Curie, were extremely keen to come to the United States for two or three months at the invitation of the foundation. In February 1946 Rapkine asked again whether Weaver had managed to "get something started" regarding the clearance by the State Department of the couple.[7] That was not going to be easy. General Groves, who had headed the Manhattan Project, and British Premier Winston Churchill feared that Joliot-Curie would spearhead an independent French nuclear program and share nuclear secrets with the Soviets. After liberation, every effort was made to keep the scientist away from his compatriots who had worked in Canada.[8] The Americans destroyed a cave containing material for a small nuclear pile, situated in what would soon be the French zone of Germany, just before the French Army reached it.[9] Joliot-Curie too made it clear that he was not going to accept the official anti-Soviet posture of Britain and the United States. He attended the meeting in Moscow to celebrate the two-hundred-twentieth anniversary of the Soviet Academy of Sciences in June 1945, along with other left-wing French and British scientists. Groves did not allow any members of the Manhattan Project to attend, and senior British physicists like Chadwick were also stopped from going. At this gathering, Joliot-Curie reputedly discussed details of the American bomb project quite openly with his hosts.[10] Weaver was accordingly careful. "We must wait a little time until the atomic dust settles," he told Rapkine, "and we can be sure that we would run into no formal difficulties in Washington, and no important limitations on his freedom to go about and visit our scientists."[11] In fact, the Joliot-Curies were refused a visa by the United States authorities for two or three years after the war, and when eventually Irène, who some people in the United States regarded as a "Communist fanatic," did go in 1948, she was detained overnight on Ellis Island by immigration authorities.

In dealing with Joliot-Curie, Weaver was concerned above all to avoid accusations by the State Department, who provided visas for foreign visitors, that the foundation was colluding in the efforts by a known Communist physicist and Soviet sympathizer to glean important information from his American colleagues. His prudence is understandable. A rather less clear-cut case arose late in 1947, when the foundation refused to become involved in the visa application made by Martin Kamen, who was invited to attend one of the conferences in France funded by Weaver's grant to the CNRS.

Irène Joliot-Curie invited Kamen, a professor of chemistry at Washington University in Saint Louis, to attend her colloquium "Isotope Exchange and Molecular Structure," in April 1948. Weaver was warned that his request for a visa might create difficulties. Linus Pauling, who was a major beneficiary of the foundation and who was close to Kamen, told the CNRS office in New York that the State Department had already refused his colleague a passport in 1947 to visit Palestine. Weaver's new assistant director, Gerard Pomerat "seem[ed] to remember" that it was also Pauling who had said that "K. had worked for the MDP [Movement for the Defense of Palestine?] and that he was a Communist."[12] In addition, "some American scientists that have been engaged in the Manhattan project" also implied that Kamen would never be given a visa by the State Department.[13] With Kamen's loyalty in question, and with the invitation coming from Irène Joliot-Curie, Pomerat and Dusty Miller refused to intervene on his behalf in Washington, preferring instead to "maintain a discreet reserve."[14] Indeed, in February 1948 Kamen's request for a passport was officially refused on the grounds that his "proposed visit abroad would not be in the best interests of the United States," a decision that he put down to the behavior of an "hysterical official."[15]

The Kamen case differs from that of Joliot-Curie in one crucial way: Joliot-Curie's Communist affiliations were a matter of public record, while the charges against Kamen were based on hearsay. The "scientists on the Manhattan project" who raised doubts about Kamen's loyalty were presumably referring to the infamous "Fish Grotto incident."[16] One evening in 1945 Martin Kamen dined with a Soviet diplomat at the Fish Grotto restaurant in San Francisco. He claimed that the meeting was purely social; the FBI insisted that he had passed on state secrets, an accusation that he consistently denied; indeed, it seems that he was unjustly accused.[17] For Pomerat and Miller, a vague recollection that Pauling had accused Kamen of being a Communist, the claim that he was suspect by some of his (apparently unidentified) colleagues, and the fact that he had been invited by Irène Joliot-Curie to France, were enough to persuade them to take no action with the State Department on his behalf.

There seems, then, to have been a qualitative change in the officers' level of concern about the Communist affiliations of scientists in the two years after the war. Joliot-Curie was working in a highly sensitive field, one in which many believed that it was essential for the United States to retain a monopoly on the science for as long as possible—a policy he openly disagreed with. One can understand Weaver wanting the "atomic dust" to settle a little. Kamen was not nearly as significant scientifically, nor was the claim that he was a Communist (sympathizer) based on any more than the opinion of a few individuals. If the foundation's officers were just as prudent about him in 1947 as they were about Joliot-Curie in 1945, it was because they were being sucked into the climate of fear regarding the loyalty and security risk of scientists, much of it based on rumor, which was beginning to take hold in the United States.

Jessica Wang has described in detail the steps taken in the late 1940s to weed out "disloyalty" and to clamp down on potential security dangers in the United States as the Cold War gained momentum and the fear of losing a nuclear advantage grew. In the first executive order signed by President Truman in March 1947, covering employees in the executive branch of the government, disloyalty was defined in terms of political engagement, or more vaguely, "sympathetic association," with groups deemed by the attorney general to be totalitarian, fascist, Communist, or subversive. Federal employees who worked with classified information had to undergo, in addition, personal security clearance. This extended also to industrial or university scientists working on secret military projects. Here, in Wang's words, "the security clearance system assumed that there existed a certain ideological type or profile that predisposed persons to commit espionage or other crimes of subversion."[18] In particular, people with liberal, leftist political leanings were all too often profiled as potential security risks, prime suspects among those willing to share sensitive information with other nations, notably the Soviet Union. As a result of these policies, by 1950, Wang tells us, some sixty thousand scientists and engineers employed by the federal government who were working in classified areas had been subject to loyalty checks and security clearance. About twenty to fifty thousand scientists, engineers, and technicians working in industry were also affected. The process disrupted research in federal, industrial, and university laboratories all over the country and threatened the livelihoods of thousands of men and women in the late 1940s and 1950s.

The reactions of Weaver, Miller, and Pomerat to the Joliot-Curie and Kamen cases were based on assessments of the danger that support for these individuals posed to the foundation's reputation in Washington. The increasingly restrictive atmosphere at home and growing tensions in Europe in 1948, though, forced the officers to formulate more general guidelines for assessing potential grant recipients and to consider taking a more aggressive stand against Communism.

Increasing Collusion, Releasing Files

The trustees and officers of the Rockefeller Foundation were deeply troubled by developments in Europe in spring 1948. In March they held a wide-ranging discussion "of problems created by the coup in Czechoslovakia and by lowering of the 'iron curtain' on other European countries where RF has had programs."[19] A meeting of the trustees followed on 6–7 April 1948 to consider the "tense situation in Europe."[20]

Three guidelines emerged after much deliberation, in which it proved difficult to arrive at an unambiguous and unanimous conclusion. The first pertained solely to countries behind the "iron curtain." Fosdick was most reluctant to sever links with such countries unless forced to, and all agreed that programs should be maintained when opportunities presented themselves. However, officers were aware that they had to be on the alert for cases in which the foundation's "continued cooperation would risk the necks of those we wished to help."[21] The situation was particularly delicate in Czechoslovakia and Romania, where, as Fosdick put it, former friends would probably "shy away from us out of considerations of their own personal safety."[22]

The other two guidelines were of a more general kind. On the one hand, the trustees felt that, with anti-Americanism on the rise, and being exploited by the Communist parties, it was "increasingly urgent" for the foundation to project a positive image of the United States abroad, "to promote goodwill and understanding among the peoples of Europe and the United States." On the other hand, they were concerned about the implications of providing support to scientists whose behavior might be deemed subversive in the United States. How best to proceed in this "very uncertain changing situation"? The foundation could not afford "a policy of over-caution,"[23] but was simply "to proceed with more caution than heretofore," as Weaver put it.[24] The officers were to keep in mind " the possibility of public reaction in this country that might damage the favorable position of the Foundation" as well as "the importance of avoiding projects that might conceivably be used to injure the United States and countries friendly to it."[25] The new situation was summed up for John Grant in the foundation's Paris office: "Bear in mind the sentiment at home. We cannot run completely counter to it. Also keep in mind that officers do not do things that will get the R. F. in trouble or undermine the Western World and the U.S.A."[26] The Communist threat was being taken seriously.

In formulating these guidelines, some of the trustees implied that they were far more likely to be needed for sensitive fields like physics or chemistry, which had a direct bearing on national security, than for something like biology. This seemed to let the Natural Sciences Division off the hook, since it generally did not support work in such areas. Weaver felt otherwise. As he put it to Fosdick a week after the meeting, "Under the conditions of modern total war, every element of scientific strength is of very great, even though admittedly not of equal, importance." In the "unspeakably horrible possibility" that a third world war, if there was one, would be biological, not atomic, Weaver went on, "then food, drugs and many other things are as essential as are the special weapons designed by the physicists and engineers."[27] The apocalyptic language is indicative of the impression that the events in Prague and Italy made on officers of the foundation, and of their fear that a third world war was imminent. The security question was now pressing and it touched every domain of science—not just those linked centrally to the A-bomb.

Officers also discussed amongst themselves the extent to which they should share their grant files with the State Department and the War Department, which had already sent representatives to their New York office to examine material. They were concerned about breaking the trust that their informants had placed in them. Fosdick was not keen to have the foundation provide information to the U.S. administration on "schools of public health, health conditions, personnel, etc. in countries where we are working." Weaver agreed and suggested that officers send a prepared statement rather than turn over original material. Fosdick had fewer qualms about fellows, however. As regards "giving information on fellows to F.B.I," he said, anything short of "full cooperation" with the authorities would do the fellow more harm than good.[28]

By April 1948, then, the trustees and officers of the Rockefeller Foundation were gradually aligning their policies with the prevailing anti-Communist climate in the United States, and some of them were willing to collaborate closely with various arms of the state apparatus to avoid being accused of colluding with "subversion" at home and abroad. This is not to say that they did so easily or that all officers were happy with the implications of the drift. Weaver, for one, was very concerned about handing over files on fellows to the security services. Writing from Copenhagen "on a very cold Sunday morning" early in May 1948, he insisted that the foundation should not "run any risk of having European scientists think that we have people, under our auspices, acting as scientific spies for the government." He specifically suggested that his division would be most reluctant about "showing records etc. to government people."[29] The foundation had built up a relation of confidence with European scientists over many years, he wrote, "confidence that we have no political connections or concerns, confidence that we are interested in the development of knowledge on an international basis, confidence that very frank comments about people and problems can be made to us without fear that this information would get out of our own hands so that the source might be subjected to embarrassment."[30] This confidence was now in jeopardy. In fact, the entire mechanism of scientific management that Weaver had put in place over the years, which depended entirely on trust in the foundation's independence and integrity and on officers forming an "intimate" relationship with researchers in the field, risked collapsing before his very eyes. The very premises on which consensual hegemony was built were being destroyed.

The director of Natural Sciences was also very concerned about allowing political considerations to affect awards. "In choosing a first rate scientist for support, [do] we have to ask about his politics," he asked

Fosdick, even if people in the person's home country respected his scientific (and political) views?[31] What about outstanding people whom the officers already knew were Communists? Should the foundation refuse support to "very distinguished British or French scientists, who [were] known to be either definitely communistic (members of the party) or, at least, Leftist, even from the point of view of the British Labor Government? [John] Haldane? [John] Bernal?"[32] Fosdick recognized that these questions posed fundamental matters of principle and suggested that another full discussion was needed in the New York office. Meantime, Weaver decided to do some research himself in Britain to find out how the foundation's image would be affected there if it allowed an applicant's politics to influence a grant decision.

Between 18 and 28 May 1948 Weaver visited a number of leading British scientists and science administrators: A. V. Hill at University College, London, Sir Ian Heilbron at Imperial College, London, Sir Howard Florey and Sir Robert Robinson at Oxford University, L. Farrer-Brown, the secretary of the Nuffield Foundation, and A. E. Trueman, the deputy chairman of the University Grants Committee.[33] His main question to them all was whether or not the foundation should take a scientist's politics into account when making awards. To a man their response was "a hearty and emphatic *no*." Heilbron, who was "exceedingly conservative" insisted that "nothing counts with them but the quality of the science." Robinson was willing to "lay my hand on my heart and say that we pay absolutely no attention to it [politics] here, and would hope you would not." Farrer-Brown was unambiguous: "The Nuffield takes no account whatsoever of the political views of its applicants." Trueman confirmed that "no Englishmen would want us [that is, the foundation] to withhold support from Haldane because H. thinks he is a communist." For Weaver's British interlocutors, in short, "good science is good science" and a person's political beliefs were altogether irrelevant in assessing eligibility for an award.

Weaver found the attitude to Communists and left-wingers far more nuanced than in his own country. Certainly no direct correlation was made between being a Communist and being disloyal or a security risk. Patrick Blackett, said Heilbron, was "very liberal, and perhaps an 'occasional fellow-traveller,' but not a communist." However, both he and Bernal were completely trustworthy, "wholly patriotic" said Florey. You knew whom to trust by looking at someone's educational background and war record. Blackett was "a Dartmouth man" said Hill, and "brought up in the traditions of a British naval officer." Bernal, Hill pointed out, had "made great contributions to Combined Operations activities under Mountbatten." The only doubt expressed was about giving Haldane access to top-secret government material. All agreed though that his political beliefs in no way interfered with his fine work in genetics. Overall, the British found the American concerns about Communism rather quaint: "We are closer to communism than you are, both geographically and in our governmental system," said Farrer-Brown, "but we do not seem to be as nervous about it as you are!" Florey banalized the issue completely. The British, he said, were "to a considerable extent, 'fed up' with the pinks and the reds, in the same sense that they [were] tired of other irritating things—bad food, criticism over doing their best in Palestine, etc."

The British position was something of an embarrassment for Weaver. It brought home to him that the foundation risked losing prestige and credibility if it "politicized" the evaluation of applications for grants. At the same time, he and his fellow officers, unlike the British he had met, were not sanguine about the nature of Communism, nor inclined to treat "pinks and reds" as just another everyday irritation. They might have deplored the hysterical anti-Communism in the United States, but they also felt that Communism was an alien and dangerous worldview and a threat to the values that they held dear. Pomerat summed up their mixed feelings after having spent time with Joliot-Curie in Paris in mid-April 1948, writing in his diary:

It would be difficult to meet someone with political views so different from our own and spend over two hours of discussion with so little feeling of resentment or irritation. Throughout the interview J. was gentle in his approach, calm in his conversation and tolerant in speaking of his cocitizens. He has a nice smile, a lovely voice, a tolerant good nature in spite of provocation which makes him an easy man to talk with. He gives one the impression that he is sincere in his conviction and honest in his reasoning. He holds views quite contrary to our own and he states them calmly in a friendly spirit of cooperation.[34]

Three years later, Pomerat would not be as sympathetic. The concerns surrounding the grant to French geneticist Boris Ephrussi encapsulate the shift in his perception, caused by the outbreak of the Korean War and new pressures on foundations from the U.S. Congress to join in the anti-Communist struggle.

Boris Ephrussi and Rockefeller Support for Genetics at Gif

Boris Ephrussi was born in Russia in 1901 and came to France at the age of nineteen. He received an International Education Board fellowship in 1926–27 and was a Rockefeller Foundation fellow in 1933 and again in 1935. At that time he was working in Paris at the Rothschild Institute for Physico-Chemical Biology, where he specialized in experimental embryology.[35] He applied to the foundation for support to come to the United States to learn more genetics, which was hardly taught in France at the time. His original plan was to go to Columbia and continue his studies on the genetics and embryology of mice. Instead, he went to work with the immensely productive group at the California Institute of Technology (Caltech) in Pasadena that had constructed the *Drosophila* fruit fly as a standard instrument for genetic mapping. There, and subsequently in Paris, he teamed up with George Beadle, a maize geneticist, to combine genetics and embryology—just the kind of hybrid research trajectory that appealed to foundation officers. They studied how genes affect pigment production and devised a means of transplanting imaginal disks from one larva into the abdominal cavity of a larval host of different genotype.[36] On the basis of this work, Ephrussi was named a *maître de recherche* of the CNRS in 1937. He was also appointed the director of a new genetics facility at the École des Hautes Études, where he set up a *Drosophila* laboratory with four full-time workers and other technicians and began to attract young French biologists to the field.

Ephrussi worked abroad during the war. On returning to France at the end of hostilities, a chair of genetics was created for him at the Sorbonne. He was also nominated director of the Institute of Genetics, to be built by the CNRS about twenty miles south of Paris on a site with "a magnificent setting on the hillside, overlooking a lovely and peaceful valley."[37] Joliot-Curie and Teissier foresaw a whole complex of laboratories at Gif-sur-Yvette. Genetics and nutrition would be the first, and they took steps in the provisional CNRS budget for 1946 to acquire a chateau, its outbuildings, and the land. The institute was to contain three separate units, one for Ephrussi (physiological genetics), one for Teissier himself (evolutionary genetics), and one for Philippe L'Héritier (formal genetics). The last two were "normaliens," who had combined their mathematical skills with an interest in biology and who had also spent some time in the United States in the 1930s with awards from the Rockefeller Foundation. All three were among the founders of modern genetics in the country.[38] Within a few years the transformation of the chateau and the surroundings at Gif was well under way. Both Weaver and Pomerat were very impressed by what they saw and felt that "from the physical standpoint it certainly gives promise of being a happy place to work in." The site, said Weaver, "could be developed into an extremely important and vital center for biological research in France."[39]

As noted in the previous chapter, the determination by the CNRS to institutionalize Mendelian genetics in France inspired Weaver. He took it to be symptomatic of the innovative spirit prevailing in the country after liberation; it confirmed the importance of the CNRS in his eyes as an institution determined to break with the past. Classical genetics had made little headway in France before the war.[40] A series of "accidents of circumstance connected with World War II" opened a window of opportunity for the geneticists.[41] Teissier's legitimacy as a former leader in the Resistance, his position in the CNRS, and his close association with chemists and physicists, including fellow-militant Joliot-Curie and the new director of higher education, Pierre Auger, provided him with the platform and the support he needed "to organize the penetration of genetics into the central institutions of France."[42]

On the face of it, we would expect Weaver to have been enthusiastic when, in July 1949, Ephrussi submitted a request for $54,000 for research equipment at the new institute. He was not. On the contrary, he was extremely cautious. Eventually, on 15 February 1950, he wrote a long letter to Ephrussi telling him that the foundation needed assurances "from the authorities of the CNRS that the men in this Institute will be free to carry out their work in the true spirit of modern universal science; and assurances from the geneticists involved that their scientific work will be uninfluenced by political considerations or party loyalties."[43] The matter was so serious that Ephrussi came to New York at once, where he was interviewed intensively for three days.[44]

Weaver's reaction was partly shaped by the decision of the foundation's new president, Chester Barnard, to tighten up on security. His new policy, first formulated in August 1949 and reiterated in March 1950, was that the foundation would not "knowingly and willingly award a fellowship to any member of a conspiratorial organization or an affiliate of such an organization whose purposes included subversive activities either in this country [that is, the United States] or any other."[45] This posed a problem for the Ephrussi grant for two reasons.

The first was the known or presumed Communist affiliations of the leading scientists at Gif, above all Teissier and L'Héritier.[46] This was now unacceptable, both at a personal and an institutional level. Indeed, Weaver and Pomerat were far more willing to act on claims that the CNRS was dominated by left-wing ideologues, including its Communist director, Teissier. Second, Weaver was concerned that Ephrussi, L'Héritier, and Teissier had not come out strongly enough against the neo-Lamarckian views of Trofim Lysenko, which had recently been elevated to the level of orthodoxy in the Soviet Union and seemed to be rapidly gaining ground in Communist circles in France. Just a few months before Ephrussi applied for his grant, a gathering of six hundred intellectuals meeting in the Salle Wagram in Paris were told by Communist leaders that they had to choose between bourgeois and proletarian science, between Lysenkoism and Mendelism. Where did Ephrussi stand? I need to break the narrative briefly to explain the dilemmas he faced.

Lysenkoism and neo-Lamarckism in France
The conflict between Lysenkoism and Mendelism had been simmering for over a decade in the Soviet Union when, in August 1948, the Soviet Academy of Agricultural Sciences decided to ban classical genetics and to institutionally cripple its proponents.[47] This move was made with Stalin's blessing and reversed earlier policies in which the Central Committee did not intervene directly to support or condemn particular scientific theories in biology. Indeed, in the first decade after the Civil War, genetics had been strongly supported by the authorities, both intellectually and institutionally, and Soviet researchers kept abreast of developments in the West, notably the United States. Through the donation of valuable *Drosophila* fruit fly stocks in 1923 by the American geneticist Herman Muller, Soviet

scientists acquired the material culture, and pursued the research lines, of the famous Morgan school, which was based first at Columbia University then at Caltech.[48] To deflect criticism, they cloaked their findings in the rhetoric required by a state whose guiding philosophy was Marxism-Leninism. In response to a severe agricultural crisis in 1929, they insisted that genetics played a practical role in the improvement of yields, and they made a major effort to have it introduced into agricultural institutes. These pragmatic moves only worked for a while, however. They could not indefinitely curb the ambitions of Trofim Lysenko.

Lysenko was an outsider to academic circles, the son of a peasant family who had no formal education. He was deeply concerned to improve Soviet agriculture, and as he moved around the country he made countless suggestions for increasing farm output.[49] His proletarian credentials and his practical bent endeared him to the authorities, who were suspicious of the "bourgeois" origins of classical geneticists and impatient with their tardiness in delivering results beneficial to agricultural output. His "Lamarckian" ideas—his rejection of the gene as a bearer of heredity and the claim that "hereditary" qualities of plants could be environmentally manipulated and transformed—struck a chord among the Soviet leadership, who valued his stress on plasticity and malleability, which resonated with their revolutionary project to build "socialist man."

Although Lysenko's supporters and the classical geneticists engaged in fierce rivalry and mutual mudslinging in the late 1930s, no side managed to crush the other. Indeed, Lysenko's star began to wane after 1945. Stalin, enthralled with what science and technology had achieved during the war, announced that the task of the Soviet Union was now to "catch up with the West." International scientific exchange briefly flourished as the scientists, encouraged by the authorities, sought to learn all they could from their erstwhile allies. Soviet geneticists, who were already internationally respected and who had strong links into the international community, obviously found much favor; Lysenko and his doctrines were in danger of being marginalized. Why then the sudden volte-face in August 1948? How was the triumph of Lysenkoism possible just months after his intellectual credibility and institutional power seemed to be declining?

The key—and I am following Nikolai Krementsov's persuasive argument here—seems to have been the formalization of the Cold War in 1947–1948, the rise of superpower rivalry, and the division of the world into "two camps."[50] In the months leading up to the Academy of Agricultural Science's meeting in August 1948, there was a steady deterioration in the relationship between the Soviet Union and the United States and its West European allies: the iron curtain fell. The Marshall Plan, the Berlin blockade, the Berlin airlift, the constitution of a separate (West) German state with its own currency, and Tito's rebellion in Yugoslavia, were matched by the signing of agreements between the Soviet Union and Hungary and Romania, the Prague coup, and a torrid election in Italy. In the fierce struggle for control of Europe, the Soviet regime shifted its strategy from one of collaboration to confrontation, from one of catching up with Western science to one of condemning everything that the West stood for. A new space was opened for those patriotic citizens who were hostile to the putatively metaphysical and impractical theories and defunct values of bourgeois science. Lysenko and his supporters seized their chance.

Classical genetics was virtually wiped out, or went underground, in the months following the August 1948 meeting. Leading scientists were summoned before party officials and forced to recant. Some major laboratories were simply closed down. Perhaps as many as three thousand biologists were fired.[51] What is more, the Soviet authorities took it upon themselves to decide whether the *content* of a (natural) scientific theory was politically and ideologically acceptable. This was a new development. In the 1930s it had been left to philosophers to discuss what theories were compatible with the tenets of dialectical materialism. In the early 1940s it was the leading members of the scientific establishment themselves who settled institutional and research policies. Now the decision lay with the Central Committee. As Krementsov puts it, "The party apparatus seized the right to judge scientific disputes and to dictate to Soviet scientists what theories to follow, what subjects to study, and what lines of research to pursue."[52] In so doing, as Julian Huxley wrote at the time, "A great scientific nation has repudiated certain basic elements of scientific method, and in so doing has repudiated the universal and supranational character of science."[53]

The decision taken by the Soviet authorities in fall 1948 to situate the disagreements between Lysenko and his opponents in the context of the Cold War struggle between "two camps" had major repercussions inside France and the French Communist Party (PCF). One could no longer be an internationalist, as many on the French left were: one had to take sides and bow to the dictates of the party on questions of scientific credibility, finding arguments to support proletarian Lysenkoism and to discredit bourgeois, classical genetics.

The Lysenko affair was introduced into France in August 1948 with an article in *Les Lettres Françaises* describing the "historic" session of the Soviet Academy at which Lysenko had triumphed; his report to the meeting was published in full in an issue of *Europe* (September-October 1948).[54] The position that the French Communist Party expected members to take was vociferously defined soon thereafter, triggered in part by statements made by Frédéric Joliot-Curie to the Anglo-American press. The first French nuclear pile, Zoé, had gone critical in mid-December, and France had entered the nuclear club. Commentators in the West were deeply concerned. Should a Communist be at the helm of the French nuclear project? Could he be trusted to keep nuclear secrets out of the hands of the Soviets? The London *Economist* mused, in an article on Christmas day 1948, that "sooner or later it seems that France will have to face the problem of purging Communists from posts affecting national security, as Britain and the United States have already done," at the risk of "jeopardizing the understanding on defence that has been reached by the western democracies."[55] Incensed, on 5 January 1949 Joliot-Curie spoke to the press about the supposed incompatibility between his official position and his political beliefs. He stressed his patriotism and his respect for the responsibilities his post carried with it. "A French Communist" he said, "like any other French citizen occupies a post to which he is appointed by his government, and could not honestly communicate to *any foreign power whatsoever* results which do not belong to him but to the collectivity which has enabled him to do his work." He was appalled by the "absurd reasoning which held that being a Communist morally relieved you of your French citizenship and automatically transformed you into a spy, working voluntarily or for compensation."[56]

Joliot-Curie's remarks were widely reported in the French and foreign press. The French Communist Party did not react publicly, however, for some time. Then, on 22 January 1949, the party newspaper, *L'Humanité*, published a speech made the day before by Jacques Duclos to celebrate the twenty-fifth anniversary of Lenin's death. Joliot-Curie was severely criticized for regarding the Soviet Union as a foreign power like any other. "Those who are fighting for socialism, for communism, all those who have their eyes turned to the future," Duclos insisted, "can say 'Each man who is progressive has two fatherlands, his own and the Soviet Union, the great socialist country.' "[57] To hammer the point home, a week later six hundred Communist intellectuals crowded into the Salle Wagram were told by Laurent Casanova that some "essential corrections" in outlook were now a matter of urgency, notably regarding Lysenkoism.[58] Criticizing "comrades (who) have begun to get angry by this intrusion of politics into science," Casanova insisted on the irreducible opposition between bourgeois and proletarian science.[59] To be a progressive French Communist intellectual was to owe one's allegiance to France *and* to the Soviet Union, whatever Joliot-Curie may say, and to promote proletarian science against bourgeois distortions. This position was elevated to an orthodoxy by the party, and anyone who dared disagree with it was branded as an enemy of the people. "The French Communist Party," wrote Communist philosopher Louis Althusser with contempt, "played a vanguard ideological and political role" in promoting not simply Lysenkoism but the "theory" of the "two sciences" from 1948 to 1952.[60]

Ephrussi reassures Weaver and Pomerat

It was against this background that, on 28 February 1950, Boris Ephrussi arrived in New York, his trip funded by the CNRS. He had a first discussion with Weaver and Pomerat that afternoon and evening, spent most of the next day and a half with them, and wound up his visit by having lunch with Rockefeller Foundation President Chester Barnard on 2 March.[61] The discussions were frank, friendly, and constructive and did much to clarify an "extremely difficult, important, confused and subtle" situation.[62] Weaver gave Ephrussi much credit for this. He suggested that no native-born Frenchman could have combined so well Ephrussi's love of France, his devotion to the Rockefeller Foundation, and his "external objectivity and understanding" for the complexities of the French situation.[63] Although much was clarified, the officers still had doubts and worries after Ephrussi had left. They and Barnard decided nevertheless to take the "reasonable risk" and recommend to the trustees that Ephrussi be funded.

One point that was quickly clarified in these conversations was that there was no particular bias in the CNRS. In his letter of 15 February to Ephrussi, Weaver noted that the institute at Gif was attached to the CNRS, and he suggested that, since this organization lacked the centuries of tradition of the University of Paris, the "scientific philosophy of its leaders and the opinion of men in its own higher levels of administration" might influence the intellectual climate in the laboratories—toward "leftist Lysenkoism" of course, though that was not said. He and Pomerat were reassured that the CNRS respected "freedom of election, freedom of appointment, and freedom of decision" as much as any other French educational body or university.[64] True, as soon as he got back to Paris, Ephrussi wrote them a rather distressed letter to say that new elections to the CNRS committees then under way might place some "very conservative elements" in the directorate, including "representatives of the still traditional French Lamarckism, more frequently associated with political ultra-conservatism."[65] Ironically they, at least as much as "leftist Lysenkoists," might pose a threat to the development of genetics in the country, since a Lysenkoist mechanism was compatible with their dislike of Darwin and their attachment to Lamarckian inheritance of acquired characteristics. This did not seem to disturb Weaver and Pomerat unduly. With an eye on the domestic situation in the United States, what mattered most to them was that Teissier had just been ousted from the CNRS directorate (on a rather feeble pretext).[66] As Pomerat wrote in his diary, "The whole problem of RF aid to the Institute of Genetics is now materially less difficult in that *Teissier* is no longer Director of the CNRS."[67] Indeed, in a more general purge of Communist scientists in senior administrative positions, Prime Minister Georges Bidault also removed Joliot-Curie from his post at the Commissariat à l'Énergie Atomique just a few weeks later, on 26 April 1950.[68]

The administration was purged, then, but what about the laboratories? In their interview with Ephrussi, Weaver and Pomerat made no bones about trying to pin down the interrelated political and intellectual orientations of the three leading geneticists at Gif. This directness, unusual until recently, was in line with Chester Barnard's new policy of not "knowingly and willingly" giving grants to "subversive" individuals or organizations. They were rapidly reassured regarding L'Héritier: "Instead of being a Communist, [he] was its exact antithesis, a rabid royalist."[69] Teissier and Ephrussi himself were not as easily dealt with.

That Teissier was affiliated with the Communist Party was well known. What concerned Weaver and Pomerat was whether he would follow the party line and promote a Lysenkoist research agenda and what Ephrussi, as director of the institute at Gif, would do about it if Teissier took this tack. Ephrussi handed the officers drafts of two of his colleague's most recent papers, which were along Morgan-Mendelian lines. He could not, however, guarantee that Teissier would not publish scientific articles with a Lysenko bias or give popular talks or write popular articles with such a bias.

Weaver and Pomerat made some extraordinary demands of Ephrussi as regards Teissier.[70] Would Ephrussi be willing to "try to force T's resignation by threatening to resign his own Directorship of the Institute" if a scientific paper along Lysenkoist lines emanated from Gif? Ephrussi refused. Would Ephrussi be willing to demand that, as director, he had to approve in advance all papers produced by his colleagues before they were submitted for publication? Ephrussi refused. Would Ephrussi be willing to ask Teissier, on behalf of the foundation, "to give written assurance that his own relation to the Institute of Genetics would not carry a Communist bias"? Ephrussi refused. Ephrussi did make one concession. Teissier gave some lectures in his course on genetics at the Sorbonne. He would ask the dean to cancel this arrangement immediately if Teissier "were to present information with a Lysenko bias" to Ephrussi's students.

At the end of the day, pragmatic considerations led Weaver and Pomerat to conclude that Teissier did not pose an undue threat. To date he had done "honest research in genetics, actually in full conformity with the Mendel-Morgan tradition."[71] He was likely to spend very little time at the institute at Gif, since he had a professorship at the Sorbonne and was very committed to the marine biology station at Roscoff. Finally, pressure on him was likely to be counterproductive and actually harm rather than help the cause of genetics in France since it might "force him to go completely over toward Communist genetics."[72]

Ephrussi himself was grilled on his political and intellectual views.[73] He told Weaver and Pomerat that he had refused to join a strike organized by his teachers' union against the Marshall Plan and that he had actually tried to force the officials to "kick him out of the union" for refusing to do so. He had agreed to lead a deputation to the minister of education to protest against the unfair dismissal of Teissier as director of the CNRS, but only on condition that "there was not a single Communist in the group." He had taken a "positive and courageous position" on Lysenkoism at a meeting with several hundred high school teachers. And he had encouraged Julian Huxley to write two articles in *Nature* on the intellectual bankruptcy of Lysenko's neo-Lamarckism and the deplorable way in which it had been elevated to orthodoxy in the Soviet Union. What he had not done, and probably would not do in the immediate future, was to "make a public statement on his attitude in the Lysenko situation." Weaver and Pomerat accepted that he had "very legitimate reasons" for not doing so.

This dismal episode is indicative of the extent to which Weaver and Pomerat had now been enrolled into the prevailing political and ideological climate at home. The reassurances they sought from Ephrussi were intended to deflect criticism of the foundation from an increasingly hysterical and hostile anti-Communist lobby in the United States, some of whose values and assumptions they shared. Those values and assumptions made little or no sense in the French context. If it was possible for them to conduct this inquiry without the foundation losing its reputation for being apolitical and independent of the U.S. administration, it was probably because, apart from Ephrussi himself, just two people apparently knew of the visit and the tenor of the discussions. One was the CNRS's acting director, Professor Pérès; the other was the president of its Committee on Foreign Relations, Professor Émile Terroine. Both of them apparently kept quiet about what transpired.

Weaver and Pomerat brought four assumptions into the room when they interviewed Ephrussi. First, they regarded Communism as an essentially subversive and politically illegitimate ideology. As Ephrussi pointed out, however, it still had considerable legitimacy and credibility in France, thanks to its links with the Resistance. This was one reason why it was impossible for Ephrussi to intimidate Teissier. He had a sense of loyalty to Teissier deriving from his courage under German occupation and, as noted, led a delegation to the minister to protest Teissier's unfair dismissal.

Weaver and Pomerat also seemed to assume that Communists, including Communist intellectuals, slavishly followed the party line and necessarily adopted Stalin's and Lysenko's critique of classical genetics. The party certainly tried to enforce discipline, but it frequently failed, being instead seriously discredited in the eyes of many intellectuals, who began to drift from its ranks. *Le Monde* explained the dissident behavior thus in 1950: "The savants, the intellectuals, whose deepest richness lies in their freedom of spirit, in their freedom of thought" could be Communists (as was Joliot-Curie) but could never accept unquestioning "fidelity and loyalty to the USSR."[74] Indeed, many years afterward L'Héritier said in an interview that, with the exception of one or two people, he did not think that any French biologist, Communist or not, ever accepted Lysenko's views. Teissier "obviously never believed in Lysenkoist theories," notwithstanding his political beliefs.[75]

The officers also assumed that theories affirming the inheritance of acquired characteristics were scientific nonsense. Matters were more complicated than this, however. For example, in 1937 L'Héritier had found that some strains of *Drosophila* flies died after being anaesthetized with carbon dioxide, and that this trait was maternally inherited and transmitted in the cell cytoplasm. He continued to study this apparent infraction of Mendelism while at Gif, where he eventually established that a virus was disturbing the hereditary behavior of his flies.[76] Ephrussi's own work on cell cytoplasm was not unambiguously incompatible with Lamarckian claims either; consequently, he was forced to admit to Weaver and Pomerat that he was "at least partially suspect" of Lysenkoist bias himself.[77]

Finally, Weaver and Pomerat assumed that there was a one-to-one correspondence between holding Lamarckian/Lysenkoist views and being a Communist: since they were nonsense, it was evident that only someone blinded by irrational political and ideological beliefs could possibly believe in them. But, as Ephrussi stressed, Lamarck's thesis had deep historical roots in France and his supporters were often politically ultraconservative. L'Héritier speculated that it was not surprising that Lamarckism flourished in a Catholic country. It provided an alternative theory of evolution to Darwin's views on natural selection, which were far more popular in the Protestant countries of northern Europe and in the Anglo-American world.[78] In general, as the experiments of these two men showed, a plausible scientific case could be made for a Lamarckist interpretation of certain results, and it was ridiculous to dismiss anyone who supported such an interpretation as being blinded by Communist bias.

To Weaver's and Pomerat's credit, the interview with Ephrussi enabled them to detach themselves from the narrow understanding of the situation in France that the domestic political scene encouraged. They were paternalistically sympathetic to his plea that modern French genetics had to be "treated as a child—protected, forgiven, not too strongly punished, treated tolerantly."[79] They understood better "the whole complicated personal, intellectual, political, and . . . emotional background against which this problem must be viewed."[80] As Ephrussi explained the situation to them both, he "grew steadily in both intellectual and moral stature" in their eyes, and Weaver felt that "a very large measure of trust and confidence" in him was justified. Pomerat summed up the situation neatly: "We must as usual place our confidence in the man who will direct the project."[81] Thus reassured, they successfully persuaded the trustees in April 1950, to award $54,000 to the CNRS to purchase equipment for the genetics institute at Gif-sur-Yvette over the next three years.[82]

In the event, this grant was never taken up.[83] The main reason for this was a bitter dispute between Ephrussi and the new director of the CNRS, Gaston Dupouy. Ephrussi was always concerned about moving his laboratories from central Paris to Gif, fearing that he would be isolated intellectually and that his staff would not be willing to move there.[84] He had secured guarantees from Teissier that as institute director he would be supplied with a car and a chauffeur, as well as a house for himself and housing for his staff on the site. Dupouy refused to meet these conditions in June 1951, when the time came to move to the new buildings.[85]

Ephrussi dug in his heels, though his case was weakened by Teissier and L'Héritier, both of whom accepted the move to Gif with a few of their staff. L'Héritier enjoyed hunting in the valley. Teissier did little more than drop in once a fortnight anyway, being preoccupied with his duties in Paris and at Roscoff.[86] With Ephrussi refusing to go, Dupouy decided to abolish the post of director of the institute and leave each of the three men responsible for their own research groups.[87] Ephrussi now began to seek employment in the United States and told the foundation in June 1952 that "he would move to another country right away if a suitable post became available."[88]

To salvage something of the situation, in March 1953, with the grant about to lapse and none of the money spent, Dupouy asked if it could not be extended and restructured so as to give $30,000 to Ephrussi for his genetics laboratory in Paris and $24,000 for L'Héritier and Teissier at Gif. Pomerat was most reluctant to pursue this solution. Ephrussi's future in France was uncertain, and there was a geneticist at Gif "who has not yet seen fit to tell the world where he stands on the Lysenko situation."[89] At a lunch with Dupouy in June 1953, Pomerat explained carefully the many reserves that he had about continuing with the grant, including that it would foreclose the foundation's giving additional help to Ephrussi, who was doing some of the best work in genetics in France.[90] He suggested, and Dupouy agreed, that the grant be rescinded forthwith.[91]

Dealing with Communist Smear Campaigns in the Early 1950s

In September 1950, five months after the Ephrussi grant had been made, a new and extremely restrictive bill was passed by the U.S. Congress. Sponsored by Pat McCarran and John Wood, the chairman of the House Un-American Activities Committee (HUAC), the Internal Security Act, or "McCarran Act," created the Subversive Activities Control Board to identify and register Communist and Communist-front organization.[92] It prohibited American citizens who were members of such organizations from obtaining U.S. passports. It also authorized the State Department to deny visas to foreign nationals who were members of Communist-front organizations or who might be deemed to engage in "activities which support the communist movement" or, more vaguely, "which would be prejudicial to the public interest" while in the United States.[93] President Truman vetoed the bill, declaring that it would be ineffective in curbing Communism, that it moved "in the direction of suppressing opinion and belief," and that it was "a long step toward totalitarianism." His veto was (easily) overridden in both houses.

The McCarran Act caused a storm of protest from scientists and immensely complicated the administration of the foundation's travel grants. Previously it had been loath to promote a visa for a man like Martin Kamen, who had been accused by some of having Communist sympathies; now it had to be careful not even to support people whom the authorities deemed "prejudicial to the public interest." What is more, in respecting the requirements of the act, which struck a lethal blow at scientific internationalism, the foundation's credibility abroad, and particularly in France, would be strained even more. Indeed, a committee of the Federation of American Scientists estimated in 1952 that at least half the foreign scientists wishing to travel to the United States had had trouble getting a visa on account of this act and that as many as 70–80 percent of requests from French scientists had been delayed or refused.[94]

The foundation's problems did not end there, however. In a speech in Congress on 1 August 1951, Representative E. E. Cox (D-GA) charged that American foundations were promoting activities that conflicted with national security. He singled out the Rockefeller Foundation among others, claiming that its "funds have been used to finance individuals and organizations whose business it has been to get communism into the private and public schools of the country, to talk down America and to play up Russia, and," he went on, by virtue of the foundation's activities in Asia it "must take its share of the blame for the swing of the professors and students in China to communism during the years preceding the successful Red revolution there. . . . Our boys are now suffering and dying in Korea, in part, because Rockefeller money encouraged trends in the Chinese colleges and schools which swung China's intelligentsia to communism."[95] It mattered little that the foundation could easily show that these accusations were misleading and mischievous.[96] About a year later, as requested by Cox, Congress established the Select Committee to Investigate Foundations and Other Organizations.[97] It was chaired by Cox, and its brief was "to conduct a full and complete investigation and study of educational and philanthropic foundations and other comparable organizations which are exempt from Federal taxation to determine which such foundations and organizations are . . . using their resources for un-American and subversive activities or for purposes not in the interest or tradition of the United States."[98]

The Rockefeller Foundation prepared a major document for the Cox committee that ran to over a hundred pages.[99] It showed that of the 14,163 grants made by the foundation since 1935 only three organizations and twenty-three individuals had been commented upon adversely or listed as subversive by several congressional committees, including the HUAC.[100] What is more, it stressed that in most of these cases the grant was made long before the organization or person had been so cited. The names of three eminent scientists were included in the foundation's list: Frédéric Joliot-Curie, for some $15,000 received between 1937 and 1941 for physics research in Paris, J. B. S. Haldane, for about $53,500 received between 1935 and 1947 for research in genetics at University College, London, and Linus Pauling, for almost $680,000 received from 1932 to 1952 for his research program coordinating physics, chemistry, and biology at Caltech.[101]

The Cox committee and its staff made an extremely thorough investigation of the activities of the major American foundations, consulting widely with the foundations themselves and eminent leaders in various fields. Its unanimous conclusion was that American foundations had played a crucial role in advancing the frontiers of knowledge. It also concluded that while some had inadvertently supported "the communist line," and a few had actually become "captives of the Communist Party," the foundation system as a whole had not weakened, undermined, or discredited the American system of free enterprise.[102] The committee found it necessary to point out that "Many of our citizens confuse the term 'social,' as applied to the discipline of the 'social sciences,' with the term 'socialism.' " It also felt it necessary to reject the accusation that the foundations were promoting "internationalism" to the detriment of U.S. interests.

Weeding Out Communists

The passage of the McCarran Act in September 1950 and Congressman Cox's attack on the foundation in August 1951 sparked another discussion inside the foundation on what to do about Communism. Lindsley Kimball, who had been appointed vice-president of the foundation in 1949, reiterated policy guidelines on 19 November 1951. His paper began with a statement whose purple prose might have been used by any senior American official fifty years later, in the aftermath of the attacks on the World Trade Center in New York:

> The people of the United States are facing a new situation under the sun. In the past, our enemies have always been well-defined and with limitations which were usually national, geographical, and temporal. We knew that our enemy was a Spaniard, an Englishman, or a German. We knew where to find him and the beginning and end of his enmity was clearly defined. In the past, we have fought and won our wars and have then pursued peace. Now our enemy is insidious. He is international. He lives in a twilight zone between war and peace. He invades not only our shores, but our society as well. And he exhibits a protective coloration, and a genius for disguise which make him difficult to combat and leads us to the verge of a national hysteria.[103]

Kimball recommended that foundation officers try to strike a balance. Some people thought that the foundation was "either unwittingly giving support to the enemies of our country or [was] itself fuzzy-minded, unrealistic, and even pinkishly inclined." These views could not be brushed aside. It was particularly necessary for the Rockefeller Foundation, as a tax-exempt organization, to "recognize public opinion as a fact." But it also had a "moral responsibility to stand between the extremes." There was a "middle ground between hysterical McCarthyism and the support of Josef Stalin." While the foundation's officers were to "accept risk rather than restrict our activities to wholly safe areas," they "certainly would not support our enemies nor contribute to subversion." In its ensuing submission to the Cox committee, the foundation reaffirmed Chester Barnard's insistence made eighteen months before that the foundation "would never knowingly participate in or support un-American or subversive activities."[104] But it went further, and assured the committee that the staff had recently been concerned "*explicitly* with the attitude of the individual or group towards Communism."[105] In short, the foundation was prepared to make a far more deliberate and determined effort than before to establish the political credentials of applicants. This was to be done mainly through *informal* intelligence gathering.

Kimball insisted that the foundation was "not an arm of the FBI and [was] not in a position itself to develop the kind of detailed knowledge on which to base infallible judgments" regarding the loyalty of individuals and organizations. Of course it would check if the name of an applicant was on an official list of potentially subversive organizations and individuals, like the attorney general's list. This approach had its limits, however.[106] Public laws and official declarations of public policy revealed a "lack of precision which itself may reflect differences about what constitutes wise policy in this field."[107] The foundation did not want to get embroiled in a murky debate over the meaning of "subversive" activities. An alternative, and in their view more reliable procedure, was to use the foundation's traditional method of establishing the suitability of applicants: by sounding out the views of peers and by discussing the grants with the candidates themselves. The trust on which the "intimate partnership" between officer and scientist was based would now be used to establish political tendency.

The grilling that Boris Ephrussi was subjected to by Weaver and Miller in March 1950 is indicative of the new spirit in the foundation to ensure that no Communist got a grant. It was part of a more general trend to probe and analyze the political views of applicants and to record the information officially. The shift in policy, and the drift toward an increasingly explicit and unabashed concern with a person's political affiliations, is evident from Gerard Pomerat's diary. In summer 1949 the assistant director of Natural Sciences was faintly embarrassed by discussions with researchers in which politics came up. Two years later he was not, and in fact he believed that it was right and proper for an officer to establish whether or not researchers were Communists and to refuse them grants on those grounds. A few examples will clarify this evolution.

In June 1949 Pomerat visited Professor and Mrs. Cordier at Lyon University.[108] Both told him that, in their view, the CNRS was "a political group which extends its aid in an entirely undemocratic manner." They spoke of the Communist politics of the Joliot-Curies and of Teissier, who, they said, had "been a convinced communist since 1920" and who

was "a much more powerful member of the party than his mild professorial manner would lead one to suppose." Pomerat was not too pleased
about being told all this and "force[d] the conclusion of this discussion
as unprofitable." He was recording it in his diary, he said, only because
he felt that the Cordiers' resentment may be typical of the gulf that still
separated Paris and the provinces (they pointed out that Teissier had
never been to Lyon, for example, even though he was the director of the
CNRS).

Pomerat's encounter with Oleg Yadoff a few months later confirms
that at this stage, while he took note of Communist accusations, he was
not pleased to hear them, believing them to be essentially irrelevant to
his job.[109] Yadoff visited Pomerat in his New York office in October
1949. The CNRS, he said, was "a communist-dominated organization
which gives aid primarily to communists . . ." or at least to "communizers." Yadoff also claimed that when the physicist Langevin was inter-
viewing candidates for CNRS fellows, he dropped their applications in
the wastebasket if they refused to go and work in the Soviet Union.
Pomerat was furious; he regarded Yadoff as completely irrational and
resented having been subjected to a "long diatribe against the CNRS."
Yadoff, he wrote, had "wash[ed] dirty French linen in an American
office to the point where it was most embarrassing," implying that the
political divisions in the community were purely an internal affair and
irrelevant to the foundation.

The passing of the McCarran Act had marked effects on Pomerat. He
visited Paris in summer 1951. On 1 June he met with Jacques Monod,
who would win a Nobel Prize in physiology or medicine in 1965.[110]
Monod was furious about the inquisitorial procedures for getting a visa
that had been imposed on foreign visitors by virtue of the act, which he
deemed totalitarian, humiliating, and perhaps a means "to trap his
friends." Pomerat had no sympathy for the man; indeed, he defended
U.S. policy. He wrote in his diary that Monod was "bitter, aggressive,
more neurotic than ever before." He also regarded him as politically
naive and ungracious, "telling M. to think a little about our war in
Korea, our help to Europe, and our enormous military preparations to
avoid a third world war by show of force." Further discussion led the
officer to conclude that though the scientist had told him that "he was no

longer a Communist," Monod in fact ended up "by following closely the Party line against the West." The fact that Monod, who was Teissier's student and brother-in-law, had immediately condemned Lysenkoism in 1948 as "senseless, monstrous, unbelievable," and had drifted away from the party was now irrelevant to Pomerat: all that mattered to him apparently was that Monod's politics were "pinkish" as seen from Washington.[111]

Pomerat was shaken by this interview and left half-way through his lunch to "wander about in the rain, and to get away from this feeling that France remains what she has been for so many recent years—a very sick country."[112] A week later he was again at the Pasteur Institute.[113] There was no pretence any longer that a scientist's politics did not matter: they were explicitly recorded in the officer's diary for circulation inside the foundation. Thus Pomerat wrote that "Lederer has for a number of years been mentioned to GRP [that is, Pomerat himself] as one of the really good chemists in France, but there is almost no question but that he is a communist." His labs were situated in the department of "Professor Aubel, who is himself definitely a communist." "[Georges] Cohen, the potential fellow, is perhaps a communist, but of this GRP is not certain." Anyway, there was no doubt at all that "at least one of C[ohen]'s immediate chiefs is a rabid communist." Pomerat gave no evidence whatsoever to substantiate these claims. He apparently judged his (unnamed) sources to be sufficiently reliable and saw no need to justify his remarks.

This visit was a turning point for Pomerat. His diary noted that in England and in Belgium too "attitudes toward aiding scientists who are communists had very definitely hardened." A further grant from the foundation to the X-ray crystallographer J. D. Bernal, who had never hidden his Marxism, would now be frowned upon; two years earlier the British had told Weaver that Bernal's war record was evidence enough of his loyalty. A grant to the Belgian biochemist Jean Brachet, who later made major contributions to understanding the role of RNA, had been "seriously questioned." Even more to the point, wrote Pomerat, an (unnamed) French scientist had told him during his trip that "if a communist scientist were to make an outstanding discovery which was of military importance he had grave doubts as to whether this communist scientist would feel free to publish the details of such a discovery without first having cleared it with the Russians"—that is, "a communist scientist was not free." Pomerat's diary entries on this visit to France closed on a somber, but decisive note. "Much against his will and much against his French love for the freedom of the individual to choose his politics and his religion and his way of life according to his conscience, *GRP now feels that the RF should not aid scientists anywhere who are for political or ideological reasons adherents of the Soviet system.*"[114] An applicant's politics, if those were too left-wing or Communist, were now a reason for refusing a grant, no matter how prestigious the scientist or how important the research topic.

Though Pomerat was convinced that Communism posed a threat to world peace and security, his opposition to it was not vulgar or populist but refracted through the prism of his professional role as a foundation officer. The basic justification for refusing an award was that a Communist or someone working in a Communist-led laboratory was not "free," meaning not free to do good science. As the foundation put it in the submission to the Cox committee, a man who was loyal to Communism was unable to "employ sound, scholarly and scientific procedure," could not "interpret his results with objectivity," and would not circulate his work "to the world of free scholarship."[115]

There are two different arguments here, and I shall consider them in reverse order. First, there was the question of publication: a loyal Communist, or someone who had to have the support of a loyal Communist to get work published, could not do so if the findings were not coherent with the party line. This made a travesty of scientific openness. It explains why both Weaver and Pomerat were so concerned to establish the chains of command in the laboratories where their fellows worked. For even if the researcher was not a Communist, his or her hierarchical superior might be and might suppress publication of original work on the grounds, say, that it was bourgeois "Morgan-Mendelian" genetics and unacceptable to the French Communist Party and to Moscow. This would stifle basic research, which thrived on the "free market place of ideas."

The second argument—that Communists violated good scientific practice and allowed ideology to blind them to truth—was used by Weaver when he first called Ephrussi to New York in February 1950. As he put it then, genetics and politics had "recently become interrelated, in some quarters, in a most confusing, a most disappointing, and indeed a most fantastic way." In the Soviet Union there was now a party line in genetics, and (quoting Huxley) "the basic scientific principle of the appeal to fact has been overridden by ideological considerations."[116] The foundation, wrote Weaver, would not support any geneticist who did not conduct research "in the true spirit of universal science" and who was not completely dedicated to the "unbiased discovery of facts—all the facts, not merely certain misleading or partial facts which conform to a predetermined code."[117] At the time, and later, Weaver was able to avoid simplistic generalizations and to decouple the ability to do good science from having left-wing or Communist sympathies. In a memo to foundation President Barnard and Vice-President Kimball early in 1952, he still insisted that people of "scientific excellence" "might have, apart from their scientific life, an undesirable political activity or reputation."[118] Others were more inflexible and generalized more readily. Officers in J. H. Willits's Social Sciences Division, for example, pointed out at the end of 1952 that the foundation expected its grantees to be intelligent and capable scientists who did not let their personal opinions, beliefs, and values affect the way they handled and presented evidence. "In terms of these criteria," one wrote, "there is, obviously, virtually no possibility that any Communist or Fascist would be judged capable of carrying out scholarly research with the Foundation's support." For adherents of these totalitarian ideologies, "the presumption of incapacity for objective scholarly research can be made."[119]

This epistemological argument neatly inverted responsibility for introducing political criteria in awarding grants. Since Communists or fascists could not, a priori, do objective, scholarly work, to finance them would be to finance not good science, but a political and ideological agenda, that is, to be "indirectly involved in supporting racial, religious, or political interests," as Weaver put it to Ephrussi.[120]

In claiming that there was a fundamental contradiction between adhering to Communist or indeed totalitarian beliefs and doing good science, the officers of the foundation were of course implementing an argument then popular in liberal circles in the United States.[121] The leitmotif of that argument was first developed by antifascist intellectuals in the late 1930s and early 1940s in response to Nazi attacks on "Jewish physics" and "non-Aryan science." They took this institutionalized intolerance as evidence that science could best flourish in a democratic society. Robert Merton, in his "Note on Science and Democracy," published in 1942, and intended as an attack on Nazism, formalized the argument by identifying a number of values essential to the conduct of science, which were institutionalized in the scientific community and which, he held, were equally sacred in democratic societies. The most important was universalism, or the view that scientific truths were not bound to local or personal circumstance and were true irrespective of the political, religious, or racial attributes of those who established them. Another value, included in Merton's original analysis but less emphasized at the time, was organized skepticism, or the conviction that only by a critical engagement of minds unfettered by preconception or prejudice could scientific truths be reliably established. Both required tolerance, openness of mind, and a willingness to follow the argument wherever it led—features that Merton felt could not flourish in a totalitarian state. These ideas, generated in self-conscious opposition to certain extreme (and actually quite isolated) assaults on physics in Nazi Germany, were reappropriated in the late 1940s by a number of American intellectuals, notably left-wing Jewish intellectuals. They sought to promote a convergence between the scientific method and the values of a liberal-democratic state. The Nazi-Soviet pact, Moscow show trials, repression in Romania and Czechoslovakia, and the Lysenko affair indicated that not just Nazism but Communism too posed a threat to both science and democracy. It was totalitarianism in all its forms that had to be opposed. David Hollinger writes that people like Robert Merton, Margaret Mead, Ernest Nagel, Richard Hofstadter, and Isidor I. Rabi "selected from the available inventory those images of science most useful to them, those serving to connect the adjective *scientific* with public rather private knowledge, with open rather closed discourses, with universal rather local standards or warrant, with democratic rather than aristocratic models of authority." These intellectuals saw a world filled with prejudice and organized efforts to stifle independent thought, which they sought to combat with free inquiry, open-mindedness, and the unfettered pursuit of truth. These were the cardinal values of the scientific community, which cohered with their deep concern for the freedom of the individual, freedom of the press, and freedom of academic pursuit.

Weaver and the other officers of the Rockefeller Foundation were situated in this intellectual field in the early 1950s. It enabled them to promote a particular conception of what science was, using it to justify withholding support from Communist or perhaps even left-wing scientists. They *built* a demarcation between science and nonscience as they struggled to define a policy that, in one move, delegitimated the intellectual credentials of applicants and excluded them as political undesirables. This "boundary-work,"[122] coupled with their informal mechanisms for identifying those whom their peers judged to be "pinkish" or "red," was in their view a far more effective protection against the dangers of "un-American" behavior than the official procedures put in place in Washington. It also provided them with a defensive shield against charges that they were supporting subversive and disloyal activities. And it allowed them to distance themselves from, and to feel superior to, a hysterical and intolerant populist anti-Communism that was the very antithesis of the liberal values that they held dear.

Epilogue: Ephrussi Again

I conclude with a brief remark regarding Ephrussi. He never did move to the United States, as he had considered doing in the early 1950s. The laboratories at Gif were expanded and, in March 1955, it was officially announced that the Faculty of Sciences of the University of Paris had the right to purchase property nearby at Orsay, a move that would reduce the isolation of Gif.[123] By May 1956 Ephrussi was reported by Pomerat to be "obviously making a real effort to be getting on with Dupouy so as to make a success of installing his laboratories at Gif." He was receiving financial support from the CNRS, the Sorbonne, and the Rothschild Foundation—and he was hoping for additional support from the Rockefeller Foundation.[124]

He was not disappointed. Meeting on 25 October 1956, the trustees agreed to award the CNRS $61,000 over three years to support Ephrussi's research effort at Gif.[125] Apparently his scientific colleagues were not asked about his political views or his views on Lysenkoism. But this did not mean that the Rockefeller Foundation was now indifferent to the political allegiance or potentially "subversive" character of grantees. It simply meant that it was using official indexes to establish culpability. Two weeks before the trustees' meeting that made the grant, a number of such lists were checked to see if Ephrussi's name appeared on them. These included the "Cumulative index to publications of the Committee on Un-American Activities," the "Cox-Reece committee index, '53–'55," the "McCarthy Committee composite index, '53," and the "McCarran Committee on aliens in U.S. '51." Ephrussi's name was not found.[126]

In March 1958 Rockefeller President Dean Rusk informed his staff that it was no longer necessary to "follow the formal procedures for checking published reports of the Congressional Committees for individuals who may be involved in prospective Foundation grants."[127] This was not because the trustees or officers were indifferent to Communism, but because the procedure followed was a waste of time and money. Only one case had arisen in the previous four years in which the information from such reports was genuinely helpful. It was clear that the officers' "normal investigations about the personal, scientific and scholarly qualities of an individual or group" were deemed so effective that there was no need for further checks using government material. In short, the informal network of contacts established by the officers, and the personal intelligence that circulated within it, was a far more efficient instrument for detecting putatively subversive and un-American attitudes than the bureaucratic structure set up in Washington.

Several factors contributed to the emergence and eventual consolidation of "molecular biology" as a distinct field of research. Nicolas Rasmussen, for example, has emphasized the role of industry, and of RCA in particular, in promoting the electron microscope as a legitimate and cognitively authoritative instrument for biological research.[128] Pnina Abir-Am has concentrated on the role of the Rockefeller Foundation, contesting Weaver's claim that his grants produced the molecular biology of the 1960s. She insists that, on the contrary, his preference for established individuals in prestigious institutions and for "technology transfer" from the physical sciences to biology blinded him to important shifts in the research frontier occurring elsewhere.[129] This chapter raises another, intriguing possibility not considered before, as far as I know: that beginning in the early 1950s, the foundation simply refused to support Communists or scientists deemed to be Communists—no matter how good their work or prestigious their institution. By 1952 the boundary work done to demarcate science from nonscience in terms of political affiliation had become a convenient rationalization for exclusion: further research is required to establish how frequently it was invoked to refuse support to scientifically deserving applicants, particularly in France.

6

The Ford Foundation, Physics, and the Intellectual Cold War in Europe

The Intellectual Cold War

Weaver and Pomerat's sensitivity to the influence of Communist intellectuals in continental Europe was part of a broader concern in the United States about the reliability of the elites across the Atlantic. Indeed, beginning in the late 1940s, policymakers in Washington were convinced that the success of the Marshall Plan—and the consolidation of U.S. hegemony in Western Europe—could not be achieved by economic and military support only. They had to be coupled with a politico-cultural offensive that would demolish doubts about U.S. motives, dispel negative stereotypes of the United States, and implant American values abroad. Soon after the National Security Council (NSC) was established early in 1948, George Kennan's Policy Planning Staff in the State Department called for the institutionalization of what it called "organized political warfare."[1] Political warfare, their memo explained, meant the "employment of all the means at a nation's command short of war, to achieve its national objectives." It was simply "the logical application of Clausewitz's doctrine in time of peace." It had helped create and sustain the British Empire. The Truman Doctrine, the Marshall Plan, and sponsorship of the Western Union (see chapter 2) were "all political warfare and should be recognized as such." But it was the Kremlin's political warfare that was "the most refined and effective of any in history." The Soviet Union was, in the words of an earlier NSC document, "conducting an intensive propaganda campaign directed primarily against the US and [was] employing coordinated psychological, political and economic measures . . . not merely to undermine the prestige of the US and the effectiveness of its national policy but to weaken and divide world opinion to a point where effective opposition to Soviet designs [was] no longer attainable by political, economic or military means."[2] The use of both overt and covert political warfare were authorized shortly thereafter by NSC10/2.[3] Military operations were specifically excluded, but the directive allowed propaganda and economic warfare as well as "subversion against hostile states, including assistance to underground resistance movements, guerrillas and refugee liberation groups, and support of indigenous anti-Communist elements in threatened countries of the free world."[4]

The Soviet Union was the prime, but by no means the only, target of political warfare; the situation in Western Europe was also extremely worrying for the U.S. administration. The French and Italian Communist parties organized militant campaigns against the Marshall Plan in the streets, in factories, and through the media.[5] In the name of an anti-imperialist, anticapitalist struggle for national independence, they derided the political parties who supported the plan as lackeys of the United States. The Soviet Union's accomplishments were praised, and its supposed desire for peace was contrasted with the United States' allegedly bellicose posture.

The ideological assault did not stop there however. As shown in the previous chapter, and as Pierre Grémion has stressed, the Communist forces did not only "attempt to organize the masses, they also sought to seduce and attract the intellectual elites."[6] Capitalizing on the fears of a nuclear holocaust, the Soviet Union embarked on a highly publicized peace crusade, which coincided with Stalin's division of the world into "two camps" and the triumph of Lysenkoism.[7] A "Stalin Peace Prize" was awarded to politicians, intellectuals, and clergymen from all over the world who were leading the struggle for "freedom and independence, for peace and progress" against "the attempts of imperialist reaction to strangle the movement of the masses against a new pillaging war, being prepared by American billionaires and millionaires," to quote a *Pravda* editorial.[8] A World Congress of Intellectuals for Peace in Wrocław Poland, in August 1948 attracted four hundred delegates from forty-five countries, including a large contingent from France (Irène Joliot-Curie, Pablo Picasso, Paul Eluard, Fernand Léger, among others). A gathering of eight hundred at the Waldorf Astoria Hotel in New York (along with a mass meeting of eighteen thousand at Madison Square Garden)[9] was followed a month later by a second World Peace Congress in Paris in April 1949 and a parallel meeting in Prague (to circumvent the French government's refusal to grant visas to some delegates). The centerpiece of the crusade was, however, the Stockholm Peace Appeal, launched on 19 March 1950. It called for a complete ban on nuclear weapons and for the condemnation of any government who used them as guilty of war crimes. All "men of goodwill" throughout the world were asked to sign the appeal, and millions did: 15 million in France, 16 million in Italy—as many as 800 million worldwide, according to some sources.[10] Pablo Picasso's now iconic *Dove of Peace* was adopted as a symbol of the movement. Frédéric Joliot-Curie was militantly engaged in these initiatives. Capitalism, he said, always had, and always would, carry war within it, as the cloud carried the storm. Joliot-Curie announced that he would never use his science to build weapons for use by "the forces of regression" against the Soviet Union. As we saw earlier, he was relieved of his post as director of the French CEA soon thereafter.[11]

For President Truman, anti-Americanism in Europe was a product of Communist propaganda. The Kremlin was "seeking to discredit the United States and its actions throughout the world" through "abuse and vilification," he wrote in June 1950. If it succeeded "to create distrust and hatred of our Government and its motives," he went on, "the gains we have recently made in Western Europe may be substantially nullified. Our material assistance, to be fully effective, must be complemented by a full-scale effort in the field of ideas."[12] Truman oversimplified the problem. It was not only the Communist Party and its supporters who distrusted the United States' motives. For many Europeans, especially the French, American initiatives in Europe were motivated less by generosity than by self-interest.

The (Léon) Blum-(James) Byrnes accords of May 1946 traded emergency U.S. aid for a lowering of economic barriers and led to an influx of both material and cultural American goods.[13] The announcement of the Marshall Plan the following year was refracted through this prism. According to a poll conducted in France at the time, 47 percent of those asked thought that the United States only wanted to improve Europe's standard of living so as to expand the market for its products and create new opportunities for American investment. The rhetoric of compassion hid the "true motives" of the United States, which were to enroll uncritical consumers and loyal allies in the struggle against the Soviet Union.[14]

The Catholic Church was as disturbed by "soulless" American materialism and mass consumerism as it was hostile to Communism.[15] Political leaders in both France and Italy balked at the loss of sovereignty that dependence on American support entailed. Having just shaken off one occupier, and the shame of defeat and collaboration, many people in France felt that another foreign power was invading their economic space with the collusion of a supine political class. Businesspeople were not convinced that mass production and an emphasis on output and consumption were appropriate to the relatively small European markets, where capital was expensive and labor cheap, and where worker-manager relationships were infused with notions of class struggle.[16] Thus an unusual anti-American alliance of church, state, and business gelled around the fear of being taken over by an alien power.

The threat to European culture was felt acutely. The assault by U.S. popular films on France's "artistic" cinema in the late 1940s, facilitated by the Blum-Byrnes agreements, undermined a key source of national pride and a bearer of cultural refinement that had partly compensated for the loss of influence and the palpable hegemony of the United States.[17] The increasing presence of mass-produced U.S. goods in French stores threatened the "Coca-Colonization" of a fragile national identity, and even to invade the deepest layers of the psyche, leading to what film director Wim Wenders called "the colonization of the European subconscious."[18]

U.S. anti-Communist measures and populist hysteria further alienated European intellectuals. Art, music, film, literature, and science: in the United States all forms of intellectual production were suspected of harboring politically undesirable elements that were a threat to national security. The House Committee on Un-American Activities (HUAC) forced the State Department to recall an exhibition of contemporary art traveling in Europe and Latin America in 1947 on the grounds that half of the artists had once sympathized with the Communist Party. A worldwide exhibition of American twentieth-century paintings had to be canceled in 1956 because too many of those represented had politically unacceptable backgrounds. An international tour of Toscanini's NBC Symphony of the Air was called off by the State Department because 4 of the 101 musicians allegedly harbored "pro-Communist attitudes." In 1953 the State Department imposed a blanket ban on the distribution of any works written by "controversial persons, Communists, fellow-travelers," witnesses who refused to testify before congressional committees, and authors who were deemed too left-wing or too critical of American values and policies. Three hundred titles were withdrawn from overseas libraries, and embassy officials even burned eleven of the books.[19] In 1947 the HUAC and the leaders of the movie industry insisted that artists who wished to work again "name names" of Communists that they had worked with. Eric Johnston, the powerful head of the Motion Picture Producers' Association, insisted that the big screen stop projecting a negative image of American capitalist society: "We'll have no more films that show the seamy side of American life. We'll have no pictures that deal with labor strikes. We'll have no pictures that deal with the banker as villain."[20]

Scientists both at home and abroad who were suspected of harboring Communist sympathies were harassed and humiliated. The State Department's refusal to issue passports to American scientists and visas to foreign scientists who wished to visit the United States in the 1950s was sometimes incredibly arbitrary and always immensely destructive of America's claim to be the bastion of individual freedom and democracy.[21] Typically, biologist Jacques Monod, who had belonged to the Communist Party from 1943 to 1945 but had completely renounced his affiliation by the time of the Lysenko debacle, was deemed to be an "inadmissible alien." He described the procedures that he would have to follow to be allowed to enter the United States—"an exceptional and temporary favor of which I am legally assumed to be unworthy"—as "extremely distasteful" and reminiscent of the kind of "inquisition" introduced by the Vichy administration in occupied France during the war. He was one of 70–80 percent of French scientists who found their requests for visas delayed, blocked, or refused in the early 1950s.[22] Even so-called independent foundations like the Rockefeller were not immune to pressures to cut off support from undesirables, as seen in the previous chapter with the Ephrussi case.

There was considerable debate about how best to achieve Washington's objectives in continental Europe (and in the Soviet Union). It was generally felt "that the creation of an American-led Western alliance would fail—that Marshall economic aid and the security structures of the North Atlantic Treaty Organization (NATO) would evaporate—if Washington's efforts were not accompanied by an equally successful assertion of American cultural hegemony."[23] But what was the best way to do this, given the deep suspicion of U.S. motives in Europe, and the exploitation of anti-American sentiment by the Soviet Union and national Communist parties? President Truman felt that the problem was essentially one of misunderstanding. The solution lay in telling the "plain, simple, unvarnished truth" about the country and its motives. Speaking to the American Society of Newspaper Editors on 20 April 1950, he called for a "great campaign of truth" that would "make ourselves known as we really are—not as the Communist propaganda pictures us."[24]

Truman's conviction that all that was needed was to provide better and (supposedly) neutral *information* about the United States was strongly contested in a report on political warfare prepared for the State Department by the Massachusetts Institute of Technology (MIT) and delivered in February 1951. Project Troy, as it was called, was a "summer study" organizationally modeled on Project Hartwell, which MIT had just finished for the navy.[25] Its original aim was to study "ways of getting information behind the iron curtain," to quote MIT President James Killian, and specifically to stop the Soviet Union's jamming of Voice of America broadcasts.[26] The scope of the study was rapidly expanded, however, to include social scientists from both MIT and Harvard—anthropologists, economists, historians, and psychologists. They argued for a richer concept of communication than simply the dissemination of "factual" information. "Political warfare" was far more subtle and sophisticated than the traditional forms of psychological warfare inherited from the Second World War and went beyond mere propaganda or the affirmation of the "truth." On the one hand, it exploited every available means to disseminate a positive image of what kind of society America was: radio certainly, but also professional journals, industrial and commercial publications, as well as intellectual exchanges, food shipments, and objects typical of American life ("drugs, flash lights, fountain pens, small radio receivers," which carried "their own messages of American skill and interest").[27] On the other hand, it had to be sensitive to the properties of the "target": "the factors—cultural, psychological, institutional, political, economic, philosophical—which affect the way people interpret the information and the way they react to it."[28] Clyde Kluckhohn, the eminent anthropologist who directed Harvard's Russian Research Center and who was studying defectors from the Soviet Union, was particularly emphatic about this. The projection of a "full and fair picture" of the United States behind the iron curtain that was intended to refute Soviet claims was "utterly inadequate as the center of our program," he wrote. It was irrelevant because it mistakenly presupposed that "the underlying sentiments, assumptions, and general frame of reference" of the target audience "were very similar to our own."[29] To destabilize Communism, one had to relate one's actions to the worldview of the target and undermine it from within. Hence the value of material and intellectual support to Tito's Yugoslavia, from food to physics (see chapter 3). As the main section of the Project Troy report put it, "Demonstration that a former Russian satellite like Yugoslavia becomes more securely prosperous when it enters the Western orbit, shakes the adherents of Russian communism throughout the continent."[30] Robert S. Morrison, the Rockefeller Foundation's associate director of Medical Sciences made a similar point. To disseminate a positive image of the United States in technologically backward regions of Asia, he said, young Americans should be encouraged to live among the people in remote villages for several years and to put in place locally adapted Western techniques of public health and agriculture. "If they were the right sort of representative Americans," Morrison went on, "they would also make use of their position to transmit *almost automatically* American ideas of cooperation in the common job, respect for individual dignity, and the free play of individual initiative."[31]

The view that the best way to promote a positive image of the United States abroad was not necessarily by countering Communist propaganda head-on with an "information" campaign, but rather by exploiting multiple channels of communication to relate American strengths and values to the needs and desires of the target became something of an orthodoxy among many liberal American intellectuals in the CIA, the State Department, and the foundations in the early 1950s. High culture—art music, literature, science—was one of the weapons used to attack anti-Americanism in Western Europe, its target: the non-Communist left-wing intelligentsia. The persuasive power of this form of cultural diplomacy, this "politics of apolitical culture" in Giles Scott-Smith's happy phrase, lay in its showcasing of individual freedom of thought and intellectual creativity as central values of the Western democratic tradition.[32] In the words of American radical publisher James Laughlin, the point was not "so much to *defeat* the leftists intellectuals in dialectical combat as to *lure* them away from their positions by aesthetic and rational persuasion." The *omission* of propaganda for "the American way" in any program intended to give European's a better appreciation for the country's intellectual achievements, wrote Harvard historian Perry Miller, "would in itself be the most important element of propaganda, in the best sense."[33] The most effective way to expose the intellectual sterility and conformity of Soviet Communism was not to preach, but to *display* the exhilarating variety that was the lifeblood of Western and U.S. democracy, to contrast the freedom of an "open" society to the suffocation of a "closed" one. (The Fulbright program set out to do just that.)[34] The secretary of state regarded this as the obligation of every American. "In a struggle where freedom is the issue," Dulles wrote, "the only adequate exponents of freedom are free people. . . . So your Government appeals for your individual demonstration, at home and abroad, of freedom so significant, so dynamic, so penetrating that it will be for all men a symbol of hope."[35]

To achieve these ends, after the outbreak of the Korean War and the consolidation of NATO, Washington set out to build, as David Ellwood puts it, "an elite body of opinion which would unite like-minded thinkers from all the [NATO] member nations around a common body of Western ideals. These would then be projected, in translation, into many different contexts by well-known opinion makers and public figures." The ultimate aim was "to foster a belief in 'Atlanticism' and a sense of shared values between Europe and America."[36] This form of "indirect propaganda," intended to change basic social, political, and economic attitudes in Europe, was achieved through organizations that downplayed or even concealed, rather than proclaimed, their American allegiances.[37] The Congress for Cultural Freedom (CCF) was a typical organization of this kind.

The CCF emerged from a meeting of the Kongress für kulturelle Freiheit that was held in Berlin in June 1950 and was a direct response to the Communist-inspired peace crusade that began with the meeting of intellectuals in Wrocław in 1948 and reached its summit in the Stockholm appeal of March 1950.[38] Just as Wrocław (formerly Breslau) by its very name symbolized the transformation of a German city into a Polish one in the Eastern bloc, so Berlin stood for an island of freedom in the encircling sea of East Germany. For a year or two, the CCF was torn by rivalry between those who wanted to turn it into an "ideological combat unit" against the Soviet Union, and whose anti-Communist fervor went so far as to condone the excesses of McCarthyism in the United States, and their opponents, who believed that this approach would alienate European intellectuals and fuel their anti-Americanism.[39] The latter faction prevailed, and the Congress became a major instrument of cultural diplomacy in Europe. It supported a number of intellectually prestigious journals—*Encounter* in Britain, *Preuves* in France, *Der Monat* in Germany, *Tempo Presente* in Italy and, beginning in 1960, Edward Shils's science policy journal *Minerva*. It organized an important arts festival in Paris in 1952, which was intended both as a challenge to "Socialist realism" and to the "disdain in which West Europeans tended to hold American culture," affirming "a common Atlantic tradition and movement in modern art, of which American composers, writers, and painters were an integral and indeed leading part."[40] In 1953 the CCF, with Michael Polanyi, held a conference in Hamburg on the theme "Science and Freedom" directed against Lysenkoism and state planning of science. Polanyi, along with the French sociologist Raymond Aron, was crucial in the planning of the next conference in Milan in 1955, where 140 intellectuals debated the "The Future of Freedom." They celebrated the "end of ideology" proclaimed by Edward Shils and Daniel Bell, deplored its dogmatic inflexibility and abstract theorizing, and advocated American-style pragmatism in the social sciences and applied research directed to social engineering.[41] The scope of the CCF's activities was also extended to include the third world, which had become the new battleground for hearts and minds against the lure of Marxism.

The CCF was an expensive operation: by 1960 its budget had climbed to about $1 million (it was double that by 1966). Much of this money came from the CIA with some help from the State Department and, notably, a number of foundations: the Fairfield Foundation, the Hoblitzelle Foundation, and the Ford Foundation.[42] Many of those who benefited from its largesse were unaware of the intelligence connection and felt deeply betrayed when it was exposed in the mid-1960s. Their aim had been merely to stimulate intellectual debate; unbeknownst to them, their work had been instrumentalized as a weapon in the intellectual Cold War. The CIA apparently made little explicit attempt to determine the content of the activities funded by the CCF—the point was, precisely, that these should represent the diversity and richness of a shared "Atlantic" culture. Yet there was deep irony in using a covert source of funding (from an agency dedicated to doing anything to preserve U.S. interests abroad) to celebrate openness, tolerance, and the critical spirit that were the hallmarks of democracy.[43]

By the second half of the 1950s, the stakes in the intellectual Cold War had changed. In February 1956, at the Twentieth Congress of the Communist Party, Soviet leader Nikita Khrushchev exposed and denounced Stalin's regime of terror, the purges, and his cult of personality. That summer, workers and students in Poland rebelled against the party leadership, and order was only restored under the threat of a Soviet invasion and an agreement that Poland would remain a member of the Warsaw Pact. Similar revolts in Hungary were suppressed a month later in November 1956, when Soviet tanks moved into Budapest, killing thousands. For all but its most dogmatic partisans, Soviet-style Communism lost whatever credibility it had ever had. The European economies began a period of sustained growth, the *trente glorieuses* as they are called in France, and while scholars still disagree as to whether or not this was thanks to the Marshall Plan, the new prosperity cut the ground away from the national Communist parties' platform of the late 1940s: as one Italian left-wing militant put it many years later, "The American myths kept their promises and won through!"[44] A new generation came of age in the 1950s, one that lived and benefited from those promises, felt ill at ease with European bourgeois culture, despised European feelings of intellectual superiority, and for which America signified modernity and the liberation from stifling European race- and class-consciousness. It was a generation that *"actively partook"* in the "stupendous success of U.S. popular culture in postwar Europe."[45] Now the major task of cultural diplomacy could be reoriented. Communism was no longer a serious intellectual or cultural threat. What was needed was to consolidate the Atlantic community around a shared body of democratic values and ideals and to weed out residual anti-Americanism and "neutralism" among the intellectual elites. That was not going to be easy. In 1959 Ford Foundation officer Waldemar Nielsen attended a meeting of writers at Lourmarin, France, that had been organized by the CCF. Nielsen noted a new self-confidence among the businessmen and government officials whom he had met on his European tour. But he deplored what he called the "sickness of European intellectuals," notably the French, as represented at the meeting. As he put it in his report to New York, many of them were "onetime Communists or fellow-travelers." Even if they were no longer formally Marxists, their souls were still plagued by intellectual and moral conflict. They did not speak of the Soviet Union or Communism any more, but "they spent a lot of time worrying and stewing and griping about the United States, about American domination, about the inferiority of American values, and so on."[46] Defeating Communism was one thing, but building an Atlantic community under American leadership with a strong scientific, technological, and ideological base was quite another. This and the next two chapters will study some of the measures taken by the Ford Foundation and by NATO (alone and sometimes in unison) to do so.

The intellectual Cold War has been more or less completely ignored by historians of the natural sciences. For many years, historians and social scientists have studied how academics in disciplines as different as anthropology, economics, sociology, and management studies were enrolled in the United States' efforts to project and consolidate its power abroad, engineering societies in its own image.[47] Historians of the natural sciences, and of physics in particular, have not participated in this debate. Of course there have been many studies made of international scientific exchange, notably between physicists, who have themselves actively promoted the idea that their circulation across national boundaries, and behind the iron curtain in particular, has been an invaluable instrument for overcoming "misunderstandings" between the superpowers and for making a dent in the Cold War.[48] The putative link between science and democracy, that is, the idea that the epistemological approach enshrined in the "scientific attitude" promotes "democratic" values of individual autonomy and critical thinking—an argument deftly used by the Rockefeller Foundation—has also received considerable scholarly attention. But while David Hollinger, in particular, has explored the use of science-as-democracy as a "weapon in *Kulturkämpfe*," he has restricted this to the culture wars *within* the United States.[49] To my knowledge, no one has fused these two perspectives to show how international scientific exchange (in physics) was an instrument in the intellectual and cultural Cold War abroad that was waged by the U.S. administration and some foundations to promote democracy in the struggle for "hearts and loyalties" and how it was intended both to build Atlanticism and to promote a positive image of the United States. In the previous chapter I looked closely at how one major American foundation dealt with the Communist threat in the early 1950s. In this chapter I consider how another used international scientific exchange at CERN and the Niels Bohr Institute in Copenhagen as an instrument in the intellectual Cold War.[50]

The Ford Foundation's First Steps toward an International Program

Henry Ford and his son Edsel established the Ford Foundation in 1936. At first its emphasis was mostly local and about $1 million was disbursed for projects that reflected the family's interests in Michigan. In 1948, following the deaths of the two founders, the trustees realized that the increased income and additional funds available to the foundation after the settlement of the two estates would enable it to embark on a greatly expanded program. Thanks to a booming stock market, the assets of the foundation increased to $492 million, and the trustees had nearly $69 million available for philanthropic activities in 1950.[51] Ford had become, almost overnight, the biggest philanthropy in the world. A study committee chaired by H. Rowan Gaither Jr. was set up to define a philosophy appropriate to the times and to the resources that the foundation expected to have at its disposal.[52]

Rowan Gaither was no stranger to the world of science or its intimate links with foreign policy and national security.[53] A lawyer by profession, in 1942 he was appointed associate director in charge of administering the MIT Radiation Laboratory. There, in the words of one commentator, he coordinated the laboratory's research activities, notably the development of radar, and acted as a liaison between its scientific staff and the armed forces. This role enabled him to build up relations of trust, mutual respect, and friendship with several leading scientists and science administrators, including laboratory director Lee DuBridge, assistant director Isidor I. Rabi, Ernest Lawrence, and J. Robert Oppenheimer. This network was not dissolved after the war, when Gaither went back to California. He assisted in converting the Rand Corporation from a branch of the Douglas Aircraft Company, performing R&D for the Army Air Forces, into an independent think tank.[54] He joined Rand's board of trustees in 1947 and served as chairman of the board almost without interruption until his death in 1961. Gaither was nominated an associate director of the Ford Foundation in 1951 and became the president of the foundation in 1953. Just before he left Ford in 1956, due to ill health, he was invited by President Eisenhower to head an ad hoc panel to evaluate the vulnerability of the United States to nuclear attack. Gaither had thus an informed and respected understanding of scientific and nuclear issues, and close contacts with both the elite of the U.S. physics community and the centers of power in Washington.

The members of the committee Gaither assembled for Ford in 1949 came from all walks of life—business, health, education, and science— the last in the person of Caltech's Charles Lauritsen, an applied nuclear physicist who had played a major role in the institute's missile development program during the war. Gaither's team consulted widely, traveling over 250,000 miles and speaking to over a thousand experts in the United States. Their *Report of the Study for the Ford Foundation on Policy and Program* was published in September 1950 to wide acclaim. To quote Francis Sutton, who spent thirty years with Ford, within the foundation it became "a kind of sacred text, scrutinized for many years by those charged with planning or justifying the Foundation's programs."[55]

The Gaither report fell squarely within the context of the Cold War. History was at a turning point, it claimed, with a choice between two paths, American-style democracy and Soviet-style tyranny. The philanthropic goal of the foundation—to advance human welfare—needed to be redefined and explicitly associated with the promotion of democracy. This was to be achieved by addressing five "problem areas": world peace, the strengthening of democracy, the strengthening of the economy, education in a democratic society, and improved scientific knowledge of individual behavior and human relations. The foundation sought essentially intellectual intervention: "The advancement of the ideals and principles of democracy," Gaither explained later, was to be achieved through education and the diffusion of knowledge. "The fundamental premise of our entire effort," as he put it, "is educational in concept and purpose."[56]

Gaither's list of problems, with one exception (the support for education as such), matched the postwar programs of the Carnegie Corporation and the Social Sciences Division of the Rockefeller Foundation. It attested to a shared conviction among these philanthropic bodies that the United States had emerged from the war as the most powerful nation in the world but with little solid and reliable knowledge about peoples and "areas" of the globe that were now central to its foreign policy and national security. The foundations would step in where the federal government and the new National Science Foundation feared to tread (partly because social research had little status, was deemed unscientific, and smelt of "socialism") and support "interdisciplinary," "problem-oriented" studies of human and social behavior.[57]

The first president appointed to implement this program was Paul Hoffman. Hoffman had been president of the Studebaker Corporation since 1935. In 1948 he became the administrator of the Economic Cooperation Administration (ECA) set up by President Truman to administer the disbursement of Marshall Plan funds (see chapter 2). When he took office at Ford in January 1951, East-West tension had so intensified—the Korean War had broken out six months before—that he consulted with Gaither about redefining the objectives of the foundation. The threat of war, said

Hoffman, "is so great that the United States and the free peoples of the world must mobilize their economic and human resources to restrain aggression, to achieve peace, and to assure victory in case of war." The foundation's programs, he said in April, should "reflect the sense of urgency of problems which confront the free world" and should prioritize "problems which arise out of the intensifying international crisis." This meant mitigating East-West tensions (for example, by providing support for refugees in Europe), strengthening the United Nations, which many still regarded as an essential instrument for maintaining world peace, and promoting understanding among peoples. For Hoffman, and many others at the time, international tension and the danger of war were fuelled by mutual distrust and ignorance. Programs encouraging "the large scale exchange of persons" were "imperative" "as a means of strengthening the free world and of promoting international understanding."[58] Robert M. Hutchins, a former chancellor of the University of Chicago and one of Hoffman's associate directors, was also convinced that "people of one country or race hate those of another because they do not know what they are really like. . . . Cultural exchange can do much to dispel these misconceptions."[59] The foundation's annual report for 1951 stressed the importance of cultural diplomacy: "The exchange of ideas, and possibly also of artistic and literary productions, [is] one of the most promising methods of fostering the development of world understanding and a sense of moral and cultural community among the peoples of the world."[60]

To implement his programs, Hoffman recruited a number of people who were familiar with the situation in Europe, notably Milton Katz, John J. McCloy, and Shepard Stone. Katz was a Harvard law professor who had been Hoffman's former deputy at the ECA office in Paris. McCloy had just completed a term of office as the U.S. high commissioner in the now partly sovereign West Germany and was elected to the foundation's board of trustees in May 1953. Stone had been McCloy's right-hand man for public affairs in the Federal Republic and had actively promoted academic and cultural initiatives in that country intended to instill democratic values among the people. Hoffman also spoke to Richard Bissell Jr., another senior Marshall Plan official who had consulted for Project Troy while at MIT. Bissell moved to the CIA in

1954 hoping to be more effective and believing that academic research and scholarly reports were unlikely to influence policy or change the course of events in any substantive way.

Hoffman's tendency to recruit people and impose programs without consulting the board and his neglect of his duties as he campaigned for Eisenhower (he was disappointed not to be nominated the new president's secretary of state) soon led to his downfall. In the short time in which he held office, however, he left an indelible mark on the overall direction of the foundation's programs. In particular, his Conditions for Peace program was regarded as one of his most important legacies. Bissell's interpretation of the program was spelled out in a major thought piece that he wrote in March 1952. For him, nuclear war with the Soviet Union was unacceptable; yet disarmament, and so genuine peace, were too much to hope for. Thus the Conditions for Peace program, he said, should have the long-term aim of finding a way to "live in the same world with the Russians without going to war with them despite profound and continuing differences of philosophy and interest."[61] Stone and McCloy began to flesh out this agenda. They concluded that it called for both a domestic effort to educate Americans on the importance of promoting democracy abroad (as a counterweight to populist anti-Communism and isolationism) and an international program aimed at improving Soviet-American relationships, encouraging educational initiatives in Asia, and strengthening democratic institutions in Western Europe.

Rowan Gaither replaced Hoffman in February 1953. Gaither spent most of his first two years in office responding to charges from the Cox and Reece committees (see chapter 5) that the foundation supported socialism and was sympathetic to Communism. Ford's international programs, which only absorbed about one-sixth of its funds, were relatively unscathed by these attacks and may even have emerged strengthened from them. As Sutton points out, they escaped the congressional scrutiny given to domestic programs. They also enjoyed strong support in Washington and so gave both officers and staff the "sense that they were serving important national interests through Ford's international programs."[62]

Gaither was not happy with the organization of Ford's international activities when he took office, as "there remained hardly a strand of policy and program related to Europe."[63] He asked Don Price, who had spent fifteen years on various Washington assignments and on the staff of the Public Administration Clearing House, to help him streamline the foundation's overseas programs. Shepard Stone was given responsibility for international affairs and laid the foundations for a major international offensive that flourished once Henry Heald replaced Rowan Gaither as Ford Foundation president in 1956.

The Foundation's Early Interest in Bohr

One of those consulted by the Gaither committee on behalf of the Ford Foundation in 1949 was Richard Courant, who suggested that Niels Bohr would be an excellent instrument to promote "endeavors in the field of international intellectual cooperation" centered on Copenhagen. Bohr, Courant pointed out, was not just "the giant of modern physics" but also a humanist and philosopher who had devoted many years to "the problems of international relations and the maintenance of peace."[64] Bohr was indeed an obvious candidate. His views on building a safer world resonated in more than one respect with the Conditions for Peace program. Bohr also shared Hoffman's faith in the importance of the UN as an instrument for world peace—his famous *Open Letter to the United Nations* of June 1950 had pleaded for "full mutual openness" in the flow of atomic information as a means to lessen "distrust and anxiety" between the superpowers.[65] Similarly, for Bohr as for Hoffman, international scientific exchange was an essential part of this process. In the words of his friend and admirer, John Archibald Wheeler, Bohr believed that "Not only control of weapons but also the even greater rewards of understanding and confidence between different branches of humanity would be secured by full freedom to travel and to exchange information and ideas."[66]

A way of using Bohr occurred to Gaither in summer 1951, when he was serving as part-time director under Hoffman. His plan was vague—in the "thoughts while shaving" category.[67] He noted that visiting scientists had so little spare money when they arrived in the United States that it was impossible for them to depart from their schedules and to travel freely. This could be overcome by financing half a dozen eminent Europeans to travel in the United States for three months—exposure that would supposedly consolidate their professional, emotional, and ideological links with America.

Gaither gave two main reasons for enrolling Bohr in this venture. First, he would be an ideal vector to disseminate democratic values. His "dedication to the principles of the free world and his efforts to improve international understanding in the interests of peace" were known to all.[68] What is more, "his influence transcends national boundary," "his following extends throughout the free countries of the world," and "his views are almost doctrine to his friends and followers."[69] What was important to Gaither was not Bohr the theoretical physicist, but Bohr the charismatic international scientific statesman, Bohr the promoter of world peace, and Bohr the sure friend of democracy and the United States. Bohr was also the director of a prestigious international center, the Niels Bohr Institute for Theoretical Physics (NBI), in Copenhagen, a location that regularly hosted outstanding scientists from all over the world.

The interest of the NBI for Gaither was twofold. First, it was a "center where scholars from all fields gather and where national interests and barriers seem to disintegrate." Second, Gaither pointed out that "several of Bohr's friends" had suggested that "Bohr and his Institute offer a strategic point for advancing international understanding, including an understanding of American objectives."[70] As Gaither put it to the trustees in a report in June 1953,

Every act of the U.S. is watched with deep concern by people everywhere, since our conduct, as well as that of the Soviet Union, holds within it the promise of peace or the danger of war. Moreover, our power has brought with it the distrust and even the dislike that the less powerful often feel towards the strong, as well as the Communist-fed posture of a nation bent on imperialist exploitation of weaker peoples. . . . We must make sure that the rest of the world sees beneath our immediate policies the real values and intentions which motivate us, the real democratic values for which we strive.[71]

Gaither believed that Bohr and the NBI were eminently suited to this task. They could provide an independent, legitimate, and influential site for undermining fundamental misconceptions and stereotypes about American foreign policy among Europe's intellectual elite.

A concrete proposal emerged in a meeting between Courant and senior foundation officer John Howard in October 1951. Courant suggested that the foundation finance an exchange program enabling a number of scholars from hard-currency areas to visit Copenhagen, extending this later to include support for additional facilities to accommodate visitors.[72] The idea inspired little enthusiasm. Ford preferred to support problem-oriented programs rather than study trips. There were also many other strong contenders in both the United States and Europe for intellectual relations programs. What was needed was something more focused—for example, a project exploring how to apply Bohr's principle of complementarity to other fields, like psychology, biology, and economics—and sponsorship by an organization that had broader intellectual interests than the NBI, like the Danish Academy.[73] A multidisciplinary, problem-based topic like this, exploring the fundamental roots of human and social behavior, could hope for something like $30,000–50,000 annually to invite people to Copenhagen.[74]

It was not until May 1952 that Courant and Bohr found the time and motivation to propose something along the lines asked for by Ford. Courant wrote to the foundation suggesting that an award of $50,000 annually for five years be made to either the Royal Danish Academy, to the University of Copenhagen, or to Bohr's own Institute. It was to be used "for the promotion of a better understanding of the impact of modern physical science on problems of biology and psychology in the light of the ideas of the principle of complementarity." By bringing together experts in several fields, Bohr hoped that the lessons of "recent progress of physical science" might "contribute essentially towards clarifying the foundations for human cooperation in many domains, and in this way contribute to the common understanding between peoples with different cultures and backgrounds."[75]

This new project was also received lukewarmly. Hans Speier, the chief of the Social Sciences Division at the Rand Corporation (which was also partly funded by Ford), was "skeptical about the success of the proposed study." As he understood the project, the foundation was supposed to finance visits to Copenhagen of specialists in biology, psychology, anthropology, and so forth—different fields were specified on different occasions, further undermining the coherence of the proposal—who were supposed to interact with the physicists in Copenhagen interested in "broader humanistic endeavors." Speier believed that Bohr was worthy of support, but suggested that if the foundation wished to help him, it should do so for just two years and commit itself only to "assistance in preparing publications," rather than financing an exchange program.[76] Discouraged, Gaither withdrew his personal support from the proposal and suggested that the final decision be left to Milton Katz.[77]

This was tantamount to killing the project. Katz, in Sutton's view, was "a cautious and skeptical officer, more given to refining the quality and design of projects and programs than to taking a strong stand himself." Worst of all for Bohr and Courant, Katz "was opposed to the idea that the foundation should have a European program at all."[78] Not surprisingly he was most reluctant to support them, and by April 1953 he was happy to treat the whole affair as "dead."[79]

Gaither's attempt to promote Bohr failed, even though the project cohered with the Conditions for Peace agenda, and even after it had been redefined to fit more explicitly with a research program in the social and human sciences. This was partly because of its somewhat abstruse intellectual character, but also because there was no obvious bureaucratic niche for a program in Europe and because at least one key officer at the time thought the foundation had no business intervening in Europe anyway. Gaither was not discouraged: his respect for Bohr and for the possibilities offered by his institute was not shaken. In summer 1954 the now-president of the Ford Foundation visited Copenhagen and invited Bohr to New York. He suggested that the physicist meet with McCloy and other officers and trustees of the foundation and arranged for Bohr to have a company car for his personal use while in the country.[80] Bohr was being slowly drawn into the orbit of the Found Foundation by dint of shared ideological and political objectives and mutual personal esteem. Thanks to Shepard Stone, the trajectory soon brought the results that both parties hoped for.

Shepard Stone

In June 1954 Gaither nominated Shepard Stone to be one of three assistant directors of international programs, with specific responsibility for Europe. Slowly he began to put a program in place. By 1956 any residual doubts in the foundation about the wisdom of taking a major initiative in Europe were dispelled. The upheavals in the Soviet bloc following on Khrushchev's denunciation of Stalin created a refugee stream that demanded humanitarian aid and proved to be an invaluable Cold War weapon. At the same time, the United States' refusal to support Britain, France, and Israel in their wish to go to war against Egypt over the nationalization of the Suez Canal strained the Western alliance.[81] European issues were once again in the limelight, and Gaither's successor, Henry Heald, was determined to have an important program in the region. Stone was made head of what was now called the International Affairs Division and given "something like plenipotentiary powers across the Atlantic."[82] His operating budget exploded from $5.7 million in 1958 to $10 million in 1963 and to almost double that by 1966, releasing considerable resources to strengthen ties inside the Atlantic community and to capitalize on cracks in Soviet hegemony in the Eastern bloc.[83]

Stone, who changed his name from Cohen in 1929, is perhaps best characterized as a left-of-center Democrat and Wilsonian internationalist, a member of a cultivated, elite, liberal Jewish establishment.[84] He completed his Ph.D. thesis, on German-Polish relations, in Berlin in November 1932, just before Hitler came to power. Subsequently he and his German wife, Charlotte, helped refugees fleeing Nazism who arrived destitute in the United States. Stone participated in the Normandy invasion in June 1944. His unit was among the first to reach the Buchenwald concentration camp. Appalled, he could still believe in "another Germany," thanks to his personal ties and to his positive cultural experience during the Weimar days. He served as an occupation officer in the American zone for a short while before returning to the United States in 1946 to take on increasingly important responsibilities at the *New York Times*. He returned to the Federal Republic in 1949 to work under James McCloy in the Office of Public Affairs in the U.S. High Commission. He was elevated to the post of director in September 1950; within a year or two he was spending $4–5 million on supporting newspapers, pamphlets, journals, radio stations, newsreels, documentaries, U.S. cultural centers, and exchange programs. His aim was, above all, to "strengthen the development of German democracy" in the country that was "the communist point of impact in the heart of Europe."[85]

Stone was always aware of the political dimension of his cultural activities. In notes prepared after a European visit in summer 1954, he stressed that the foundation's support for basic research on social and economic problems, for schools of business administration, and for the social sciences were essential for strengthening "a free and democratic society in Europe."[86] As he put it afterward in an interview, the postwar reconstruction of Europe was not simply a question of economics and politics, but also of "education, science, culture and others. Let's face it, so many of these things do have political implications."[87] For Stone, cultural diplomacy was an essential component of the reconstruction of Western Europe and the consolidation of an Atlantic alliance under American leadership.

Stone's priorities reflected the changed international climate after Stalin's death and the less aggressive approach to the West being advocated by his successor. The tasks were now two: to maintain a dialog with the Soviet Union as a "condition for peace" and to consolidate the Atlantic alliance, above all by abolishing the weak link in the West's armory—the "neutralists," those intellectuals who were disaffected by the Soviet Union but who were unwilling to align themselves with the United States. Stone shared the growing consensus in the United States, canonized in the 1956 book by Carl J. Friedrich and Zbigniew Brzezinski, that Stalinism, like Nazism (or Communism, like National Socialism), was just another form of totalitarian dictatorship that had to be opposed. For him, working along with the U.S. administration to keep Soviet pretensions in check and to support dissidents from the Eastern bloc was the only intellectually and politically responsible thing to do. At the same time, Stone shared the contempt that many European intellectuals had for McCarthyism and for the mass American culture that invaded Europe in the 1950s as exemplified by vulgar Hollywood movies and rock-and-roll music. For him, though, this did not merit being anti-American, rudderless, overwhelmed by cultural pessimism, and prey to neutralism. There was another America—a democratic, open society whose contemporary art, literature, and music bore witness to its respect for the creative individual. Stone used his position at Ford to "defuse and depoliticize the vulgarities of American mass-culture" by conveying to Europeans the importance of the United States as a partner in the Atlantic community, with shared political and ideological agendas.[88]

This stance explains Stone's support for the Congress for Cultural Freedom, the first meeting of which was held under the aegis of McCloy and Stone in Berlin in June 1950. In 1954 Stone wrote that the CCF was "the most effective organization in Europe working among political, intellectual and cultural leaders."[89] Soviet repression in the Eastern bloc along with the Congress's promotion of high culture was the best antidote to European "myopia" and the best demonstration of the true nature of Communism and the common cultural heritage of the United States and Europe. By the early 1960s the Ford Foundation, through Stone, was contributing as much as half the annual budget of the CCF ($550,000 in 1960, for example).[90] A congressional committee investigating the charge that foundations were being used for tax evasion exposed its sources of revenue in 1967; in so doing, the committee also exposed the CIA's involvement. In the ensuing scandal, the federal government withdrew its support for the CCF. McGeorge Bundy, who had recently taken over the presidency of the foundation from Heald, decided to establish a new structure almost entirely funded by Ford, the International Association for Cultural Freedom. He also abolished the International Affairs Division inside the Ford Foundation. Shepard Stone left the foundation and, in 1967, became the first president of the revamped CCF. Tarnished by its past associations, his new organization never enjoyed the success of its predecessor, and within a decade it had effectively disappeared.

Stone was always comfortable in the company of powerful and influential people, and Bohr was no exception. Early in June 1955 he visited Copenhagen as Bohr's guest. His host's hospitality and stimulating conversation left a lasting impression on Stone. "The high point of my trip," he wrote on returning home, "was the two days as your guest in Copenhagen, talking and smoking a pipe with you."[91] Bohr, for his part, seems to have had a genuine affection for Stone, who became a close friend of the family, a relationship Stone judged to be "one of the stimulating and rewarding experiences of my life." After Bohr's death in 1962, the family invited Stone to a memorial service in his honor, an event that was, as he wrote to Aage Bohr afterward, an "occasion of personal significance for me."[92] It was this esteem for Bohr, coupled with the belief that international exchange was an invaluable instrument of cultural diplomacy, that led to the fellowship grants that the Ford Foundation made to CERN and to Bohr's institute in 1956.

The Fellowship Grants of 1956

The origins of the fellowship schemes at CERN and the NBI can be traced back to October 1954, the month after CERN officially came into being. Jean Willems, a key Belgian science administrator who had promoted the Geneva laboratory to his national authorities, visited Gaither to see if the foundation could support his Institut Interuniversitaire des Sciences Nucléaires, a nonprofit organization interested in the development of atomic energy for peaceful purposes. On returning home, Willems suggested that Gaither might be interested in supporting an application for a "Ford Foundation Visiting Professorship" at CERN, which would enable five outstanding American nuclear scientists to spend a year each at Geneva. This would help improve the laboratory's international standing and prestige and expose European scientists to "the knowledge and the enthusiasm" of their American colleagues.[93]

Willems's initiative provided Stone with an opportunity to see "what the Ford Foundation might do in the nuclear area in Europe."[94] The International Conference on the Peaceful Uses of Atomic Energy, held in Geneva from 8 to 20 August 1955, provided him with a helpful window on the situation. The meeting was one component of President Eisenhower's Atoms for Peace initiative, announced to the United Nations in December 1953. For the United States it served as an opportunity to promote civilian nuclear power, to enable third world countries to have access to research reactor technology, to get an insight into the state of the art in the Soviet Union, and to project a positive, nonbellicose image of the country abroad, winning hearts and minds in the global struggle for influence.[95] The Ford Foundation asked R. G. Gustavson, the director of Resources for the Future, to report on his impressions of the meeting and to inquire about CERN. Gustavson visited the construction site of the laboratory in the suburbs of Geneva and discussed prospects with its director general, Cornelis Bakker, and his predecessor, Felix Bloch.[96] He also spoke with a galaxy of physicists: Niels Bohr, John Cockroft, Manne Siegbahn, and The Svedberg from Europe; Ernest Lawrence, Walter

Libby, Glenn Seaborg, and Walter Zinn from the United States. Several important points emerged from these encounters, as reported by Gustavson.[97]

First, there was widespread enthusiasm about CERN and in favor of Ford Foundation support for the laboratory. Gustavson made much of Bohr's claim that CERN was "of the greatest importance not only to Europe but to the world because it will enable the European scientific fraternity to be reborn and Europe to once again make the contribution to basic science which it made in the past." All the Americans he consulted, with one anonymous exception, were also positive about the new Geneva laboratory.[98] In one of the distinguished evening lectures at the Atoms for Peace conference, Ernest Lawrence spoke of CERN as embodying the "finest flowering" of the kind of international cooperation the gathering was intended to embody.[99] Gustavson reported that nearly all of those he spoke to mentioned "the tremendous debt which America owes to the European scientific community. Without that community we undoubtedly would never have had the opportunity to become the leaders in atomic science which we are today."[100]

European scientists also insisted that the foundation should not help finance the basic construction and operation of the laboratory. That was the responsibility of the member states that had set it up, and nothing should be done that would allow them to renege on their responsibility. There were things that governments could not do, though, and here Ford could help. In particular, it could fund a fellowship program for scientists in countries who were not member states of the laboratory. Bohr mentioned "some of the poor countries" that "were unable to support CERN financially," and also Austria and Finland as potential candidates.[101] Bakker and Bloch agreed—but thinking of what Ford could do for the laboratory, rather than what the laboratory could do for the dispossessed, they also insisted that any fellowship program should make allowance for visits from leading American scientists, as Willems had originally suggested.

These positive reports suggested that not only Bohr's institute but also CERN might be suitable candidates for Ford funding. After returning home from the Atoms for Peace conference, Bohr wrote to Stone with a definite proposal that resonated with the overall U.S. aims for that event.[102] He emphasized the key role that his institute had played in fostering international collaboration and "atomic science" and the close contacts his group had had with "atomic energy research centers in the U.S.A. and various other countries." These activities, Bohr added, were limited in that usually he could only invite people to Copenhagen if they had their own funding. A grant from Ford could be used to support "senior experts" and "younger promising physicists" who would "receive further guidance in studies ranging from the epistemological foundations to specific nuclear problems."

Bohr carefully distinguished this request from potential Ford support for CERN. The two were complementary, he said, not competitive. The laboratory in Geneva was interested in studying the "very high energy" region, comparable to that of cosmic rays. By contrast, the "experimental as well as theoretical researches" at his institute were "more concerned with the lower energy nuclear phenomena, basic for atomic energy developments." By supporting both centers, the foundation could enlarge its "research programme in atomic science" in "a harmonious manner."

With two concrete suggestions before them, the officers of the foundation needed to work out how, if at all, they could justify awards to Copenhagen and Geneva. An internal meeting was held during the first part of December to define policy and, "for obvious reasons," to clear the details of the project "with responsible people in the State Department, AEC and other agencies."[103] Those consulted included Walter Whitman, who had served as the secretary general of the Atoms for Peace conference, Lewis Strauss, the chairman of the Atomic Energy Commission, and physicists Lawrence and Rabi.[104] Strauss confirmed that Bakker was "a staunch friend of the United States and of democratic principles generally," as well as being "a good administrator."[105] Others in Washington, including in the State Department, "all expressed strong support for the projects and said that American interests would be served by closer contacts with these institutions."[106] Lawrence told Gaither that "he could think of no government policy which would be inconsistent with the proposed grant [to CERN] and indeed thought that it was well within the boundaries of the U.S. Government's policies and purposes with respect to the international encouragement of the peaceful uses of atomic energy."[107] Thus assured that the proposals were intellectually, politically, and ideologically sound, and in America's scientific and strategic interests, on 8 December 1955, Stone drafted an outline budget after consultation with Bakker, Bohr, and Willems.[108]

Stone proposed that Ford award $400,000 to the European laboratory and $200,000 to Bohr's institute, both grants to cover five years. CERN's money would support five senior American scientists ($125,000— Willems's original proposal), and fifty fellowships of $4,000 each ($200,000). The balance, $75,000, was set aside to pay the costs of meetings and conferences "designed to increase contacts with non-European scientists." Half of the support for Bohr provided five $4,000 fellowships a year for "senior experts" and "outstanding young physicists," as Bohr had requested; $25,000 was earmarked for facilities at the institute. The remaining $75,000 was for "projects related to Prof. Bohr's ideas to bring nuclear science, philosophy and humanities closer together"—in fact, the original Courant proposal.

The board of trustees accepted Stone's recommendations. In detailed discussions with Bohr and Bakker over implementation, the parties stressed again the need to finance researchers who were not nationals of the home institutions.[109] On 27 April 1956, President Gaither officially approved the award to Copenhagen to support "study and consultation by young and senior physicists from outside Denmark."[110] The award to CERN that Gaither agreed to six weeks later similarly stipulated that it was for "cooperation with research primarily in U.S.A. and other non-CERN countries."[111] The foundation treated the grants as a single award with two institutional beneficiaries.

Ford, Physics, and the Cultural Cold War

The Ford Foundation did not regard the awards to the Niels Bohr Institute and to CERN as support for physics research. Indeed, Stone took pains to reassure Gaither that they were made "specifically for the expansion of international activities and not for the support of physical science as such."[112] The importance of this distinction in the minds of the foundation's officers is clear from their treatment of two other requests for funding from Bohr. In 1958 he asked for $500,000 for the construction in-house of a tandem electrostatic generator to do nuclear physics research. His request was refused on the grounds that it was irrelevant to promoting international relations.[113] By contrast, when Bohr, who immediately changed tack, asked if Ford might not like to put the money instead into a new building "to provide improved facilities for international co-operation in atomic physics," the initial reactions were more positive. In the event, Danish sources supported the extension of the institute's buildings, and Ford renewed its existing program instead.[114] But the point is clear. Ford gave money to the NBI and CERN to encourage the international exchange of persons, skills, and ideas, not for research equipment.

Eisenhower's Atoms for Peace program, and the outstandingly successful conference in Geneva in summer 1955, provided the overall rationale for Ford's intervention. They shaped the rhetoric surrounding the justifications written by Stone for the trustees, and they also affected the nationalities of the people who were supposed to benefit from the awards. Stone's formal request for support was headed "Atomic Development," and it situated the grant squarely in the nuclear energy field.[115] It began by emphasizing that nuclear science in Europe, which had previously made important contributions to American science, had fallen behind the United States due to setbacks suffered in the "Hitler period" and during World War II. "The problem is, thus, to renew this contribution by supporting the development of the peaceful uses of nuclear energy through increasing cooperation among American and European nuclear scientists." Stone stressed that before the war, Bohr's Institute had helped train an "extraordinary number of scientists . . . who later helped to produce the atom and hydrogen bombs." It spoke of CERN as "one of the major centers of nuclear research in the world," as the "central research institution in Europe for nuclear energy," and as a place that was "becoming a meeting place for scientists and non-scientists interested in the future implications of nuclear energy."[116] The official program action form gave the name of the grantee as the "European Organization for Nuclear *Energy* (CERN)." The error was corrected by canceling the last word and typing the word "Research" above it.[117]

The research in Copenhagen and above all in Geneva was mostly remote from civilian nuclear energy. But Bohr and Bakker understood that a foundation like Ford, uninterested in financing science per se, needed to relate its programs to foundation and State Department policies. Bohr, possibly prompted by Stone, wrote to him in September 1955 that the Ford Foundation might like to support the NBI "in connection with their [foundation] plans of promoting peaceful co-operation on atomic developments."[118] Bakker had to be more prudent, as he spoke as the director of a multinational high-energy physics laboratory that was under the watchful eye of the member state delegations. Fortunately he could trade on a widespread misrepresentation of the laboratory's aims: as Gunnar Randers, head of Norway's Joint Establishment for Nuclear Energy Research, put it, "In my work on international cooperation concerning atomic energy, I meet nearly in hundred percent of the cases the misconception that the CERN work is an European atomic energy development centre."[119] Bakker's grant application to Ford, which spoke simply of CERN as doing "pure nuclear physics research" without going into detail, was suitably circumspect, and surely did not disabuse anyone of the misconception bemoaned by Randers.[120]

The international scope of the grants was also supposed to respect the aims of the Atoms for Peace program to reach out to poorer countries. When the awards were first discussed in-house in the fall of 1955, officers stressed that researchers from underdeveloped countries would be offered fellowships. "It seems to me," wrote one of Stone's colleagues in November 1955, "that anything that is done by responsible organizations to advance the competency of people from the underdeveloped countries in the field of nuclear science is on the plus side."[121] Ernest Lawrence agreed that "Asian scientists need assistance and training and CERN provides an excellent way to take care of this need."[122] Before Gaither gave his final approval, Stone assured him that "the purpose of our grants was to make it possible for Indian and other non-European scientists to receive advanced training both in Copenhagen and in Geneva."[123] The document submitted to the trustees argued that support for the two centers was recommended specifically both because "European-American exchange in the nuclear area is of deep significance to the United States" and because "aid to scientists from the underdeveloped areas will benefit the Foundations program in those areas," notably Asia. In the event, over half the fellowships were given to American scientists, while India and Pakistan were poorly served. CERN used the first grant to support just one scientist from each. Bohr provided fellowships for four scientists from India and one from Pakistan.[124]

The "internationalism" that Stone and the foundation hoped to promote was restricted by geopolitical realities. As an early draft of the grant proposal put it, the awards were to promote "research work and consultation by young and senior physicists *from the free world.*"[125] This was not mere rhetoric; Stone explicitly urged the policy on Bakker and Bohr, who went along with it. Bakker and Stone came to a "gentlemen's agreement" that "CERN would not use the Foundation's grant funds to finance visits by scientists from the Communist countries."[126] The same arrangement was made with Bohr. In a letter to Copenhagen of 30 April 1956, Stone wrote that, to assure the consent of the trustees, "It would be very helpful to us if Ford Foundation funds were not used to support persons from Soviet Russia, Communist China, and the so-called Satellite countries."[127] Bohr agreed, even though this policy was in stark contradiction with his public stance in favor of open, international scientific exchange, including with colleagues from iron curtain countries.[128]

The geopolitical reach of the fellowship program changed with changing foundation policies toward the Communist world. After the tumultuous events of 1956 in the Soviet bloc, Stone relaxed the constraints on the national origins of Ford fellows at the NBI and CERN in line with the foundation's policy in the region. In April 1957 the foundation announced an award of $500,000 for intellectual exchanges between Poland and Western Europe and the United States. Stone was the architect of the program. No sooner was it official than he informed Bakker (and presumably Bohr) that the foundation would have no objection if CERN (and Copenhagen) invited Polish scientists as Ford Fellows.[129] A few months later Price told Heald that "State department officials and Allen Dulles, head of CIA, are urging the foundation to continue and expand the program in Poland, and also in Yugoslavia, Czechoslovakia, Roumania, and possibly the Soviet Union."[130] By 1958 international exchanges with the Soviet bloc became normal Ford Foundation policy, and the constraints on Bakker and Bohr as regards offering fellowships to nationals from the entire region were lifted.[131]

Communist China posed a different problem. In 1958 T. D. Lee and C. Yang, Chinese nationals resident in the U.S. who had just won the Nobel

Prize for physics and were seeking American citizenship, proposed that Bohr invite a few physicists from their homeland to Copenhagen, where Lee and Yang would meet them. Bohr checked first with Ford to see if he was allowed to use foundation money for this purpose. Stone consulted the authorities. The State Department was divided but, and this was crucial for the foundation, both Robert Amory, the CIA's chief of intelligence, and Allen Dulles "strongly favored such an action," on the grounds, according to Stone, that "the USA needs such information." Bohr's request was officially approved at a meeting between Stone, Price, and President Heald, and Bohr was told that he was free to invite whomever he chose from Mao's China to Copenhagen.[132]

In Bohr's global report covering the grant awarded to him in 1956, he stated that between 1956 and 1961 no less than 138 scientists from outside Denmark had participated "for longer periods" in the research work done at the Institute with funding from the Ford Foundation and other sources. Among them were four from Poland, six from the USSR, and six from Yugoslavia: no other Eastern European country was mentioned. The report stated that further approaches toward the Communist world were under way, including attempts "to establish co-operation with Chinese physicists." A subsequent report, written three years later, mentioned 222 non-Danish visitors who had spent more than three months at the NBI. Seven were from China, nineteen from the Soviet Union, eleven from Poland, five from Czechoslovakia, and a few each from East Germany, Hungary, Romania, and Yugoslavia.[133] East Germany apart, the geopolitical distribution of the countries on this list maps neatly onto that proposed by the CIA in September 1957.

It is clear from these episodes that Stone used the NBI and CERN as platforms for advancing the foundation's Cold War agenda in Western Europe rather than as centers for promoting the Atoms for Peace program, as his grant application implied. This was done both by encouraging exchanges between scientists from the "free world," in particular by enabling American physicists to spend extended periods in Copenhagen and in Geneva, and by offering fellowships to scientists from the Soviet bloc and Communist China. To ensure that the foundation and its grantees acted in line with American foreign policy and strategic interests, no step was taken without close consultation with the administration in

Washington. Stone "made a point" of talking "with some of the best minds I knew in the Federal government" and visited the capital three or four times a year to confirm that the awards he was making did not run afoul of U.S. interests.[134] The interactions across the iron and bamboo curtains made possible by his grants to the NBI and CERN were intended to spread an appreciation for Western democratic (and American) values and to enable informal intelligence gathering on behalf of the CIA. In fact it seems that in the early 1950s the CIA hoped to recruit Ford Foundation officers and some of its fellows as undercover agents for just this purpose: by one account, John McCloy was "brokering a clandestine relationship" between the foundation and the agency.[135] It was suggested that foundation officers and fellows could be asked to gather information about the situation in the Communist world. The scheme apparently came to naught, though in the late 1950s, it seems that "the names of Foundation fellows whose grants had already expired were regularly forwarded to the CIA" by Ford and by "dozens of other foundations."[136] Some grantees, including those of the Rockefeller Foundation, may have been approached and even recruited by agency officers to provide informal intelligence about the situation in the Communist world.[137]

Stone obviously had no hesitation about working with the CIA, whether by helping fund activities like the Congress for Cultural Freedom or by aligning his international fellowship program with its geopolitical strategies. A sensational recent history of the CCF by Frances Stonor Saunders describes in detail how British intellectuals, including the eminent poet Stephen Spender, who edited the journal *Encounter,* became enrolled in the activities of the CCF without apparently knowing that the CIA provided much of its funding.[138] Was Stone less than honest with a man like Bohr about the close associations he had with senior people in the CIA?

Definitely not. Stone never concealed his close, privileged personal links with senior officials in the U.S. administration, including the CIA. For example, in February 1958, he took Bohr to Washington, where they "had some fine talks with Allen Dulles [director of the CIA] and his boys, Senator Clinton Anderson [chairman of the Joint Committee on Atomic Energy from 1954 to 1956 and again in 1959], Senator Flanders [who led a campaign in the Senate to discredit McCarthy], and other dignitaries."[139] Later that year Stone introduced the Danish physicist to the secretary general of the CCF in Paris, Nicolas Nabokov, and apprised him of the congress's activities.[140] Both the CCF and the Royal Danish Academy of Sciences and Letters, whose president was Bohr, supported Hungarian refugee intellectuals and artists.[141] In short, Bohr was fully enrolled in the efforts by the foundation to support refugees from Communist regimes and was aware that Stone had close personal relationships with senior CIA officials. Whether or not he passed on this information to his friends, colleagues, and the fellows who came to the NBI—thus allowing them to refuse to collude in this political agenda—we will probably never know.

Niels Bohr's and Cornelis Bakker's grants from the Ford Foundation were not simply based on individual merit and on the prestige of the institutions they directed. They were also in recognition of their Cold War political allegiances: if these had been suspect, the funds would surely not have been forthcoming. Bohr and Bakker were not simply nodes in a network of international scientific exchange, they were also witting instruments of American foreign policy. There is a price to pay for patronage—"nothing so simple as coercion, though coercion at some levels may have been involved, but something more like the inevitable relations between employer and employee in which the wishes of the former become implicit in the acts of the latter."[142] The coproduction of hegemony requires the consensus of those who work with the United States to promote democracy, a consensus that sometimes involves making difficult compromises, not to speak of a certain amount of (self-) deception.

The Extension of the Grants

The foundation's awards to Geneva and Copenhagen were each extended twice. The trustees approved a second grant to CERN of $500,000 in December 1959 and a final grant of $250,000 in December 1962, bringing the total award to the laboratory to $1,150,000. Bohr's Institute for Theoretical Physics received $300,000 in 1959 and $150,000 in 1966, bringing its total to $650,000.[143] Atoms for Peace and civilian nuclear energy were no longer used to justify these grants. The argument for the award to CERN in 1959 spoke unambiguously of it doing research in "high energy physics."[144] The justification for the third and final grant spoke of CERN as "Europe's main center for fundamental particle research." One reason why CERN received just $250,000 and not the $750,000 it asked for was that "funds for research in the field of high energy physics could be obtained from many sources, particularly from governments."[145]

Along with the decoupling of the institutions from "Atomic Development" went an emphasis on the quality of the basic research done in Geneva and in Copenhagen. Victor Weisskopf, outstanding MIT physicist, Ford Foundation fellow at CERN, and CERN director general in the early 1960s, told Stone in 1959, even before its main accelerator had been commissioned, that the laboratory "had developed as a research center far beyond expectations. It was one of the most exciting, if not the most exciting, centers of nuclear research in the world." As for the NBI, Weisskopf described it as "the Mecca of all physicists. . . . Every American nuclear structure physicist of any standing must have spent a year at that place."[146] John Wheeler's immersion in weapons work (he played a major role in the development of the H-bomb) supplied him with the metaphor he needed to express his enthusiasm: "If a Martian were trying to degrade mankind's ability to make progress in this field [nuclear structure research] to the maximum extent and had only one bomb in his arsenal to accomplish this purpose," wrote Wheeler, "he would surely drop it on Copenhagen."[147] These evaluations were repeated in the officers' submissions to the Ford Foundation trustees.[148]

The success of CERN and the NBI in fostering international collaboration was again stressed. The former "constitutes a model for significant international co-operation in both scientific and non-scientific efforts," wrote the officers in their plea for the third grant.[149] The NBI served "increasingly to make its members, from whatever country, feel part of the same family of man—and act that way. . . . If it did not exist it would have to be invented."[150] The East-West connection was singled out for praise. Ford's grants to Bohr's institute, wrote Weisskopf, were supporting "one of the tenuous bridges between West and East which will be so essential in the future to preserve some unity among the nations." Its links with "Red China and other Asiatic communist countries under Chinese influence" were "a unique phenomena" that had to be maintained at all costs. The scientific relations between "Russia and the West" that Ford had fostered, Weisskopf claimed, had "helped to improve the political situation between these two countries" and had probably "prevented the outbreak of a major war."[151] The foundation's officers backed off from these exaggerated claims in their written submissions. But they made mention of CERN's importance for promoting "the Foundation's objective of advancing the development of East-West scientific exchange."[152] And they stressed the "special position" that the NBI had for forging "high-level links," particularly between scientists from the West and those from China and Eastern Europe. These made it possible for American researchers to develop contacts that would not otherwise be available.[153]

One new argument appeared in the documents justifying the grants to the trustees: the importance of the two centers as platforms to strengthen "scientific efforts in the Atlantic community in general, and in Europe in particular."[154] This was a key plank in U.S. foreign policy in the region and an issue that mattered deeply to Stone. Defending a grant for Jean Monnet's "Action Committee for the United States of Europe" just before the Treaties of Rome came into force on 1 January 1958, Stone told the trustees that "A strong European Community, closely associated with the United States, is of great importance to American national interest, and support for educational and research activities related to the European community is a principal element of the Foundation's European program."[155] By 1959 CERN, which had "proved to be a most successful institutional symbol of European unity and of Atlantic partnership," had become an ideal candidate for a grant.[156]

Concluding Remarks

The Ford Foundation, and Shepard Stone, were fully engaged in the cultural and ideological component of Kennan's program of "political warfare." In consultation with the State Department, the CIA, and scientific experts, notably Rabi and Lawrence, Stone joined the battle to strengthen democratic values in the Free World wherever they were threatened by the Communist menace; to build a positive image of the United States as open, innovative, and giving full scope to individual expression; and to strengthen the Atlantic alliance. The idea that science could be used as an instrument of foreign policy had been part of the State Department's thinking since 1950, when Berkner delivered his report on "Science and Foreign Relations" (see chapter 3). International scientific exchange was one of the instruments Berkner singled out, both for giving U.S. scientists access to new knowledge produced abroad and for intelligence gathering. For Ford Foundation President Paul Hoffman it was also essential to fostering world peace in a highly dangerous nuclear age. In embarking on this program, Stone had a freer hand than did any government agency. The presumed autonomy of the foundation gave him considerable legitimacy among the European intelligentsia (many of whom felt betrayed when Stone's close ties to the CIA were publicly revealed) and enabled him to finance a politically inspired agenda under the cover of an "apolitical" program.

The Niels Bohr's Institute and CERN were ideal loci for these programs. The unusual combination of esoteric basic research with the social weight and high-level contacts of many of the physicists who worked and visited there made them crucial vectors for keeping open lines of communication between nation-states and power blocs otherwise sharply at odds with one another. By using the NBI and CERN as hosts for international scientific exchange, Stone hoped to consolidate a transnational elite in the nuclear field whose members were not only drawn from the United States and Western Europe, but also from the Soviet Union and its satellites and from Mao's China.

CERN and the NBI's location on the geopolitical map helped to depoliticize the programs even more. They constituted supranational spaces not just metaphorically and symbolically, but also in bricks and mortar. One of the main reasons why the founders of CERN insisted that it be built on neutral soil in Switzerland was to indicate the strictly civilian nature of what was done there and its detachment from the immediate national industrial or military interests of any of the member states. Yugoslavia was one of its founding members, and any qualified scientist with an acceptable proposal, including someone from the Soviet Union or Communist China, could use its facilities. CERN was thus at once squarely in the Western camp and an international laboratory, open to visitors from the Communist bloc. Similarly, as John Wheeler wrote to

Shepard Stone in 1967, "the Niels Bohr Institute, located as it is in a small country, provides an 'international free port of ideas,' dissociated from the political overtones of a kind that no center in a large country, however good, can match."[157]

These three levels of relative autonomy from national interest—financial, scientific, and geopolitical—were harmoniously fused in the Ford Foundation's fellowship grants to CERN and the NBI. Situated in a denationalized space, Stone's program to advance the cause of peace and to demolish negative stereotypes of American society was stripped of explicit political and ideological overtones, and its cohesion with the aims of the State Department and the CIA was (temporarily) masked.

7

Providing "Trained Manpower for Freedom": NATO, the Ford Foundation, and MIT

The Soviet bloc, *not* the Atlantic Community, had taken man's first step into outer space. A country barely one generation into the industrial age had surpassed us.
U.S. Senator Henry Jackson, "Science and Freedom" (1958)

We do not need war. Peaceful competition is enough. . . . We will bury you.
Nikita Khrushchev

The Ford Foundation's support for European physics was not simply justified by the desire to promote scientific internationalism as an instrument of world peace. The original grants to CERN and to Bohr's institute made in 1955 were triggered by the ideological climate provided by the Atoms for Peace conference that summer in Geneva. But their renewal in 1959 was justified in quite different terms, as seen in the previous chapter: the two research centers were to be focal points to strengthen "scientific efforts in the Atlantic community in general, and in Europe in particular."[1] In fact, they were now part of an emergency effort by the Ford Foundation, along with NATO, to increase the pool of trained scientific and engineering "manpower" in Western Europe in the face of a new kind of threat from the Soviet bloc.

The launch of the Sputniks in October and November 1957 gave added weight to those who had been saying for some time that the output of trained scientists and engineers in the Soviet Union would soon outstrip that in the United States and its allies. Before these visible and highly publicized blows to U.S. pride, prestige, and national security, many in the West still had little regard for Soviet scientific and technological capabilities. The obstacles to the exchange of information imposed by the regime made it difficult to evaluate the state of Soviet science and technology.

Ignorance was fueled by propaganda, by naive assumptions about the incompatibility between science and Communism, and by the belief that Soviet scientific and technological achievements owed much to foreign help and to spies: Edward Teller was convinced that Klaus Fuchs's treachery had advanced the Soviet atomic bomb project by ten years, a ludicrous exaggeration.[2] Perceptions began to shift in some quarters at the Atoms for Peace meeting in Geneva, where the U.S. delegation was "surprised" by "the highly technical competence of Russian scientists and engineers generally and the large numbers of students in training in universities and technical schools."[3] Even then, until the Soviets launched Sputnik, General Medaris remembered afterward, many in the missile community regarded the Russians as "retarded folk who depended mainly on a few captured German scientists for their achievements, if any. And since the cream of the German planners had surrendered to the Americans, so the argument ran, there was nothing to worry about."[4] With Soviet intercontinental missile capability confirmed, complacency gave way to hysteria born of a sudden sense of vulnerability. There was a renewed and widespread emphasis on the need for improving the quantity and quality of highly trained scientists and engineers both at home and abroad.

This chapter and the next will discuss the steps taken by NATO, and by NATO and the Ford Foundation in concert, to increase the qualified pool of scientists and engineers in the Atlantic community to satisfy the growing needs of Western democracies and to challenge an alleged Soviet superiority. The launch of Sputnik was not at the origin of these anxieties, but it catapulted them to prominence for key decision makers in both the foundation and NATO. A sense of panic permeated Shepard Stone's argument to the trustees in December 1957 for a block grant of $150,000 to encourage "Scientific Activities in the Atlantic Community."[5] While the sum of money was far too limited to make a "broad or fundamental attack on the educational and scientific deficiencies of the Atlantic community," he pleaded, "these problems are so urgent that it seems important to provide limited and stop-gap help to prevent a weakening of determination during the coming few months."[6] The North Atlantic Council (NAC), also meeting in December 1957, was so concerned about the Soviet achievements that it established a new NATO committee to strengthen science and technology. President Eisenhower himself sponsored the idea: his science adviser, and president of the Massachusetts Institute of Technology (MIT), James R. Killian, was a member of the American delegation at the council meeting. Their "fundamental motivation" was that science's "strong international and cooperative nature would serve as an additional bond to strengthen Nato."[7] With the United States leading the way, millions of dollars were soon released for training and research, notably through the NATO fellowship program, but also by funding summer schools and research projects in militarily relevant fields like oceanography and operations research.

In this chapter I first unravel the historical origins of the NATO Science Committee, chart the intentions of those who served on it, and describe its overall programs. The level of documentation in the NATO archives (whose post-1965 holdings are closed to researchers) does not allow one to probe beyond numbers, however, or to study in detail the limits of the committee's efforts to "bring U.S. scientific sophistication to Europe," as Isidor I. Rabi put it. To that end, I also analyze, using other sources, the dramatic failure of Killian's radical attempt at institutional reform—the establishment, under the sponsorship of the Ford Foundation and NATO, of an Atlantic, or international, institute of science and technology modeled on MIT. In chapter 8, I probe even deeper, considering proposals to reform the content of the training required in NATO and detailing the activities and setbacks of the Advisory Panel on Operations Research, chaired by Philip Morse, also of MIT. The immense difficulty of exporting American educational structures into a Europe that was now economically and scientifically independent and able to assert itself more self-confidently against the United States, limiting the reach and penetration of empire, will inform my concluding remarks in chapter 9, where I draw together the threads of the overall argument in this book.

The "Manpower Gap" in the 1950s

After World War II, scientists were transformed into a coveted national resource every bit as important as oil or steel, thanks to the invaluable "gadgets" that they had helped develop, to the determination to maintain the West's scientific and technological supremacy, and to the rearmament triggered by the Korean War and consolidated during the Cold War. Speaking to the American Association for the Advancement of Science in 1951, Henry DeWolf Smyth, who had previously headed the Physics Department at Princeton and was then a member of the Atomic Energy Commission (AEC), said that "scientific manpower" was a "major war asset" that needed to be "stockpiled" and "rationed" just like any other scarce resource.[8] This manpower obsession, as David Kaiser has called it, led to a flurry of reports on the production of scientists and engineers from almost every quarter, including the Office of Defense Mobilization, the Office of Naval Research, the newly established National Science Foundation, and the American Institute of Physics.[9]

Physicists were particularly sought after by the Department of Defense, the AEC, and the major defense contractors.[10] Nuclear physics and solid-state physics flourished, being deemed particularly relevant to national security. To meet the increased demand for qualified manpower, graduate education was reconfigured. "Training" came to be valued more highly than research, and, as Kaiser has argued persuasively, there was a new "*pedagogical* emphasis upon *efficient, repeatable*—and thereby *trainable*— techniques of calculation, thus solidifying a prewar instrumentalist trend."[11] Physics was transformed from a calling into a career, in which one circulated smoothly (and without scruple) among academia, industry, and the military. The popular image of the physicist was reconstructed to fit with a new social climate and a new social role that prized conformity, and to destroy the pervasive view that scientists were intellectually and politically subversive.[12] To improve recruitment, MIT's Jerome Zacharias, along with a stellar panel that included Vannevar Bush, James Killian, Polaroid founder Edward Land, and Nobel Prize winners Edward Purcell and Isidor I. Rabi, set out to "Americanize" the image of the physicist. Their goal was "to show that a physicist was *not* a Hungarian with a briefcase talking broken English but the Ed Purcell's of this world and the J. R. Zacharias's, somebody who spoke English with no accent, who was one of the boys." Teachers were instructed to let students know that physicists were very much "like other Americans," and that most of them "marry, have children and belong to PTA's; some play golf and bridge and watch westerns on TV."[13] And indeed, as Kaiser has shown, it was just this image of affluent suburban "ordinariness" that was aspired to by many graduates in the late 1950s and that was invoked in promotional material by employers.[14]

Thanks to the huge influx of students into universities made possible by the G.I. Bill—when its effects peaked in 1947, 49 percent of college enrollments were veterans—and to the transformation in the identity and training of the physicist, "output" from efficient, if overcrowded, graduate programs soared.[15] Calculation techniques like Feynman diagrams were used in "Taylorist, factory-production style" to educate a new generation of socially respectable and politically reliable "manpower."[16] From about 250 in 1950, the annual number of Ph.D.'s awarded in physics by U.S. institutions grew to 500 in the mid-1950s, and increased sharply again thanks to Sputnik, reaching 1,000 by 1965.[17] Notwithstanding the expansion, demand still outstripped supply, at least during these two decades. Nor was the manpower crisis restricted to physics. M. H. Trytten of the U.S. Office of Scientific Personnel made the point forcibly in February 1955; speaking at the "Military Industrial Conference on Technical Manpower" in Chicago, Trytten needed no detailed statistics to support his case.[18] For him, it was self-evident that there had been a qualitative change in the demand for trained scientists and engineers due to the "spectacular growth of applied science, now reaching maturity in this mid-twentieth century. The personnel demands of our technological industries have doubled every ten years since the beginning of the century, with the exception of the last ten years, when the rate seems to have accelerated." This acceleration was due to "billion dollar research budgets for military purposes," and to the "birth of the massive new technological area of nuclear research and development." "The laboratory and the plant," Trytten suggested, "are no longer merely military suppliers. They are the very nerve centers in many respects of total strategy."[19]

Trytten pointed out that the United States had produced only 4,300 Ph.D.'s in all the sciences in 1954 and that this figure was likely to drop off. This was not enough. Indeed, Trytten said, "There has never been so much concern expressed on any one point in my experience as on this matter of adequate manpower in science and technology for a wide variety of defense-connected research projects in which present difficulties exist because of lack of personnel."[20] There was no surplus available to fuel an expansion in case of a national emergency. What is more, the demands of nonmilitary sectors were expanding rapidly. (Trytten remarked that Eisenhower's recently announced ten-year road building program, for example, would create a demand for 32,000 additional engineers.)[21]

Trytten was not merely concerned to draw his audience's attention to the difficulties faced by U.S. civilian and military industry. He also wanted to stress that the shortfall in qualified scientists and engineers posed a grave threat to national security. For America was dealing with a formidable foe, who believed firmly in the importance of science and technology for economic and military power, and who organized and directed training and research to meet national priorities. Trytten quoted the famous nuclear scientist Peter Kapitza to the effect that science under socialism needed to be "more systematic and conscious of its aims than it is in capitalistic countries." In the Soviet Union, Kapitza added, "the bonds between science and life must be closer and deeper" and not left to chance as in laissez-faire systems. Trytten insisted that Soviet education bore out this philosophy, arriving at the "startling fact" that in 1954 the Soviet Union produced about twice as many Ph.D.'s in the sciences, of probably comparable quality, as did the United States.[22] In Trytten's view, the United States risked "fritter[ing] away its 'technological superiority' " if drastic measures were not taken to address the growing "manpower gap" with its archenemy.[23]

The more people peeked behind the iron curtain, the more anxious they became. Returning from the Atoms for Peace conference in August 1955, AEC Commissioner Lewis Strauss remarked that what he had learned about the quality of Soviet science was "sufficient to shatter any complacency we may have enjoyed in regard to our own imagination and ability." What is more, the Soviets were not only training scientists as good as the best in the America, they were training more of them. "It is evident," he wrote, "that we are rapidly falling behind Soviet Russia in the training of engineers and scientists. The level of our reservoir of trained brainpower is dropping, in relation to demand," Strauss went on. "They, on the other hand, are striving to fill theirs to capacity as rapidly as they can."[24] Melvin Price, the chairman of the subcommittee of the U.S. Congress's Joint Committee on Atomic Energy that was responsible for research and development was equally alarmist. The country, he wrote in March 1956, was not producing enough qualified scientists and engineers for both the peaceful and military nuclear programs. This shortage of scientific manpower, he said, was "becoming increasingly urgent in view of the massive strides being taken by the Soviet world" in this regard. "When the Committee attended the Geneva conference last summer," Price added, "it gained a firsthand impression of this alarming fact." Immediate and strenuous measures were needed to resolve the situation: "at stake," said Price, "is not only our national defense and well-being but our ability to compete with the Soviets in the struggle for men's minds throughout the free world."[25]

Several attempts were made to quantify the manpower gap in the mid-1950s. Disquieting data were presented to a meeting of Eisenhower's cabinet on 30 April 1954. A brief report attached to the minutes pointed out that the Soviet Union was "training a body of scientists and technicians which is increasing in size and quality and approaching comparability with that in the United States."[26] A graph comparing the number of graduates per year in all scientific fields in both superpowers suggested that while the United States had far more trained scientists and engineers at its disposal at the time, the Soviet Union would soon challenge its leadership because it was training far more people annually to graduate level—in fact, twice as many (140,000/year versus 70,000/year) by 1956 if present trends were maintained. This graph was reproduced in part in the *New York Times* in November. A lengthy article on the front page of the Sunday edition drew the "inescapable" conclusion "that the United States is not educating a sufficient number of scientists, engineers, and technical personnel" that were "essential for survival in the atomic age." "The Soviet world," the *Times* warned, was "bending all its energies to win the race for technological supremacy."[27]

More reliable data were available in 1955, when Nicholas De Witt at Harvard University's Russian Research Center published the first substantial, well-documented account of the organization, functioning, and size of the Soviet educational system.[28] His report, entitled *Soviet Professional Manpower,* was commissioned by the National Academy of Sciences and the National Science Foundation. De Witt's figures were somewhat lower than those made available to the *New York Times* the year before, but the trends were similar. If one lined up figures available from his book with those from the U.S. Scientific Manpower Commission, it emerged that between 1956 and 1960 the United States would train annually about 57,000 scientists and engineers; the equivalent number for the Soviet Union was about 136,000. If the Soviets maintained that momentum, by 1960 the gap between the two countries would have been eliminated—there would be about one million trained scientists and engineers in both countries. Moreover, it was not simply a "manpower gap": already in 1954 about 50 percent of the people taking professional degrees in the Soviet Union were women.[29]

One obvious way of meeting this Soviet challenge was to improve the output of trained woman- and manpower in the Western alliance as a whole. Eisenhower and Secretary of State John Foster Dulles both strongly supported the moves, gaining momentum in the mid-1950s, toward the establishment of a European Economic Community (EEC) and its associated civil nuclear energy authority, Euratom. The latter, in particular, with its implications for a European nuclear capability, was linked in their minds to the notion of Western Europe as an independent "focus of power," of a "united Europe as a third great force in the World." "Were Western European integration to take place," Dulles remarked in this context, "this could remove the burden of Europe from the back of the United States, draw France and Germany together, and constitute a unified pool of power to balance the USSR."[30] Improving European scientific and technological capabilities in the NATO framework would make an essential contribution to "burden-sharing" in the defense of the West: it would create a stockpile of trained personnel superior in quantity and quality to that in the Soviet Union and its satellites.

During the early 1950s the Manpower Committee of the Organization for European Economic Cooperation (OEEC) had accumulated a considerable amount of data on the training of scientists and engineers in its member countries.[31] Building a quantitative picture of the relative strengths of science and engineering in NATO and the USSR using this data revealed that the gap between the Soviet Union and the European members of NATO was even greater than that between the two superpowers (see table 7.1).

In 1955 the European members of NATO were training only about 100 scientists and engineers to bachelor's level per million of the popula-

Table 7.1
Approximate Number of People Taking Bachelor's Degrees in 1955

	Population	Scientists	Engineers	Total
NATO Europe	255.9 million	12,000 (47)	15,200 (59)	27,200 (106)
NATO North America	176.1 million	23,600 (134)	24,100 (137)	47,700 (271)
Total NATO	432 million	35,600 (82)	39,300 (91)	74,900 (173)
USSR	200 million	15,000 (75)	60,000 (300)	75,000 (375)

Source: "Recruitment and Training of Scientists, Engineers and Technicians in NATO Countries and the Soviet Union," Report by Robert Major, Consultant to the Committee of Three, NATO, Document C-M(56)128, 26 November 1956. Figures in brackets are per million of the population.

tion, far fewer than the United States (271) or the Soviet Union (375). To retain parity between the NATO pool and the Soviets, a major effort was called for, not merely in America, but even more so among its NATO allies. In November 1956 U.S. Senator Henry ("Scoop") Jackson, a major figure in atomic energy and national security affairs, was asked to suggest appropriate measures.

The Jackson Report, "Trained Manpower for Freedom"

The bureaucratic origins of the Jackson report can be traced back to one of NATO's responses to the changed situation in the alliance in the mid-1950s. The division of Germany into two separate states was accepted as temporarily unavoidable, and each was integrated definitively into opposed geopolitical constellations of power. Most constraints on West Germany's sovereignty were lifted and the Federal Republic entered NATO in May 1955. Soviet Premier Nikita Khrushchev responded by establishing the Warsaw Pact, which included East Germany: the boundaries between two major blocs living in "peaceful coexistence" were consolidated. A fundamental reassessment of the nature of the Communist threat and how to address it was called for. In May 1956 the NAC asked the foreign ministers of Italy (Gaetano Martino) and Norway (Halvard Lange), along with Canada's Lester Pearson, to look into "ways and means to improve and extend NATO co-operation in non-military fields and to develop greater unity within the Atlantic Community." The Three Wise Men, as they came to be called, delivered their report in January 1957. It made several points that are pertinent here.[32]

First, the report stressed that "large-scale all out military aggression" by the Soviet Union against Western Europe had become unlikely. The USSR had come to realize that such an act would be met by "a sure, swift and devastating retaliation, and that there could be no victory in a war of this kind with nuclear weapons on both sides." In this climate of nuclear stalemate, the Cold War would be fought on other fronts by "non-military or paramilitary methods," involving "penetration under the guise of coexistence, with its emphasis on conflict without catastrophe." NATO had to recognize that it was faced with "an additional challenge in which the emphasis is largely non-military in character." What they called civil security could no longer be kept distinct from military security: there was an "essential inter-relationship" between the two.

Second, to assure civil security, one needed to breathe new life into the sense of belonging to an Atlantic partnership. In the context of a nuclear stalemate and the receding fear of a nuclear conflagration, what would hold together the loose association of states that was NATO? To prevent "the centrifugal forces of opposition or indifference from weakening the Alliance," one needed to strengthen the feeling of belonging to "an Atlantic community whose roots [were] deeper even than the necessity for common defence." Indeed, as the Three Wise Men stressed, when NATO was established in 1949, it was intended to be more than just a military alliance: it was supposed that "common cultural traditions, free institutions and democratic concepts . . . were things which should bring the NATO nations closer together, not only for their defense but for their development." NATO (in the words of its secretary general almost fifty years later) was "above all, . . . a community of values—values that are not negotiable, values that will be protected whenever they are under threat."[33]

Third, the panel stressed that science and technology had an important role to play in this process, reinforcing civil security and contributing to the "sense of Atlantic community." The Three Wise Men concluded from the comparative manpower data for NATO and the Soviet Union mentioned above that there was "an especially urgent need to improve the quality and increase the number of scientists, engineers and technicians in NATO countries."

Two separate initiatives were taken to implement these recommendations. In June 1957 the NAC established a task force chaired by Joseph B. Koepfli, a chemist, assistant to the president of the California Institute of Technology, and science adviser to the State Department from 1951 to 1953, to plan a conference on scientific and technological cooperation in the NATO zone. In parallel, in November 1956 the Second Annual Conference of Members of Parliament from the NATO countries asked Senator Jackson to chair a Special Committee on Scientific and Technical Personnel. Its brief was to survey the present situation regarding the training of scientists and engineers, compare the situation in NATO with that in the Soviet Union, and suggest measures for improving matters. Jackson asked John Archibald Wheeler to join other distinguished American scientists, educators, and industrialists on a committee that would help him formulate his proposals—men like Detlev Bronk (president of the National Academy of Sciences), Richard Courant, Edward Teller, James Killian, Howard Meyerhoff (executive of the Scientific Manpower Commission) and David Sarnoff (chairman of the board of RCA).[34] Wheeler was enthusiastic about his involvement: although overcommitted already, he told Jackson that he would "go into bankruptcy to accept your invitation." Eventually Wheeler was to chair the committee.[35]

Wheeler, introduced in the previous chapter, was a leading plasma physicist at Princeton University and had been worried for years about the technological capabilities of the Soviet Union. He was convinced that if it was ever allowed to gain military superiority, it would "become adventurous, seeking expansion and conquest, possibly triggering World War III."[36] Wheeler had helped produce plutonium in the Manhattan Project; he had lost many friends by deciding to work closely with Edward Teller on thermonuclear weapons in 1950–1953. And he was concerned that many of the brightest and the best in the United States were no longer prepared to do defense-related work. In his eyes "The Struggle to Preserve a Free World" required a strong Europe. In one of a number of (undated) back-of-the-envelope notes he prepared for the committee's discussions he stated: "1. Only 1 man in 20 in the world is an american. 2. We can't go it alone. 3. Europe's traditions and strength make it our strongest ally.

4. 200,000,000 European versus 160,000,000 Americans. 5. Europe's weakness our deepest concern." "Many of Europe's Weaknesses Stem from Shortage of Scientific and Technical Manpower," Wheeler felt, and to remedy them he drew up the outlines of a "NATO Defense Technology Training Program" to kick off in summer 1957.[37]

Jackson's report, entitled "Trained Manpower for Freedom," was submitted to the third conference of parliamentarians from NATO countries five weeks after the launch of Sputnik.[38] NATO faced "a genuine crisis in the form of serious shortages of skilled scientific and technological manpower" owing to what the report called the "scientific revolution" of the postwar period and its technological consequences. Industrial progress and economic well-being in the Atlantic nations, and continued exportation of science and technology to the "under-developed regions of the world," demanded more and better skills—the Soviet Union had already won the first round of the struggle for the hearts of minds of recently decolonized peoples. In light of the increase in quantity and quality of Soviet scientists and engineers, there was "a real danger that the Soviet bloc may surpass the NATO Community in science and technology. This shift in the balance of scientific power could lead to a shift in the balance of military power."[39] With these considerations in mind, "Trained Manpower for Freedom" recommended that NATO, for example, finance a talent development program to produce at least 500 doctoral degrees annually in fields of economic and military significance; take steps to make science in secondary school more attractive; actively promote more American-style summer schools in Europe; expand international exchanges of scientific and technical personnel; and expand cooperative training and research projects in fields that would lend themselves to international action.

The report by the Koepfli task force to the NAC was also available in November 1957.[40] It went far beyond its original brief, its most substantive recommendation being that NATO establish a science committee and that it appoint a science adviser. This was to be a person of high scientific caliber and prestige who would chair the committee and oversee the day-to-day implementation of its recommendations. The Koepfli report also devoted a special section to the need to train more women scientists and engineers in the Western bloc as way of meeting the Soviet threat.

In the wake of the Sputnik shocks, the NATO Council, exceptionally, met at the level of the heads of government on 16 and 17 December 1957. In a stirring speech to the assembly, President Eisenhower affirmed, "We are here to re-dedicate ourselves to the task of dispelling the shadows that are being cast upon the free world. We are here to take store of our assets—in men, in minds, in materials. We are here to find the ways and means to apply our undoubted strengths to the building of an ample and safer home for mankind here on earth. This is a time for greatness."[41] The council agreed on a number of measures to reinforce the military might of NATO in Europe. It also established the NATO Science Committee and the post of science adviser to the secretary general, affirming that "the full development of our science and technology is essential to the culture, to the economy and to the political and military strength of the Atlantic Community."[42]

The Activities of the NATO Science Committee

The NATO Science Committee first met from 26 to 28 March 1958. Its members, although formally representing their governments, were chosen principally for their scientific merit and seniority in their national scientific communities. They were not expected to have security clearance, and they were not given access to classified NATO material. Unclassified basic science that, at most, straddled the civilian/military divide, was the committee's preferred instrument to strengthen capabilities in Europe through a range of programs that fostered international scientific exchange. This also secured its legitimacy among the European scientific community, many of whom would otherwise have balked at taking money from a defense-related establishment. Membership was heavily biased in favor of physicists and astronomers; just one biologist (from Denmark) and one chemist (from Italy) were among the twelve that gathered at the first meeting. During the early years, the proceedings were dominated by the American and British representatives, Isidor I. Rabi and Sir Solly Zuckerman, chief scientific adviser to the British government.[43]

The first five science advisers to the NATO Council, and chairmen of the committee, were Americans. All of them were physicists, and all but one was in academia. The first was Norman Ramsey, who had done his Ph.D. under Rabi at Columbia University in the 1930s, then on leave of absence from Harvard University. Frederick Seitz, a solid-state physicist from the University of Illinois, filled the post in June 1959. William Nierenberg, professor of physics at the University of California in Berkeley, took on the job from 1960 to 1962. He was followed by William Allis, a plasma physicist at MIT. In 1964, in a deliberate change of policy inside NATO, J. L. McLucas took over. McLucas also had a Ph.D. in physics, but he was the chairman of an electronics firm who had just completed two years of service with the Department of Defense, working closely with the military's R&D program.

Two educational programs absorbed the bulk of the Science Committee's budget, both inspired by the Jackson report: a fellowship program, administered by the national authorities, and summer schools. During the first year of operation, the fellowship scheme had $1 million at its disposal, while $150,000 was set aside for the summer schools. By 1965 the budget for the fellowship scheme had reached $2.5 million, while that for the Advanced Study Institutes, as they were now called, was $650,000.[44]

These programs took off quickly, thanks to U.S. money. The Ford Foundation was the first to fill the needs of the summer school program, before the Science Committee was organized. Shepard Stone worked closely with Senator Jackson, so closely in fact that his December 1957 appeal to the trustees for the block grant of $150,000 to strengthen science in the Atlantic community lifted the kinds of programs worthy of support directly from "Trained Manpower for Freedom." Ford funded summer schools at Les Houches, in the French Alps, and Varenna, in Italy, in 1958 (NATO took them over in 1959), as well as a summer school in solid-state physics for Pierre Aigrain in Paris in 1958.[45] It was Nevill Mott's turn in Cambridge, England, in 1959.[46] The United States paid 50 percent of the NATO Science Committee's fellowship and summer school budget in 1959 and 1960 (just over double what the United States should have paid on the standard NATO cost-sharing formula). It reduced its contribution gradually from 1961 onward, until it was down to the regulation 24 percent by 1965.[47] This was compatible with its overall philosophy: pump-priming to get a program started, followed by an increased burden-sharing by the other allies.

The most thorny policy issue facing the NATO Science Committee was its relationship to defense research. The committee defined its role as primarily educational; however, should it not also support mission-oriented basic research of military significance? The French thought so, but they quickly discovered to their chagrin that not everyone agreed. Their delegate, André Danjon, director of the Paris Observatory, suggested in 1958 that, this being a committee of NATO, it should privilege research in fields deemed of "particular importance for strengthening the defenses of the West"—electronics, radio-astronomy, physics of the solid state, oceanography, and so on.[48] To this end, the French proposed that NATO sponsor a separate foundation, endowed with $10 million of start-up funding, to place contracts with European industries and research centers. Danjon insisted that if Western Europe was going to face the Soviet threat, it should set priorities, as they did, "building up the same common approach to the development of science as existed in the east, instead of discussing piecemeal measures which did not go to the heart of the problem." This was a fatal blunder. Such *dirigisme,* said Sir Solly Zuckerman, might be common in Communist countries, but "it would be impossible in Western democracies to direct scientists in the way suggested by the French representative"—a convenient pretext, for the French took their cue from the approach followed by the U.S. Armed Forces, who steered research agendas in Europe by placing contracts for mission-oriented research.[49] Anyway, Sir Solly went on, his government would never relinquish control over its defense research priorities to the NATO Science Committee (nor, by implication, help the French strengthen their national defense capability in areas where they felt vulnerable). In short, as long as there was not closer defense cooperation and integration within NATO itself—impossible given the nationalistic military ambitions of major countries like the United States, Britain, and France—it was utopian to think that a coordinated, defense-related R&D effort could be organized in the way Danjon had suggested.[50]

Though the Science Committee rejected the French effort to centralize planning, it did set up a research grants program to promote scientific fields that straddled the military/civilian divide and that were suitable for international collaboration. These grants were managed by independent panels that brought together leading experts in the field to promote a particular specialty. For example, Håkon Mosby of the Geophysics Institute of the University of Bergen, chaired an oceanography panel that did a good deal of research in strategically sensitive areas of the North Sea, as well as in the Straits of Gibraltar and the Turkish Straits.[51] Philip Morse, who ran a pathbreaking graduate program on operations research at MIT, chaired an advisory panel in this area from 1960 to 1965. The work of this panel, and the setbacks Morse suffered, are so interesting in their own right, and pertinent to the topic of this book, that the next chapter is entirely devoted to them.

In 1963 the NATO Council believed that the time had come for a more concerted effort to promote an integrated military research program. It set up an "Exploratory Group," chaired by Pierre Aigrain, to look into the matter. Aigrain had the ideal profile: an expert in semiconductors, he was a professor of physics at the École Normale Supérieure and a research director in the "Direction des Recherches et Moyens d'Essais" at the Defense Ministry.[52] The Aigrain report confirmed that the Science Committee had done very little directly to stimulate military-related science.[53]

Aigrain's group analyzed how this had happened. It noted that, while the original intention of the Koepfli task force had been "to see the major part of the scientific resources of NATO organized in support of the military," the launch of Sputnik had reoriented priorities and "concentrated NATO's attention on the importance of increasing the number of scientists being trained in the West." The net result was that when the terms of reference of the Science Committee were drawn up, scientific education was given more prominence than military technology. In effect it had been assumed that "by simply adding a civilian science adviser and a scientific committee to the existing NATO machinery, [a] flow of scientific advice to the civil and military authorities within NATO on defense matters would automatically be forthcoming." This failed to happen not only because the science adviser was not engaged in the military decision-making process itself, but also because he had been chosen "on the basis of high academic qualifications, but with little concern for defense experience."[54] As a result, his office had been almost totally divorced from the problem of defense, which was, after all, the fundamental justification for the alliance.

The solution to the problem proposed by the Aigrain group was to give more weight to the existing ad hoc Defence Research Directors Committee. It was to become a standing committee of the council, and its members were to be senior scientists or engineers responsible for integrating R&D into military preparedness at the highest level in their own countries.[55] The Science Committee would maintain its basic mission, "continu[ing] to be responsible for the promotion of science and technology, and for the stimulation of cooperation in science between member countries."[56] Henceforth, however, its chairman would need to be someone whose defense planning and policy- making experience counted for as much as his scientific achievements. In summer 1964 Allis returned to Princeton to continue plasma physics research, and McLucas, who had direct experience of the U.S. military's R&D program, replaced him.

By 1965 the total annual budget of the NATO Science Committee was $4 million, over 60 percent of it for the fellowship program.[57] Since 1958, 4,600 NATO fellows, almost three-quarters of them under the age of thirty-one, had received advanced training, the majority in universities that were not in their home country. About half of them had studied physics and chemistry, but engineering, mathematics, the medical sciences, biology, and geology were also well represented. The summer school, or Advanced Study Institute program was now arranging about fifty meetings annually, with from forty to more than a hundred scientists attending each meeting. One of the program's major achievements was to encourage interdisciplinary approaches that were largely absent from university curricula in Europe. The budget for research grants was $735,000. Operations research, which had been funded by the Department of Defense in 1960 and 1961, before the other member nations gradually chipped in, had its own budget line, with $115,000 in 1965.

The impact of the NATO Science Committee cannot only be measured by numbers like these: it was both more intangible and perhaps more significant. In a brief note written for the American authorities around 1964, Rabi explained that "the general objective of the Committee [had been] to raise the effectiveness of Western Science by encouraging cooperation between the scientists of the NATO countries and to use the prestige and universality of science as a means of strengthening the bonds between the members of the Alliance, as well as raising the strength and vigor."[58] The committee had also sought to break down the antipathy in intellectual circles in Europe to NATO itself and to working on defense-related science. At its very first meeting, Rabi made a point of praising the summer schools as something that "would certainly add to NATO's dignity": they were a "reversion to the strong tradition of oral intercourse between scientists, and would diminish the antipathy to NATO by associating it with higher learning."[59] By 1964 he felt that this objective had been achieved: NATO had "become a respected name in scientific circles whereas before it was regarded with suspicion." It was "almost unprecedented," Rabi went on, and a sign of how important the allied scientific communities and their governments deemed the work of the committee, that the United States had been able to steadily decrease its percentage contribution to the standard NATO share even as the budget of the committee increased almost fourfold. Nor was it only NATO that benefited. The various programs of the Science Committee had "been of the greatest importance in bringing U.S. scientific sophistication to Europe" and in raising "the prestige of the U.S. in intellectual and scientific circles." It was now deemed an honor for a European to have studied at an American university or to have worked in a major research laboratory in the United States. A decade before the committee of Three Wise Men had stressed that "common cultural traditions, free institutions and democratic concepts . . . were things which should bring the NATO nations closer together, not only for their defense but for their development." For Rabi at least, one of the great achievements of the NATO Science Committee was to have used basic science and international scientific exchange within the alliance to strengthen a shared cultural tradition and to bring its intellectual communities together around a core set of values under American leadership.

The Failed Attempt to Establish a "European MIT"

While Rabi's enthusiasm for the science program promoted by NATO was probably warranted, he did not mention one of its most spectacular failures: the attempt to create an Atlantic, or international, institute of science and technology (IIST) somewhere in Europe—preferably near Paris—modeled on MIT.[60] This initiative is worth studying in depth for several reasons.

First, it created much interest in some scientific quarters at the time: enthusiastic supporters included Sir John Cockcroft, Hendrik Casimir (the research director at Philips in the Netherlands), and Raymond Latarjet (director of the biology section at the Curie Institute in Paris). Killian himself spoke of it as a proposal "that stirs one's imagination deeply because of its fresh potentialities, its importance, its intellectual spirit, and its appropriateness at a particular moment in time."[61]

Second, the IIST project engaged the highest levels of military and governmental officials in the NATO countries: Secretary of State Dean Rusk in the United States; Lord Hailsham, the British minister for science and technology; Pierre Piganiol, responsible for scientific and technological research in France; as well as President de Gaulle himself, to name just a few. Even ex-President Eisenhower weighed in, commending Killian for his efforts: "Everything we do along this line cannot fail to promote better understanding among peoples," Eisenhower wrote in 1962, "and, mutual understanding, while it may not insure peace, is still one element without which no real peace can be sustained."[62]

Third, the attempt to establish an IIST is directly pertinent to the argument of this book: this was a conscious effort by scientific and political leaders on both sides of the Atlantic—with strong support from the Ford Foundation and NATO—to export an American style of graduate education in science and engineering to Europe, with a view to reforming a university system they deemed obsolete and ill-adapted to the demands of the postwar military-industrial state. There are echoes here of Warren Weaver's perceptions of the French scientific community in the late 1940s (see chapter 4). More to the point, in the late 1950s Shepard Stone "shared the view, common at the time, that European institutions, particularly the universities, needed 'modernization'" if they were to contribute to democratic progress.[63] At the end of 1959 Stone persuaded the trustees to contribute $1 million over five years to the establishment of a new college in Cambridge intended "to do honor to a great Englishman by acting to meet one of Britain's and the free World's central needs—the advancement of science and technology." Churchill College would "provide for the increase in the output of superior scientists and technologists, more especially for an increased output of the leaders and teachers in these fields," enhancing "Britain's survival as a modern industrial nation, . . . its exports, and its role as a partner in the Atlantic community."[64] That same year, the foundation also authorized a thumping $9 million grant to Gordon Brown, dean of engineering, to develop a "science-based engineering curriculum" at MIT.[65] Brown was an enthusiastic proponent of the "MIT-Idea," a concept that, if somewhat fuzzy overall, did embody "major characteristics" that, in Brown's view, made MIT "different from other institutions of technology" and that were an "exportable quantity."[66] In short, this project was an attempt to remodel European education along lines that presupposed that the American way, and the MIT way in particular, were ideal types to be emulated by any state or region with pretensions to modernity. Its failure was also a failure of the hegemonic enterprise in this domain.

The seeds of the project were sown in a report publicly released by NATO in October 1960. "Increasing the Effectiveness of Western Science" was the result of almost a year's work by a study group established by the Science Committee and chaired by Louis Armand, former president of the French Railways and of Euratom, and president of the École Polytechnique at the time.[67] He was helped by about a dozen eminent scientists and science administrators, including Rabi and Seitz from the United States, Cockcroft and Zuckerman from the UK, Casimir from the Netherlands, Danjon from France, and Paul Bourgeois, the director of the Royal Observatory in Belgium.[68] Jean Willems, the director of the "Fondation Universitaire" in Brussels, also made an invaluable contribution. He took over as acting chairman from Armand in April and May 1960 and arranged for the translation of the report into French. Most crucially, though, he administered the study group's budget, made up of equal shares of $75,000 from NATO and the Ford Foundation. Willems had served on the Koepfli committee, so he was well aware of NATO's concerns in this area, and immediately accepted when Stone asked his friend if, to simplify bureaucratic procedures, the "Fondation" could act as the fiscal agent for the study group.[69]

"Improving the Effectiveness of Western Science" was dominated by the concern that the alliance was not making proper use of its scientific and technological talent both to meet "a new and massive challenge . . . from the Soviet bloc," and to satisfy the demands of its populations "for a higher standard of living, for relief from toil and drudgery, for increasing health of body and mind." The Soviet Union's threat to "their security and, indeed to their culture" derived essentially from "a massive and, on the whole, effective organization of scientific research, development and education" that enabled them to direct limited resources to specific goals under the control of a centralized system. This approach was impossible in democratic societies. Implicitly drawing on a teleological model of capitalist economic expansion—when unhindered, it would be driven forward by its own internal logic—the report suggested that intervention should be aimed not at "directing" science, but at removing "the obstacles which at present retard [its] growth." There followed a routine catalogue of impediments to the development of science: lack of funds, of buildings, of qualified science teachers, of trained technicians and craftsmen, as well as restrictions on international travel and scientific exchange, to mention just a few. Some of the suggested improvements were equally mundane, as the journal *Nature* pointed out ("the report merely adding its authority to proposals already advanced"), like improving the availability of documentation.[70] One proposal stood out from the rest, however: the founding of an international institute of science and technology.[71]

The study group and the Science Committee saw several advantages in this proposal. First, they were convinced that short-term palliatives to the current crisis were not enough. From the outset Armand insisted that the group needed to suggest "a few major undertakings so far-reaching and novel" that they would precipitate changes in national habits that were obstructing the optimum use of science in the West.[72] Casimir reinforced the chairman's line of argument. He brought to the table a proposal for an Atlantic university, which, he said, had already been discussed on several occasions. Such a center, Casimir wrote, "should be a challenge to existing universities and should shake them out of their complacency."[73] Increasing the effectiveness of Western science meant using shock tactics to remove irrational resistance to the effective training and exploitation of scientists and engineers.

The international university (or institute) would also consolidate Atlantic solidarity, promoting, Armand noted, "Western cohesion at the scientific level." For Zuckerman, Rabi, and Seitz this was the main rationale for setting up the study group in the first place. Indeed, Sir Solly would have preferred the report to be entitled "Increasing the Effectiveness of the West *through* Science." Seitz agreed, appealing to the arguments used by the Three Wise Men for expanding the scope of NATO to encompass civil security.[74] For Armand and Casimir an international university would do just that. "If the Western world is to become more of a cultural and economic unity," the Philips research director wrote, "there is evidently a great demand for leaders who combine professional excellence with knowledge and understanding of Western culture, knowledge of languages, etc. This is the type of person, that one should try to train at an Atlantic university."[75]

The Communist card was also played. The group noted that Moscow had just created a "University of the Friendship of Nations" for people from Africa, Asia, and Latin America (more than twenty new states were created between 1945 and 1959; more than thirty between 1959 and 1969).[76] "The West cannot afford to be outdone by the Soviet Union in this area," the report insisted.[77] Jackson made much of this in promoting the project to NATO parliamentarians. "One third of the world—populated by the billion people who live in the underdeveloped nations—is now in ferment and turmoil," he wrote. "These people are bent on reaping the harvest of scientific advance. . . . They are not content with the timetable of gradualism." If the Atlantic community did not help them, Moscow and Beijing would. He ended his report with an impassioned plea for the NAC to "speed studies leading to the earliest practical establishment" of an IIST in Western Europe.[78]

America pointed the way. Institutions like MIT and the California Institute of Technology had had "a tremendous influence on the development of science and technology"—a reference to the Route 128 corridor and Silicon Valley, where many dynamic industries manufacturing highly sophisticated equipment had grown up in the 1950s. In addition, such institutes made provision for rapid upward mobility based on a professorial system that "offers substantial rewards and freedom to young members of the faculty."[79] By contrast, existing institutions in Europe had too few professorial chairs in important fields and too few rewarding posts for people under the age of thirty-five. Resistance there would be: Casimir reported that just recently a rectors' conference at Dijon had been unenthusiastic about European or international universities.[80] On the positive side, it was "doubtful whether any single European country has the resources in manpower and money to proceed on its own," while "Western Europe as a whole could easily muster the required talent and equipment to found an International Institute comparable to the Massachusetts Institute of Science and Technology in size, scope and quality."[81]

Things moved fast once the Armand group's proposal was released. The new NATO science adviser, William Nierenberg, officially recommended it as being "concrete, straightforward and bold," "an order of magnitude step in approach," and having the "the very greatest possibilities for strengthening the NATO Alliance and the general union of the free world."[82] At the end of October 1960, Cockcroft spoke about the idea when he formally opened the new physics building at the Imperial College of Science and Technology in London. Sir John was reported in both *Nature* and the London *Times* as saying that Britain should take the lead and have the new institute established there as part of its university expansion program.[83] On 2 November the NAC authorized the NATO secretary general to appoint a small, high-level working group to make a feasibility study of an international institute of science and technology. On 9 January, Nierenberg sent Stone a list of the people, headed by James Killian, who would make up the committee: Casimir and Cockcroft, from the Armand team; Piero Caldirola, the director of the Institute of Physics in Milan; A. Rucker, from the Technische Hochschule in Munich; and Piganiol, the "Délégué Général à la Recherche Scientifique et Technique" in France, an extremely senior official in the new machinery for science policy put in place by President de Gaulle when he came to power in 1958, and who reported directly to the prime minister.[84] All five men attended the centenary celebrations of MIT in Cambridge, Massachusetts, a few weeks later. The chairman of the corporation told them that he thought that the IIST proposed in the Armand report was, as Piganiol put it, the "best possible solution that could be foreseen."[85] The group delivered its report to NATO on 10 October 1961. It was given a low level of classification and eventually released to the general public in November 1962.

Before Killian and his study group got to work, Nierenberg gave an upbeat account of the first reactions to the Armand proposal. The science adviser had been surprised by the "immense verbal support" from "responsible members of nearly all NATO governments" for an IIST.[86] They liked the idea of setting up an interdisciplinary postgraduate program for a gifted elite in a wide variety of fields: this was something governments could not do individually, either because it was too expensive or because it was too difficult to break down the entrenched disciplinary structure of traditional European universities. The institute would also inspire and challenge talented youth, both because of the limited career prospects available to them in academia as organized and because its degrees would be recognized in all participating countries—not then the case in Europe—thus increasing their mobility. Finally, "a great technical university of this kind" would be tightly linked to the European community as a whole, notably European industry: in this regard what "M.I.T. and Cal Tech have achieved in the U.S. must be reproduced in Europe."

That said, Nierenberg admitted that "the most serious split of opinion is as to whether such an Institute be consolidated into one huge campus or whether it be a collection of widespread departments with a central administration." Critics argued that it would be more practical to build on existing structures, which could be expanded and internationalized— say Louis Néel's institute in Grenoble, or the National Physical Laboratory in Teddington, England. The French were particularly keen on this solution. A general paper written in February 1961 for the Délégation Générale à la Recherche Scientifique et Technique (DGRST) made this point as soon as it was familiar with the contents of the Armand report: the "radical" proposal for an IIST ignored the fact that Europe already had many centers that could be used for achieving the aims of the institute, "and that its creation from scratch without considering what is already available is not only a waste of money but will also have serious repercussions on recruitment."[87] Piganiol (and Rucker) made a similar point when they first met Killian in Cambridge. This did not particularly deflect Nierenberg or Killian: they felt that some compromise with the French would be possible.

The institute foreseen in the Killian feasibility study had five interdisciplinary centers and a center for advanced study, modeled on that at Princeton, to host visiting fellows of outstanding intellectual achievement or promise.[88] These centers would not duplicate the research foci of existing national and international institutes; rather, they would do research ranging from basic science to applications in emerging areas that required large-scale facilities not readily available elsewhere. The five centers were applied mathematics (where Killian judged there was a particularly serious "manpower" shortage)[89] and theoretical physics; technological processes and systems, specifically concerned with "shortening the interval between scientific discoveries and their application"; materials research, which would have a large concentration of small items of equipment; earth sciences, which included geophysics, meteorology, oceanography, and the space sciences; and life sciences, including, of course, molecular biology, but also "viruses, bacteria, nervous systems, and the transmission of biological information." Overall capital investment was put at a little more than $55 million; annual running costs for a target population of 1,000 students, 400 faculty, and 1,000 support staff was about $16 million. If, for political or financial reasons, the institute had to be set up piecemeal, the group felt that it should begin with the first two interdisciplinary centers and the center for advanced study. The list of centers was not definitive anyway. "The cultivation of the humanities and social sciences as partners of science and engineering" should be foreseen in the longer term. The institute should also have "the courage to drop centres and programmes if more important programmes emerge that should replace them. Education in science and technology must constantly be modernized," the Killian group insisted.[90]

Three main features would set the IIST apart from existing European institutions: it would encourage flexibility, it would be embedded in society, and it would be a source of inspiration. The graduates of the IIST, having been trained in interdisciplinary centers that did not respect traditional academic boundaries, would be creative, versatile, and adaptable. They would not only possess "a mastery of the latest developments and techniques, but also a deep understanding of the underlying disciplines"; they would be willing to "master new ideas and new technologies throughout their working careers."[91] Furthermore, they would be sensitive to linking science and technology, new knowledge and its applications.

The IIST would be fully embedded in society and responsive to the changing demands that the modern world placed on scientists and engineers. The institute would foster strong links with industry, as well as with government and nonprofit research organizations. Faculty would be recruited not only from academia, but also from industrial and national laboratories. Visiting professorships or exchanges would be arranged to enable scientists and engineers in industry to participate in research and teaching. External consulting and industrially sponsored research would be encouraged. Summer courses providing industrial refresher and management training would be held. An advisory council to the president would include leading industrialists among its members. The age of the "ivory tower," devoted to the pursuit of knowledge for its own sake, was past.

Finally, the IIST would be "a symbol and a precedent" with a meaning "transcending its direct contributions to education and research." Apart from being a "fillip for Western morale," it would be a "dramatic demonstration of the intellectual strength, vigour and unity of Western nations working together"—a demonstration "that only in the Free World is it possible to achieve a truly international educational institution attractive to first-rate scholars." This institute would do more than simply encourage existing universities to "modify their systems and traditions to accord with modern conditions."[92] More positively, in its internal organization, but also in its relations to industry and to other universities, it would "set a pattern" that would "encourage the withering of old and sterile practices and lead to new and creative organizations and relationships."[93] It was intended, then, to be a model to be looked up to and copied, an inspiration not just for Western Europe but also for newly developing countries as they sought to chart their path into a new age. In Killian's striking rhetoric, the IIST would be a "transforming institution" that would establish "the tradition of the future."[94]

How did the working group counter the objection that it would be preferable to exploit available European resources rather than to build something from scratch, thus ensuring that the IIST did not cream off the best talents from national institutions? The report refused to regard it as a major obstacle. Europe was embarking on a major program of university expansion. The new institute would be one among many and would "supplement," rather than duplicate or compete with, national universities and research institutions, which could be "associated" with it. Its effects on staffing would be minor, because the numbers concerned would be small and the pool was large, because growth would be gradual, and because a special effort would be made to recruit faculty from outside academia, notably from industrial and government laboratories.[95]

One major concession was made to pacify the French. As noted above, Danjon had unsuccessfully tried to persuade the NATO Science Committee to establish a foundation to encourage directed research in Europe. This proposal remained a cornerstone of French policy. The Killian report discussed the foundation, all the while insisting that it was outside its terms of reference. It recorded "strong support" for the idea of establishing a foundation that encouraged interdisciplinary research and international exchange using "research contracts, but also scholarships and subsidies" and suggested that the idea be studied further. It did not, however, present it as an alternative to establishing an IIST.[96] As Eugene Skolnikoff, on the White House staff, put it, commenting on the negotiations over the issue inside the Killian working group, "The primary obstacle we encountered up to the last meeting in June [1961], was the French position that an international science foundation, performing functions similar to those of our National Science Foundation, is more important than an international graduate institute. The French now agree, however, that the two concepts are not in conflict."[97]

The Reception of the Killian Project in Europe

In a synthetic account like this, it is difficult to do justice to the diversity, nuances, and evolution of the reactions in the many countries in Europe that, through NATO, debated the Killian report from 1961 to 1963. Limited access to "internal" archival sources in which national responses were discussed—from London to Athens, Paris to Rome, and The Hague to Bonn—amplifies the problem. One also has to distinguish between the positions taken by representatives of the scientific community, on the one hand, and science administrators and government officials (including the military), on the other. These constraints must be born in mind when reading the account that follows, which is based mostly on NATO documents, Killian's own papers, and a small but superb collection of official French documents.

Support for the creation of the IIST came from several quarters. Small countries strongly favored it (with the exception of Belgium, which tended to follow the French line in European affairs). Scotland and Luxembourg offered one or more sites for the institute. The Netherlands was prominent and aggressive in its support for setting up a centralized institute, and in 1963 Hendrik Casimir played a key role in trying to keep the idea alive against all the odds. Italy was also strongly in favor. A meeting in January 1962 of two dozen rectors of Italian universities, three directors of polytechnics, and sixty members of the scientific committees of the Italian Research Council unanimously supported the concept. Sites were offered near Rome and at the University of Milan, though it was stressed that Italy's participation in the project was not conditional on the institute being located in the country.[98] This approach was in line with Italian foreign policy, which was consistently in favor of European-wide initiatives, here sweetened by an American engagement. From the perspective of researchers and academics, the scheme had the added advantage that the government would be formally locked into its financial obligations by international treaty. This protected the funding stream from the vagaries of the national political process and the changing policies of a fragile and unstable state system. Finally, NATO countries that were scientifically, technologically, and educationally lagging behind the rest of Europe, like Turkey and Greece, were also enthusiastic. The Greek delegation to NATO made an impassioned plea for siting the institute in Athens, and leading figures in the country were actively engaged in 1963, along with NATO officials and Casimir, in trying to maintain the project's momentum.[99]

All of this was encouraging, of course, but it was not enough. The official American policy, as spelled out by the State Department in January 1962, was that, subject to the provision of necessary funds by the Congress, the United States was prepared to take an active part in the establishment and operation of the IIST. This of course had to be in a single location: "a great part of the political motivations which inspired the Killian recommendations would be sacrificed if the institute were to be decentralised," as one NATO official put it.[100] The United States agreed with those who felt that the IIST, once established, should be independent of any existing international body, in particular of NATO or the Organization for Economic Cooperation and Development (OECD).[101] As for funding, the NATO formula (24 percent paid by the United States) was not being considered. Rather, the State Department would seek "maximum contributions from private foundations" and, crucially, "approximate equality of government contributions among US, UK, France and FRG, with other countries making up balance as required."[102] Put differently, it was imperative, in the view of the State Department, that these three big European countries support the Killian project and be willing to share much of the cost. That proved to be a tall order.

What little we know from NATO reports about the position in Germany is that the project was not enthusiastically received. The Forschungsgemeinschaft was opposed to a centralized institute, and the Max Planck Gesellschaft even more so. The organization representing university rectors (Rektorenkonferenz) was lukewarm. At the political level, the foreign ministry was negative, if not downright hostile. After visiting the country with Rucker in summer 1963, Allis wrote that, although the situation there had improved a little (particularly after removing the misconception that the IIST would be a NATO center), it was difficult "to find people in Germany who [were] sufficiently interested in the project, and who [were] in a position to exert some influence." As he was leaving for home the state secretary of one of the Länder, with whom Allis had struck up a cordial relationship, told the NATO science adviser that he did not think that Germany would, in fact, ever participate in an IIST.[103]

Senior representatives of the British scientific community expressed their opposition to the scheme almost as soon as Cockcroft explained it to them. In December 1961 the Advisory Council on Scientific Policy (ACSP), after consultation with the University Grants Committee, formalized their recommendation. They agreed that it was important to strengthen international links in science and to encourage interdisciplinarity, but they did not think that setting up a centralized IIST was the best way to do this. Instead, they proposed that "international agreement should be sought on the selection in Europe of individual centres in existing universities etc. in particular fields of science and technology which could readily, and economically, be developed into a nexus of the kind of international centres envisaged." This solution might enable Europe to build "truly international universities" on national soil, chosen on the basis of the people who were there already and "without violence to the natural evolution of existing institutions." "We believe," wrote the ACSP, "that in the case of most fields of science the nucleus around which work of the highest quality can be developed is the exceptional individual, and not a research facility." This approach would also avoid the disproportionate benefits in favor of the host country brought by a central international institute.

The ACSP's recommendation put the British government in a difficult position. It needed to find a way to respect both the views of its distinguished scientists—which were fully coherent with the position taken by the French from the outset—and those of the Americans, who, along with Killian and the majority of the working group, had insisted on brushing this alternative aside and on calling for a single, centralized institute. Two foreign policy objectives met head on. In summer 1961 Prime Minister Macmillan formally deposited Britain's request to join the European Economic Community—a supranational arrangement that he had originally shunned and even tried to sabotage by leading a rival free-trade organization. During 1962, while "the six" were still debating the pros and cons of British membership, plans for important joint technological projects were being discussed, notably the development of an Anglo-French supersonic airliner (Concorde) and of a multistage satellite launcher (Europa).[104] While it was important not to upset the French, London also wanted to protect its "special relationship" with Washington and to take advantage of the American offer to provide significant financial support for the IIST. After a high-level tripartite meeting in England in July, the chairman of the ACSP, Lord Todd, put forward an alternative proposal that he hoped would be acceptable to all parties. It was a classic British compromise. The IIST would have a "core" consisting of two of the five centers proposed in the Killian report. These "headquarters" would be contiguous with an appropriate existing institution in Europe. They would be "a central symbol," a "live and vigorous focus," which would help the Institute "have a visible moral effect." The other three units of the institute would be located in different countries and associated geographically, though not governed by, an existing major educational institution.[105] Todd suggested that CERN could serve as the nucleus around which the other lesser bodies circled: the ACSP immediately quashed that scheme.[106] The whole plan seems to have drifted in limbo until May 1963, when the British government formalized a decision allegedly taken some months earlier to retreat to the original ACSP proposal for a fully decentralized structure. Cockcroft's and Zuckerman's efforts to persuade Lord Hailsham to think again were met with a "very chilly reception" by the minister for science and technology.[107]

The position adopted by the French had much in common with that of the British. Both sought decentralization. The French, however, persisted in their view that a foundation, established by international treaty, should finance the selected centers for interdisciplinary studies in the different participating countries. It would also define joint research programs and targeted projects (*actions concertées*) and facilitate international exchange. The majority of the governing board of the foundation would be from participating European countries. It was essential, a high-level advisory committee said, that any new organization maintain the scientific and cultural independence of the European countries, whose traditions and diversity were a source of intellectual richness and creative power.

This last requirement was indicative of another feature that distinguished the French from the British: the French were far more sensitive to the fact that the United States was engaged in the venture. Thus they noted that the IIST would provide a bridgehead for the United States into European research capabilities. This had become all the more important as increasing national investments in science and education retarded the flow of top-quality scientists across the Atlantic. The advantage that the United States had in exploiting potential research results was also of some concern to the French. Assuming that any discovery made jointly by researchers from different countries was open to joint national exploitation, they feared that the United States, with its superior technological and industrial power, would be able to shut out the Europeans.

That said, the French were also deeply concerned about how best to modernize their institutions and make them more flexible and open; while they shared the doubts of many about the Killian proposals, they were also fascinated and attracted by them. This was the case of chief of staff, General Jean Olie. Pierre Guillaumat, the minister responsible for atomic energy at the time,[108] also believed that an international institute could be a valuable stimulus to the reform of obsolete and ill-adapted structures. Raymond Latarjet, who had spent a year at Cold Spring Harbor, made a spirited appeal for a central institution. It would break science in Western Europe out of its closed world and accelerate the slow evolution of existing organizations. People who opposed it did so because they saw it as an instrument of American "interests"—a view that failed to understand the United States—or because they feared competition—a view that admitted the inferiority of existing structures to those proposed by the Killian group. The IIST was an open-ended experiment, something that would evolve and adapt itself to the European milieu over time, said Latarjet, a product of the empiricist Anglo-Saxon spirit so different from the *esprit français*. Despite their pleas, Latarjet and those who thought like him did not prevail.

The collapse of the IIST is generally attributed to the French, and to de Gaulle in particular. Seitz, in his autobiography, tells how, at a tripartite meeting in London, Pierre Aigrain was unexpectedly informed by a messenger that the French President had vetoed the plan for an IIST.[109] In several letters written in 1964, James Killian also held de Gaulle primarily responsible for the demise of the project.[110] At one level this is obviously a gross oversimplification: the project was opposed by strong factions in Britain and in Germany too. What truth there is in it lies in the fact that sometime in 1963 de Gaulle decided that if there was to be an IIST in Europe, it was to be without the United States.[111] This was not acceptable to Washington. In June 1963 Killian wrote to Jerome Wiesner, Kennedy's science adviser, telling him that the British felt that there might still be some point in the United States' taking the initiative to reopen discussions on the IIST. Wiesner's desultory reply suggested that Killian come to see him when he was next in Washington; the letter is annotated: "JRK tried to see JW while in Wash. 9/5 but JW was not there." In September 1963 the IIST was dead.

Concluding Remarks

The most important question to ask about the IIST is not, Why did the project collapse? (and to blame de Gaulle for that), but, How could serious scientific statesmen have been so optimistic that it would ever take root? (which shifts the locus of culpability onto Armand, Killian, their working groups, and NATO). Indeed, given the prevailing situation in Europe at the time, it is a wonder that anyone believed that the project could get off the ground at all. Knowing what we now do, we can only assume that the State Department was completely in the dark about the measures being taken to transform European higher education in 1960 when it singled out Britain, France, and Germany as key partners in a centralized IIST.

The hostile reactions of the scientific community were predictable. Indeed, in the early 1960s there was considerable opposition in Europe to *any* new international scientific project that might drain national resources at the very moment when they were being built up. Consider, for example, molecular biology, one of those emerging fields planned for in the Killian proposals. There were major new initiatives to develop national capabilities in France (spearheaded by the Pasteur Institute), in Germany (where a new genetics institute was opened in Cologne), in England (in 1962 the famous Laboratory for Molecular Biology moved from the Cavendish into a new modern four-story building), and in Switzerland.[112] As I have shown elsewhere, an attempt by John Kendrew, flush with his recent Nobel Prize, to set up an international laboratory for molecular biology, using arguments that were carbon copies of those presented by the proponents of an IIST, foundered on criticisms that were carbon copies of those heard here, namely, that it would "tend to remove many people from Universities, and thus tend to kill the goose that lays the golden eggs"—and if Kendrew could not see this, it was because of the "self-satisfiedly imbecile" attitude at Cambridge that blinded him to anything happening in biology outside the four walls of his own laboratory.[113] The same reproach could be made of the view of the world from Cambridge, Massachusetts.[114]

Intense opposition from the academic establishment was also predictable. The debates over establishing an IIST were conducted in parallel with an ongoing and tortuous discussion on establishing a European university for the member countries of the EEC. Early in 1960 an interim committee set up by Euratom proposed that this university have six departments: law, economics, social and political science, history and the development of civilizations, as well as mathematics and theoretical physics. The idea was to build an interdisciplinary program to train an elite in fields that were essential to the construction of the community foreseen in the 1957 Treaties of Rome. In this respect the university (which was eventually established in Florence in 1976 without the physical sciences) was, in the words of its biographer, "called on to play a pilot role, which would be one of its original features, and ipso facto guarantee its success."[115] In inspiration it was thus very like the IIST—and indeed, Armand and Willems were engaged in the negotiations for the university, the Ford Foundation offered support for the scheme in 1958, and Oppenheimer and Lilienthal were consulted in 1959 on how it might be implemented.[116] The opposition to the European university also mirrored that to the IIST—"all the academics saw it as a future rival that risked outshining their national universities," attracting the best professors and draining resources. The German rectors were so opposed that they even threatened to sabotage the entire scheme.[117] In January 1960, at a meeting of the Armand group and the Science Committee, Willems reported that all discussions on the European university had been broken off *sine die* while waiting for the (probably hostile) response from the cultural ministries of the German Länder.[118] Notwithstanding these strong negative signals, Armand and then Killian went ahead with their dream of an IIST, undaunted.

The opposition of governments was predictable, above all in those countries where key stakeholders in graduate training and education were against the scheme, namely, Britain, France, and Germany. It would be extremely difficult to support a project that was largely opposed by the academic and research establishments. De Gaulle's foreign policy added additional constraints. Soon after he came to power, he began to affirm the independence of France inside NATO: the French fleet in the Mediterranean was withdrawn from NATO command in March 1959, the stationing of American nuclear weapons on French soil was banned, and French air defense forces were returned to national command. This affirmation of French autonomy was not simply "anti-American." In an important press conference in September 1960, the French president also challenged the whole notion of a supranational Europe, which he saw as diluting the sovereign power of the nation state. The negotiations over the European university fell prey to this policy; so too did the negotiations over the IIST, where the leading role played by the United States simply added an irritant to what was already seen as an unsatisfactory supranational solution to a pressing problem that each country had to resolve itself.

Immediately after getting back from his first meeting with Killian in Cambridge in 1961, Pierre Piganiol reported that "The United States have decided to provide Europe with an MIT-like Institute to serve as a rallying point for the elite of underdeveloped countries [*des pays sous-développés*]. They would like this English-speaking institute to be situated near Paris."[119] The sarcasm was surely intentional. Seen from MIT, European higher education was sclerotic, hidebound by tradition, and unable to adapt to the modern world: it risked becoming "underdeveloped in science," to quote NATO Secretary General Dirk Stikker.[120] His metaphor was suffused with all the arrogance of the West and of a Cold War–driven imperative to reform countries that were regarded not only as backward but also as incapable of defining their own futures. It also resonated with Gordon Brown's and James Killian's determination to export the MIT-idea abroad. It made little difference whether the target country was "under-developed" Europe, India, or Iran. As Stuart W. Leslie and Robert Kargon have pointed out, "Brown and his colleagues believed that modern engineering, like modern capitalism, was essentially global and linear."[121] This belief partly explains why its advocates in the United States and Europe plunged forward with their plan for an IIST, indifferent to local specificities and to existing European strengths: for them, "MIT was a universal panacea."[122] It was not. "A university was built in the sociological context of a country," said the French mathematical physicist and educational reformer André Lichnerowicz. He "did not believe in transporting a 'pre-fabricated' model in its entirety from one continent to another."[123] Those who wanted to export (or import) an MIT to Europe failed because they were blind to the achievements, the needs, and the limits of the local communities whose condition they were supposed to ameliorate. It was a lesson too that Philip Morse would learn as he enthusiastically set out in 1960 to propagate his view of operations research through NATO.

8

"Carrying American Ideas to the Unconverted": Philip Morse's Promotion of Operations Research in NATO

The success of operations research (OR) in improving the fighting capability of the British and American forces in World War II ensured its place in the thinking of NATO strategists.[1] OR was one of the first activities promoted by the émigré Hungarian engineer Theodore von Kármán through the Advisory Group for Aeronautical Research and Development (AGARD), which he established in 1952.[2] By the mid-1950s NATO's Supreme Headquarters Allied Powers in Europe (SHAPE) had consultants from the Rand Corporation engaged in OR, SHAPE's Air Defence Technical Centre, at The Hague, had it own international OR team, and NATO had arranged classified conferences for OR specialists.[3] In a glowing preface to the proceedings of a conference organized by AGARD in 1957, General Norstad, the Supreme Allied Commander in Europe, also emphasized the civilian possibilities of OR. Norstad affirmed that operations research was making "a significant contribution to the military potential of the Atlantic Alliance," whose strength depended not only on its manpower, its skills, and its material, but also on "what we do with what we have"—the now standard phrase justifying the importance of OR.[4] He especially drew attention to the promise of OR in nonmilitary fields, suggesting that it could "stimulate and strengthen the economies of the NATO nations."[5]

The Jackson report, "Trained Manpower for Freedom," paid particular attention to operations research. It recommended that NATO establish an institute for OR in which "mathematicians, physicists, economists, and representatives of other fields of science" provided "rational and objective answers within an acceptable time period" to problems arising as the conduct of war became more complex.[6] Not surprisingly, OR was one of the first tasks brought to the attention of the NATO Science Committee. Prodded by the United States, the committee established an Advisory Panel on Operational Research (APOR), which was chaired by MIT's Philip Morse from 1960 to 1965.

In the previous chapter I described the educational programs of the Science Committee and analyzed in detail one initiative that it sponsored with leading figures at MIT. In this chapter I continue my analysis of attempts to export an MIT-like model across the Atlantic, probing beyond the institutional dimension into the actual content of the proposed program. Morse's concept of OR was strongly influenced by his wartime experience and had both pedagogical and social dimensions. He believed that, to be effective, "independent" OR scientists had to work closely with people who had executive authority, using advanced mathematical techniques and powerful computers to tackle problems that could help them to make decisions more effectively. This was the vision that he wanted to import into NATO. He failed because he did not realize that his conception of OR jarred with that which then predominated in Brussels. Nor did he grasp that the social configuration of education in Europe, and above all the relationship between academia and the military, differed considerably across the Atlantic and even among European countries. In the five years that he led the APOR, Morse saw his conception of OR gradually transformed into something quite different, both in content and in social organization, from what he had wanted. What he learned, to his chagrin—as did Killian with his IIST project—was that one could not easily transfer an American educational model abroad, particularly when it was embedded in the set of social relations that was the military-industrial-academic complex in Cold War America.

A Brief Introduction to OR

The origins of operations research are generally traced back to British attempts to optimize the use of radar beginning in the late 1930s.[7] Civilian scientists working in multidisciplinary teams (mostly people with training in physics and mathematics) were commissioned to study how best to use radar in combat. One of their first major successes was to improve the positioning of the radar sets and gun batteries defending

London from the Luftwaffe's night bombing raids. The scope of their work was extended to improve defense against German U-boats that were devastating convoys shipping food and war matériel from the United States. Its credibility established, OR became an integral part of the British war effort, assisting the armed forces to optimize their personnel and equipment.

Patrick Blackett, who won the Nobel Prize for physics in 1948 for his work on cosmic rays, was the driving force behind these early British OR efforts. From 1942 onward, he reported directly to the Admiralty's vice chief of the naval staff. He and his team worked mostly on antisubmarine warfare in the Bay of Biscay and the North Atlantic, as well as on improving the effectiveness of bombing raids.[8]

Radar was also the innovation that gave civilian scientists interested in operations research a foothold in the U.S. military, though only with some difficulty. By the 1930s the technical bureaus of the armed services had well-equipped laboratories, with (rather poorly funded) R&D programs.[9] Their research horizons were defined by generals who believed that new weapons should be developed in response to specific tactical needs. This attitude limited technological development under the aegis of the military to incremental improvement in performance and discouraged weapons innovation. Enter Vannevar Bush and his friends: MIT President Karl Compton, Harvard University President James B. Conant, and National Academy President Frank Jewett (also the vice-president of AT&T). With their support, Bush persuaded President Roosevelt in June 1940 that a federal agency should be established to mobilize scientists in universities and industrial corporations to work under contract on weapons research. Bush was nominated director of the National Defense Research Council (NDRC). Shortly thereafter, a British team led by Sir Henry Tizard showed the Americans the latest radar device they had developed for detecting enemy aircraft—the resonant cavity magnetron. This vacuum tube could generate microwave (cm-length) radiation with an intensity a thousand times greater than anything yet available in the United States, giving it a detection capability far in excess of any known device. The NDRC immediately gave MIT a contract to establish a radiation laboratory to explore the possibilities of microwave radar. The laboratory's future was guaranteed by the attack on Pearl Harbor in

December 1941 and by the development of a new, highly effective air-to–surface vessel (ASV) radar. The NRDC was subsumed in January 1942 under Bush's more powerful Office of Scientific Research and Development (OSRD), with responsibility to develop weapons "from the germinal to the production stage."[10] The military, now persuaded that university-based research into new weapons could change the course of the war, populated MIT with officers who kept in touch with the latest developments. By the end of 1942 the Rad Lab's monthly budget was over $1 million, and its staff numbered about two thousand. This figure had doubled by 1945 and included about a thousand academics, almost half of whom were physicists.

Bush believed that effective weapons development called for a two-way flow of information between designers and users. The laboratories would let the armed services know of any promising developments; the military would tell the scientists what their requirements were.[11] Bush was against civilian scientists being too closely integrated into the military command structure, fearing that their creative talents would be cramped and their objectivity impugned by the overriding requirement to respect senior officers. His approach did not satisfy Edward L. Bowles, professor in electrical engineering at MIT, pioneer in microwave technology, and newly appointed consultant on radar to Secretary of War Stimson. Bowles felt that to fully exploit the potential of new weapons, one needed to embed them in "a command and operational framework expressly and objectively set up to make them 'click.' "[12] He noticed the findings of the Anti-Submarine Warfare unit in Boston which, inspired by operations research being done in Britain, concluded that radar-equipped aircraft rather than navy destroyers provided the best defense against German U-boats. Shifting the focus of submarine warfare to aerial search-and-destroy was not easily achieved, however. As German Admiral Karl Dönitz recognized, the most important advantage of the submarine was the difficulty of being located. For U.S. Admiral Ernest J. King, scouring the vast expanses of the ocean for U-boats from the air was about as useful as "the proverbial 'search for a needle in a haystack.' " He emphasized "that anti-submarine warfare—for the remainder of 1943 at least—must concern itself primarily with escort of convoys."[13] King's opposition gradually crumbled after heavy losses in the North Atlantic and the immense success of the Royal Air Force in destroying submarines in the Bay of Biscay. This was partly thanks to the deployment of a number of devices developed by scientists under contract to the OSRD, including ASV radar. The admiral reluctantly conceded that hunter-killer groups could turn the tide of war, and that if naval officers and scientific researchers pooled their professional talents, they could make optimal use of the newest technologies coming out of the labs. By summer 1943 the Boston operations research team, led by the MIT mathematical physicist Philip Morse, was integrated into a new Tenth Fleet that grouped together all the antisubmarine units in the Atlantic. The Army Air Force also began expanding its dependence on OR. By V-J day, writes Daniel Kevles, "from Africa to Southeast Asia and on to the Aleutians, civilian scientists were in vogue as strategic and operational advisers to a degree without precedent in the annals of American military history."[14]

Philip Morse and OR

Philip Morse did graduate work in physics at Princeton University in the late 1920s and spent 1930–31 in Munich and at the Cavendish Laboratory in Cambridge.[15] On returning home, he was recruited by Compton to join a new physics department being set up under John Slater, his task being to build a first-rate acoustics program at MIT. Morse's first major contribution to the war effort was to develop a decoy for acoustic mines that used the sound of a ship's propellers to detect and demolish the vessel. He saw little interest in building gadgets for the military, however. By the end of 1941, as he put it in his autobiography, "Some of us wondered whether the only thing trained scientists were good for in war was to do the measurements and design work thought up for them by the supply departments of the armed services."[16] The pace of technological change, and the need to react rapidly to it in war, created different demands on scientists, as well as new opportunities for them to influence policy. His chance came when, in April 1942, he was authorized to set up the navy's Anti-Submarine Warfare Operations Research Group (ASWORG), later a subgroup of the Operations Research Group (ORG).[17] Building the team was not easy. Morse was competing for scarce personnel against other rapidly expanding and prestigious projects. Within six months he had about a dozen and a half people working with him, mostly physicists and mathematicians. By the end of the war, ORG numbered close to a hundred analysts. They had to persuade senior military officers that new and unfamiliar weapons being developed substantially improved their capacity to wage war. Technically knowledgeable civilian scientists had to work closely with users in the field to explain how devices worked—and to debug them. OR researchers had to personally collect reliable data on which to base their analyses, rather than get it secondhand from archives or from others at the front.[18] Finally, they had to work closely with commanders responsible for deployment to ensure that new or enhanced technologies were used to greatest effect. As one report put it after the war, "In some cases it pays far more to improve the tactics than to improve the weapons, and it is almost always easier, faster and cheaper to do so."[19] The military, writes Andrew Pickering *"enfolded* civilian science, reconfiguring its internal structure to embrace OR practitioners and reconfiguring its operations around their findings and recommendations."[20]

The successful management of the civilian/military interface was the key to the wartime success of OR. Morse used a revealing metaphor to describe how he saw the relationship between the civilian expert and the military officer. In a speech in May 1962 celebrating the twentieth anniversary of the ORG (now the OEG—the Operations Evaluation Group), he suggested that it was akin to the relationship between a doctor and a patient.[21] It had two key components: independence of decision and privacy of disclosure. The navy ("patient") had to give the OR researcher ("doctor") access to all the facts of the problem it faced. The physician "had to be outside the usual chain of command, in order to maintain an impersonal point of view," and "free to use [his] own methods of analysis, to reach a diagnosis." The only constraint on this independence of decision was that the expert's recommendations had to be realistic and could be applied. In return, the "doctor" had to respect the "patient's" privacy, that is, respect "the principle of privileged communication, the principle that the reports of our work belonged to the Navy alone and could only be disclosed to others with the Navy's consent." By adhering to these guidelines, "We could be loyal and responsive to the Navy's aims in spite of our independence of procedures and assignments." Mutual trust was indispensable: "Without this trust we might as well have gone back to the laboratory."[22]

Building a Program at MIT after the War

The results that OR had achieved during the war ensured that the U.S. military would continue to support it afterward. A Weapons Systems Evaluation Group (WSEG) advising the Joint Chiefs of Staff was established in 1948 as part of Truman's new National Military Establishment (see chapter 2). Morse was its first technical director. His influence grew when, in the mid-1950s, MIT was asked to set up a private, nonprofit body to advise the WSEG, the Institute for Defense Analysis (IDA).[23] Morse was also appointed to the board of trustees of the Rand Corporation when it was separated off from the Douglas Aircraft Company in 1948.[24] He stayed only briefly with Rand before taking up his duties at the WSEG; nevertheless, he cemented his relationship with the think tank when he returned to MIT in 1950 and remained for the rest of the decade on the board of Rand, where he became fully engaged in its war games simulation program.[25]

While the links with the military, derived from its success in the field, gave OR credibility and endowed its practitioners with prestige, influence, and access to resources, Morse's ambitions for the field were far broader. In 1951 he and George Kimball, his wartime deputy in ASWORG, published *The Methods of Operations Research*—its core was the declassified, "theoretical" fragment of a three-volume tome written originally for the navy. They defined their topic as "a scientific method of providing executive departments with a quantitative basis for decisions regarding the operations under their control."[26] This generalized definition was deliberate. In fact, Morse was a tireless promoter of OR as a tool for rational, quantitative, science-based planning in every sphere of social life. As he said, people like himself who came out of the war with experience in OR "weren't seeking support for big laboratories. We weren't out to learn more about the nucleus or the solar system." We had "just been through a concentrated course in how to generate political action" and "we believed we had learned some techniques which could help the decision process itself, could improve the speed and effectiveness of military decisions, or industrial decisions, or even perhaps decisions in public affairs."[27] To fulfill this aim, however, required the education of new generations of researchers who believed in the value of the technique and the formation of a group with the usual trappings of a professional society. The first meeting of the Operations Research Society of America was held on 26 May 1952, under the auspices of Columbia University. It was attended by about seventy people, over half of them from the military or from think tanks like Rand that did contract work for the armed services. Morse was elected its first president and a year later could announce that membership had grown to more than five hundred, thanks above all to industries (aircraft, electrical, oil, railroad, insurance, department stores, and so forth) that had either established groups in-house or bought OR consultants.[28] National journals and an international federation followed in quick succession.

The interpretative flexibility of OR[29] made of it an ideal passage point for the social sciences, notably economics, to gain public respectability, scientific legitimacy, and military funding after the war.[30] In 1951 Robert Solow, a distinguished economist at MIT, was skeptical about the possibilities of extending OR beyond the military domain. Five years later he was enthusiastically extolling its benefits in the pages of *Fortune*. "Longhairs—typically PhDs with no business training or experience—are getting into business more and more," he wrote, "supplying not only the technology of machines or processes, but also a new general technology linked to decisions."[31] Indeed, according to Philip Mirowski, "The reason that OR was so important after the war was not due to any technical innovation or discovery: rather, it was the workshop where the postwar relationship between the natural scientists and the state was reconfigured, and the locus where economics was integrated into this scientific approach to government, corporate management and society." John von Neumann was crucial to this reconfiguration in the United States. He introduced his game theory "as an adjunct to OR, lending mathematical firepower and intellectual clout to the wartime concern over 'strategy': it was he who forged the lasting links with the computer as tool and as exemplar for organizational rationalization; it was he who endorsed the promise of linear programming to various military organizations; and it was he who supplied the bridge to information theory."[32]

The OR curriculum that Morse developed at MIT, and sought to export abroad in the late 1950s, reflected these postwar developments in the United States. To establish the program, Morse needed to subvert traditional disciplinary boundaries, producing "a more general kind of scientist" who, while having a "thorough training in the scientific method," was also interested in the broader interactions between new technical developments and the social system.[33] Several things facilitated his task. First, the MIT faculty and administration alike began to realize that rigid boundaries between academic departments were no longer sustainable. The Research Laboratory in Electronics (established in 1946) set the precedent and, to quote its first director, Julius Stratton, "was destined to influence the development of interdepartmental centers at the Institute over the next two decades," centers that took "account of the fact that newly emerging fields of science commonly cut across the conventional disciplinary boundaries."[34] Second, there was the administration's determination to integrate the social sciences into the curriculum. In 1949 Stratton, then provost, shaped a major report on the postwar educational challenge to MIT, which insisted, inter alia, that the social sciences and the humanities be strengthened.[35] The new president, James Killian, whose vision for MIT was "a university polarized around science, engineering, and the arts," shared Stratton's ambitions and actively supported Morse's efforts.[36] Finally, there was a willingness to accept federal (as opposed to industrial) patronage, notably from the military, to the extent that Alvin Weinberg famously remarked in 1962 that it was increasingly difficult to tell "whether the Massachusetts Institute of Technology is a university with many government research laboratories appended to it, or a cluster of government research laboratories with a very good educational institution attached to it."[37] These three features of the new MIT that emerged from the war—a commitment to interdisciplinary work, a determination to integrate the social sciences into the engineering curriculum, and a faculty that was anything but coy about accepting federal, particularly military, patronage—were crucial to Morse's success in getting his teaching program in OR off the ground.

The field began to be taught at MIT in 1948, but it only really became institutionalized when Morse returned to campus in 1950. In 1952 he was nominated chairman of the interdepartmental Institute Committee on Operations Research, which included faculty representatives from the departments of mathematics and of civil, electrical, and mechanical engineering, as well as from the new Sloan School of Industrial Management.[38] The committee instituted a weekly seminar and launched a summer program, first held in 1953. Seventy people signed on for the course, about forty of them from industry and fourteen from military organizations. All were expected to have basic knowledge in calculus and differential equations. In the morning, Morse and mathematician George Wadsworth gave theory lectures dealing with topics such as the fundamentals of probability, mathematical modeling, queuing theory, Monte Carlo techniques, game theory, linear programming, and feedback-network equations. The afternoons were devoted to practical studies with visiting staff that had field experience in industrial and military groups. Firms and government agencies soon began sponsoring research at MIT, and people were trained in mathematical techniques that called for high-speed computers.[39] (In 1957 Morse became director of the MIT Computation Center, which was equipped with an IBM 704, quickly replaced by an IBM 709.) By the time Morse had become seriously involved with training Europeans through NATO in the late 1950s, he had already built an undergraduate and graduate program demanding advanced scientific knowledge and computing skills. This was the educational model that he hoped to export abroad.

Morse's First Close Encounter with OR in Continental Europe

Morse was convinced that NATO's capacity to function as an integrated military alliance depended not only on having a capable central command structure, but also on having competent OR groups in every participating nation. He elaborated on this to a navy audience in 1962.[40] The United States, he said, was having trouble communicating with its allies in NATO. This was not simply a matter of training: it was primarily due to the unique way in which civilian scientists doing OR had been integrated into the armed services in America for about twenty years: "The interplay between scientists and military staffs, between weapons designers and weapons users, is so close in this country that it is affecting our vocabulary as well as our planning. Officers talk like physicists and acousticians argue tactics." America's NATO allies were not used to "this interpenetration of scientific and military thinking." This made it extremely difficult for countries without strong military OR groups on the American model to coordinate their planning with the United States. Here then was an urgent task for NATO, and Morse was obviously the best man to take it on. He had already established a summer program in OR at MIT, and he had the staff who could teach it abroad. He had close links to the defense establishment and knew whom to tap for money. And he had already worked closely with NATO's first two science advisers, Norman Ramsey and Fred Seitz.[41]

At the Science Committee's second meeting in July 1958, the United States offered to pay the living costs and tuition fees of European scientists from NATO countries, funded in part by the Department of Defense, to attend Morse's annual summer school in Cambridge. A contingent of twenty-seven, mostly from France, Germany, and Italy, along with about a dozen people from other countries, went to the United States at once.[42] The American delegation also distributed a detailed document describing a number of concrete steps that the U.S. was prepared to take to promote OR in the alliance. This was, in fact, a blueprint for a major U.S.-inspired and U.S.-funded training program to strengthen operations research in NATO.[43]

Morse took the initiative by contacting Ramsey during his visit to the Boston area in fall 1958.[44] Ramsey explained that any OR activity in the NATO centers would have to be approved by the Mutual Weapons Development Program (MWDP). In December 1958 Morse and his wartime colleague Bernard Koopman went to see General Larkin, the head of the MWDP. They proposed combining a variant of the MIT summer school in Europe with brief visits to several countries to survey the national scenes firsthand. Larkin liked their plan to "carry . . . American ideas to the unconverted," as he put it.[45] The U.S. Air Force quickly agreed to place a contract with MIT to "Provide Consulting Services to Aid in Establishing Programs in Operations Research in NATO Countries."[46]

About 120 people participated in Morse's first course in Europe, which was held at the Training Centre for Experimental Aerodynamics, near Brussels, from 17 to 28 August 1959.[47] Half were from the military,

30 percent were from industry, and the remainder was from universities. Belgium, France, and Germany each sent about twenty-five people, followed by Italy (sixteen), Norway (ten), and the Netherlands (nine). Great Britain was not represented: ironically, as we shall see, Morse believed that the field there was so far ahead of OR on the Continent that his training programs were unnecessary. A survey of the participants showed that, while everybody agreed that the course was well-prepared and intellectually stimulating, many felt that less attention should have been given to theory and more to the discussion of practical, notably military, problems.[48] Notwithstanding its success, Morse and his team concluded that the exercise should not be repeated. "A 'broad-brush' program, such as the one given in Brussels . . . can be of some help in getting things started, but further NATO seminars and programs will be of limited help." The single most important reason for this decision was that the participating countries had different levels of development in OR. Since each had different needs, help had to be "tailored to the specific needs of the different countries, as determined at the technical level."[49]

The Diverse Forms of OR in Europe

The variety of situations prevailing in the alliance was brought home forcibly to Morse's team when they made their site tours.[50] Turkey was at one extreme. It was not simply that the level of work in OR was not particularly high there—it was more or less irrelevant. The American military advisers on the spot "emphasized that, while a great deal of equipment has been placed in Turkey, nobody knows where it is, how much there is, and what condition it is in." It was obviously ridiculous to concentrate OR efforts on, say, methods of equipment supply, a standard topic in more industrially advanced nations. As for trained manpower, and those teams of civilian scientists who were supposed to work with the military on OR problems—there was widespread illiteracy in the country, although the government had just passed a law that every child should have six years of schooling. "This means," the Morse team reported sanguinely, "that in about one or two generations things should look better."[51]

France was at the other extreme. The visiting Americans were deeply impressed by the breadth and depth of OR in the country. Even though it had been started barely six years before, they judged the level of French operations research to have "a place close behind the U.S. and the U.K."[52] Each branch of the military—the army, the navy, and the air force—had its own OR sections, as did an interservice group, the Centre Interarmée de Recherche Opérationnelle (CIRO). There was also a large and growing civilian group headed by Charles Salzmann, the Centre Français de Recherche Opérationnelle (CFRO). When Morse and some of the other team members visited the CFRO in 1959 it had fifteen scientific members and about half a dozen civilian consultants, mostly in academia. Sixty percent of its work was for the military, 30 percent was for private industry and 10 percent was for government concerns.

Morse and his colleagues waxed lyrical about the superb education of the French military, including their knowledge of mathematics and probability, and the fertile ground for operations research in a country that, they felt, had a "rationalist outlook on life." Their only concern was that "brilliantly educated as the military members of the O.R. groups may be," they did not have that "extra 'degree of freedom' of imagination" of the civilian scientist in basic research. Being relatively low in rank, their talents would be stunted by having to report to and respect their superior officers. What was lacking in France, then, was that "interpenetration" of the civil and military, of academia and the armed forces, that Morse and his colleagues prized so much: indeed of the twenty-five French participants in the summer school in Brussels, only one was from a university.

The visit to Germany was disheartening. Morse felt that there was virtually no communication at all among the universities, industry, and the military, nor any desire to cross boundaries. Young men saw the need for cooperation, the American visitors wrote, but did not know how to accomplish it. Industry and the military, for their part, seemed to have "no appreciation of the benefits which could come from university cooperation." The military in fact were downright hostile to "allowing any 'university types' to have much part in military planning," apparently because they saw them as being too academic and incapable of tackling concrete questions. As Morse put it in his autobiography, "there seemed to be no place in the military organization chart for a civilian group, particularly one containing academicians."[53]

Norway lifted their spirits. Morse and his colleagues gave talks at the University of Oslo and to the Norwegian OR society. Speaking to people at the Norwegian Defense Research Establishment (NDRE) and to the Joint Chiefs of Staff, they found "considerably more understanding of the problem and more development of possibilities . . . than in Germany." They thought highly of the experimental and theoretical quality of the OR studies made at the NDRE, which was combining "weapon development with operations research and weapons evaluation for all three Norwegian services."[54] After his discouraging visit to Bonn, Morse was relieved to find that in Oslo, unlike "countries farther south," "differences in culture and thinking" were negligible. In Norway "we found we laughed at the same jokes, and we felt no need to pull punches in an argument."[55]

These impressions, based as they were on a few days of face-to-face interaction, obviously need to be interpreted carefully and read through the grid of a thorough understanding of the trajectories of OR in different countries after the war—a task I cannot undertake here. However, they do suggest a somewhat different interpretation of European diversity than that put forward by Mirowski. As noted, Mirowski has claimed that in postwar America the disciplines of OR and economics became intertwined, with economics providing a sense of coherence, legitimacy, and identity to a field that lacked them. He has extrapolated this interdependence beyond the boundaries of the United States, claiming that while operations research "looked different when constituted in Britain, or America, or on the European continent, or in the former Soviet Union . . . much of this can be traced to the vicissitudes of World War II, divergent state policies towards science, and ensuing local interactions with culturally variant conceptions of economics."[56] The locus of Mirowski's analysis is governments, corporations, and "society"; he pays no particular attention to the military. Morse and his colleagues did, and they found that the way in which academia was organized and the social structures that embedded advanced education in national military systems had a profound effect on a country's interest in OR and in the kind of OR that was practiced.

Consider Germany, which did not integrate civilian scientists closely into high-level military command structures during the war, and whose military establishment only really began to flex its muscles after the country entered NATO in 1955. Here the field was only beginning to become formally recognized in the early 1960s (the German society for OR was founded in September 1961). One of the consultants hired by Morse found that "important segments of policy-making groups in the military, in public administration, and in industry were unaware (or perhaps unconvinced) of the general applicability of scientific methods to the solution of important command or executive problems."[57] The relations between the universities and the military were mostly confrontational. Academic structures were rigid and discipline-bound: a colleague from MIT reported in 1965 that German professors had a "tunnel vision," distrusting any interdisciplinary activity on the grounds that "contact with another domain [was] by definition dilettantism, hence unprofessional" and "contaminating."[58]

In Norway, by contrast, Morse found operations research flourishing in the NDRE, which was closely collaborating with the University of Oslo. Established in 1946, the NDRE drew on the skills of Norwegians who had worked in Allied laboratories during the war, following German occupation in 1940.[59] Throughout the 1950s it was heavily funded by the government, generally more than even the strong atomic energy program, and far more generously than industrial research institutes (indeed, Morse found industrial OR "just beginning" in the country). More to the point, the NDRE was heavily engaged in the MWDP, devoting considerable resources to the development, with the United States, of an antisubmarine weapons system, Terne III, in addition to working on proximity fuses and on radio communications in the polar region. It was also responsible for the installation and testing of parts of a strategic communications system for NATO headquarters, then in Paris. A Ferranti Mercury computer was installed at the establishment in summer 1957. A team led by Kristen Nygaard, who had been with the NDRE since 1948 and who had just completed his master's thesis in mathematics at the University of Oslo ("Theoretical Aspects of Monte Carlo Methods"), began writing computer simulation programs with it.[60] Nygaard, along with Ole-Johan Dahl went on to develop the first object-oriented program languages. In Norway, Morse saw a porous barrier between the civilian and the military, and the integration of advanced academic research in OR in a military establishment that was very much in line with the American model.

The military also provided a key site for the development of OR in France, but the sharp division in the educational system between the universities and the grandes écoles, dedicated to training cadres and an officer class, produced yet another configuration.[61] The military sponsored theoretical studies in OR. In 1963 Professor Malavard described to NATO the kind of training he arranged at the Centre Interarmée de Recherche Opérationnelle.[62] The CIRO taught advanced mathematical techniques to students of a suitable level from the grandes écoles or having a bachelor's degree from a university. This lasted seven months and was followed by two-and-a-half months of project work on a concrete problem. The basic training could be supplemented by attending a series of week-long specialized courses in particular topics—stochastic processes, dynamic programming, simulation, and so forth. This system, while integrating advanced scientific theory into the military, did not also "enfold" *civilian* students and researchers into military OR to the extent that Morse hoped for. In short, insofar as Morse's perceptions are significant, they suggest that Mirowski's insistence on the close link between the rise, even survival, of OR and changing "conceptions of economics" needs to be refined to give due weight to the very different social structures in which OR was conducted in Europe and, in particular, to the relationships between academia and the military, which were sometimes very different to those prevailing at MIT.

Philip Morse and his colleagues came to continental Europe in the summer of 1959 to train people in operations research. They did so because they believed that the security of the West depended on combining the skills of civilian scientists with those of military officers to make the best use of military equipment, and that an alliance could only be effective if top officers in each of its member nations knew how to optimize resources to achieve a common goal. They realized that Europe was a diverse patchwork of nations at different stages of development, but they believed that the kind of program they had successfully taught at MIT for more than five years would provide the necessary basis for welding the alliance into an efficient fighting unit. The program they proposed combined theory with practice and, crucially, also embodied their views on how OR groups should be constituted and run. In particular, they stressed that an OR group could only be successful if it included civilian scientists who were free from the constraints of a military hierarchy and whose relationship to the military was analogous to the doctor-patient relationship. This was at the core of the training. Morse emphasized it on several occasions at the summer school and in various discussions with attendees. It was also "made a central topic in the talks to the various nations visited during the second stage of the course."[63] And it was endorsed, in broad outline, by the U.S. Department of Defense: operations research, according to a 1961 statement, had "been most effective where analysts (a) have worked in strong interdisciplinary groups (b) on substantive problems (c) at those administrative levels at which the problems occur and (d) with direct access to the responsible staffs and executives."[64] OR was not just a clutch of theories and not just a set of computer-based models "applied" to practical problems; it was also a set of social relations.

The MIT team's three weeks in Europe gave substance to the platitude that diversity and specificity characterized the situation across the Atlantic. The Europeans, particularly the French, were theoretically strong, and many groups on the Continent were putting OR to work in military and sometimes industrial contexts. If exporting the MIT model was going to be difficult, it was not because people in Europe lacked the skills or the interest in OR. Rather, it was because the structure of social relations into which the practice of OR was to be embedded was quite different, not only between the United States and Europe, but also among European nations themselves. It was not simply that implanting OR in some countries was meaningless or that the field had reached very different levels of development. An interdisciplinary program in an emerging discipline that was intended to combine the skills of civilian experts with military executives was not novel in an institution like MIT, situated at the heart of the military-industrial-academic complex; it was barely conceivable in most European countries. Remodeling the Old World to fit Morse and Koopman's paradigm did not simply involve "carrying American ideas to the unconverted," as General Larkin thought; it required restructuring the social fabric itself.

Defining What Training in OR Would Be in NATO

During the course in Brussels, Morse went down to NATO headquarters in Paris to speak with Seitz about sustaining the momentum of the two summer schools. Seitz suggested that his American colleague set up an advisory panel on operational research of five or six of the best scientists in the field from different NATO countries under the auspices of the Science Committee.[65] In the months that followed, a project was developed that drew on the recommendations made by Morse's team after their summer tour in Europe, on suggestions made by Koopman, now based in London on behalf of the IDA, and on input from the Department of Defense, which was going to put up quite a bit of the money.[66]

Morse's panel met for the first time on 28 March 1960 in Seitz's office.[67] Sir Charles Goodeve, director of the British Iron and Steel Research Association and a man who played a major role in promoting civilian OR in Britain, was there, as was Charles Salzmann, from France.[68] So too were Erik Klippenberg from the NDRE, whom Morse had met during his visit to Oslo a few months earlier; Giuseppe Pompilj, director of the Institute of Calculus and Probability at Rome University, who had been inspired by the Brussels meeting; and Henry Görtler, a professor of economics at Freiburg University, added somewhat "tentatively" by Morse after his bad experience in Germany.[69] Four draft recommendations were drawn up, revised, and accepted by the Science Committee meeting on 6 and 7 April. Ten years later Morse tried to belittle the importance of these revisions to his original scheme.[70] In fact, they were the first step in an increasingly determined challenge to his conception of how best to strengthen OR in the alliance.

The draft recommendations of 28 March 1960 placed university education at the heart of the educational process. The revised recommendations, accepted by the Science Committee, no longer gave university education the pride of place in the training program. Equal if not greater weight was given to more technically oriented training and to practical experience in the field.[71] Even Morse's proposal that a recruit had to spend a year in an academic OR program before being apprenticed to an existing OR group for a further nine to twelve months was watered down. The committee suggested that the "basic scientific education provided by standard universities and technical schools" provided enough background for a fellow, who could use it for further "academic study in OR *and/or* for work with well-established O/R groups."[72] In short, the Science Committee felt that an OR field-worker did not really need an academic training in OR as such: all that was required was a general scientific background.

There are two versions of what happened at the meeting of the NATO Science Committee when this paper was discussed for the first time—one by Morse, and one in the official minutes. In his autobiography, published in 1977, Morse wrote that Isidor I. Rabi and Sir Solly Zuckerman were hostile to the proposal. Rabi feared that it would draw money away from the fellowship scheme, which was the heart of the education and training program the Science Committee was promoting. Once Seitz had reassured them that the MWDP would bear the major financial burden, and after an impassioned appeal for support by Finn Lied, director of the NDRE, "the tide was turned," wrote Morse. "The proposals were approved; they would be reviewed after three years' experience."[73]

There is no trace whatsoever of this in the official minutes, dominated by the comments of a Mr. Johnson, the assistant scientific adviser to SHAPE.[74] Johnson made three main points. First, SHAPE was only interested in the military aspect of operations research. Second, practical studies were valuable; theoretical research was not. Third, the boundaries between civilian and military OR should be drawn sharply for security reasons; international collaboration should also be discouraged for the same reason.

SHAPE's demands flew in the face of Morse's and Seitz's ambitions and surely influenced the redrafting of the original proposals. When the science adviser had sent out invitations to join the APOR, he was careful to stress that "our Office is first and foremost interested in the *science* of the field and not necessarily in its military applications even though, we will, of course, not ignore the applications to any appropriate field."[75] This was coherent with the academic make-up of the committee and with its overall aim to strengthen Western science. It also defined a well-trained OR worker as someone who had university training in OR, who had acquired a range of mathematical skills that could be deployed indifferently in military or civilian contexts, and who could aspire to becoming part of an international scientific community. Morse's original set of proposals embodied this goal. It now emerged that this was not a goal shared by the military authorities in NATO. To defend Morse's approach from Johnson's criticism, Seitz was obliged to affirm at the meeting that the APOR was "not aiming at directly influencing the work of the military authorities, but rather at establishing reputable institutions capable of training competent scientists with a view to service in the practical field."[76] Seitz and Morse wanted to embed OR in European training programs that produced scientists with "universal" skills, detached from local requirements and specificities but applicable to specific practical problems. It was not what the NATO authorities wanted.

The Defeat of the Model

With his proposals officially accepted, Morse and the APOR went ahead with "indoctrination" courses in OR at several universities (courses intended "to arouse interest and demonstrate how regular instruction in operational research can be established") and in training courses in various NATO centers.[77] To take account of the very different national situations in the Continental NATO member states, consultants were recruited to work for six months to a year in specific countries, adapting their programs to the local context. David Stoller was released from the Rand Corporation to spend six months in Italy; Maurice Sasieni, from Case Western, went to Norway; and Erwin Baumgarten (from the U.S. Navy's OEG) went to Germany for a year.[78] By special request of the Federal Ministry of Defense, his leave of absence was extended for a further six months at German expense. For the first two years the only cost of these activities to the Science Committee was the expenses of the British and French lecturers who helped with the teaching.[79] Morse was supported under a contract with the IDA, which also paid most of the cost of the consultants that he recruited.

All three consultants played an important role in catalyzing interest in OR in their host countries.[80] All three interacted productively with OR groups doing practical work there. All of them reported having difficulty maintaining student interest in OR due to the rigidities of the national educational structures and existing demands on students' time. Stoller, who taught at the University of Rome, was most successful: his graduate course in queuing theory started with about thirty students, though attendance then fell off to about half that. But he had an ideal link in the person of Pompilj, and many of the students in his Institute for Calculus and Probability were air force and army officers. Sasieni, who spent 50 percent of his time teaching two courses at the University of Oslo, found his numbers shrink by two-thirds, from fifty to twelve in his introductory course, and to four or five in his advanced course. Baumgarten taught a number of introductory and advanced courses at the Universities of Bonn and Freiburg. The courses were elective, and attendance was irregular. Baumgarten saw hopeful signs in Germany— an excellent OR capability in the defense establishment, a chair of econometrics established at Bonn, the formation of a national OR society, and the identification of OR as a "Schwerpunkt-programm" by the Deutsche Forschungsgemeinschaft, thereby privileging operations research for financial support. All the same, the extension of his contract for six months at the expense of the Federal Ministry of Defense was not to stimulate education and training: it was for briefing and consultation with the military establishment.

The experience of the consultants showed clearly that European universities were structured quite differently than an institution like MIT and that many of them were situated in a quite different social space. The postwar structure of MIT was articulated around interdepartmental and interdisciplinary programs, and the mutual interpenetration of the civil and the military was a matter of course. In much of Europe, as noted in the previous chapter, the disciplinary departmental structure was still deeply entrenched in universities, and links with the military were delicate: at least in appearance, and often in practice, these were institutions for teaching and research that shied away from doing classified research.

In 1962 Morse lost control over the evolution of his program, previously made possible through his substantial contract with the IDA. The NATO Science Committee began to pay an increasingly large share of training for OR from its own budget, and the British delegation, which had never been enthusiastic about the program, immediately called for changes. In a note dated 2 February 1962 the British presented the findings of a recent "critical review" made in the UK of the NATO Science Committee's "Operational Research Training Programme."[81] It was a devastating indictment of Morse's concept of operations research.

Operational research, the British wrote, was "an exceedingly practical matter." It was "not to so much a science as a scientific attitude of mind on the part of its practitioners," and "on-the-job training was the only way in which the proper attitude can be inculcated into young and inexperienced operational research workers." Recruits could have a wide variety of basic training—from botany to zoology. Furthermore, the British did not believe "that instruction in a random selection of mathematical techniques can convert a newly-graduated scientist into a fully-trained operational research worker." For them, learning was by doing—topped off with "short specialist courses" as necessary. As the British paper put it, rather ponderously and with all due caution, "During the course of his work the average operational research worker will from time to time find it desirable to use a technique with which he is unfamiliar; if his attempts to gain enlightenment from a colleague, textbooks and other technical literature should fail, recourse would be had to attendance, as the occasion arose, at a suitable specialist course." Advanced intellectual training was a last resort, then, not a top priority, and had to be highly focused.

The British approach stripped operations research of the intellectual credibility it derived from sophisticated mathematical techniques and computers. It also deprived it of the legitimacy it gained by being a subject worthy of a master's or Ph.D. program at a prestigious university. This had immediate practical consequences. Late in 1961 the Science Committee decided to abolish the system of NATO consultants, intended to introduce OR into academia. Instead, NATO would spend its money for OR on summer schools and two-year graduate apprenticeships.[82] What is more, the British wanted *both* years of the apprenticeship to be used for on-the-job training, not just the second year, as Morse had planned. As for encouraging "post-graduate degrees in operational research"—the kind of thing that one found at MIT—the British delegation believed that "the experience there *would not be sufficiently broad* to justify the grant of a Graduate Apprenticeship."[83] Those who needed to enrich their understanding could do so by attending specialist short courses and seminars dedicated to a specific topic that included discussions of case histories and experimental work. They could also attend biennial conferences at NATO headquarters having a definite theme—civil, military, or both.[84]

The pressure to prioritize the practical, service dimension of operations research at the expense of the advanced training of scientific researchers increased in 1963. The NATO Science Division appointed Alan L. Oliver to serve as an executive officer to the APOR to liaise with all military operational research organizations in the alliance and to suggest how better to apply OR to military problems.[85] Morse's new right-hand man, who took up his post in April 1963 and who was certainly British, had a view of OR that clashed head-on with those of his chairman. Oliver insisted that the panel had to reorient its goals if it wanted to survive in the framework of NATO. The Science Committee quickly endorsed this view.[86] His emphasis on problems of importance to the military was coupled with skepticism regarding the need for advanced theory—"much worthwhile operational research is done with simple schoolroom mathematics."[87] If consultants were again recruited at the national level—and there was considerable demand for this from some NATO countries—they would not be primarily linked to academia, as Morse had done. Oliver felt that the consultant's "primary emphasis" should be working with "the national operational research groups" rather than with academics, and that a local OR society should sponsor them. The apprenticeship scheme, as conceived by Oliver, lost any vestige of Morse's original view of it as work experience to supplement postgraduate training at a university. Its aim was "to provide staff for the operational research units of Government Departments" in nations that needed them, that is, to produce civil servants.

In 1964 the Department of Defense decided that it could no longer provide Morse with financial support.[88] He had already prepared the ground for his departure. In January 1963 he chaired a meeting of a Group of Experts on Operational Research for the Organization for Economic Cooperation and Development that wanted to promote postgraduate training in OR teams and to link an OR approach with macroeconomic analyses.[89] By 1965 Morse had left the NATO framework to pursue his vision in Europe through that organization.

The Peculiarities of the British

Philip Morse and his team encountered multiple social forms and material practices of operations research in Europe. But it was the British who left an indelible mark on the structure of OR as promoted by the Science Committee, particularly with the recruitment of Oliver. That mark was symptomatic of the country's postwar history of civil OR, which the British imported into the military world of NATO. It was defined by two main features.

First, OR had little credibility in British government circles, where it was regarded as politically suspect and not much more than applied common sense.[90] The idea of using OR to rationalize and optimize the postwar economy was never embraced with much enthusiasm by the first Labour government, particularly after the onset of the Cold War, when centralized state planning smacked of totalitarianism. The left-wing credentials of many proponents of OR, notably Blackett and Zuckerman, added to government suspicions. Treasury officials reported that the more they read about operations research, the more it seemed to them to be little more than a fancy name for "a whole range of sensible activity (already known, studied and applied under other names)," and they effectively decoupled its practice from its wartime association with physics and mathematics. Until the mid-1960s, in fact, civil OR in Britain was largely restricted to two nationalized industries, thanks only to the personal commitment of a few key people in the National Coal Board and the British Iron and Steel Industry Research Association (BISRA), where Sir Charles Goodeve played a major role.

The view that OR was little more than educated common sense implied, of course, that it was not the kind of subject one needed to learn at university. The practitioners who had developed OR for the British military during the war, men like Blackett and Zuckerman, did not, like Morse, try to introduce it into academia: they went back to their prewar activities once hostilities were over. As a result, whereas by the end of the 1950s there were at least six American universities that had graduate programs leading to postgraduate degrees in operations research, OR in Britain remained "practitioner-dominated," and "apart from one or two short courses at individual universities, British postgraduate OR programmes were notable for their absence."[91] Again it was only in the 1960s that the situation began to change, as new universities and business schools were built that were willing to experiment with new curricula. Knowing this, we can better appreciate the scathing attack of the British in the NATO Science Committee on Morse's educational ambitions in February 1962, their contemptuous dismissal of his advanced educational agenda, and the practical, vocational dimension of OR that they, and the science adviser to SHAPE, promoted so strongly.

It would be wrong to conclude that Morse's attempt to train operations researchers in the NATO countries failed. On the contrary, by 1965 about five hundred people had participated in APOR conferences and symposia. Almost thirty graduate apprentices had been placed in various military (SHAPE, Canadian Defence Research Board) and civilian (BISRA, French CFRO) organizations.[92] Oliver was increasingly successful at getting national military OR groups to provide enough information on what they were doing, without violating security restrictions, so that they could collaborate across national boundaries.[93] The APOR did not fail to improve the level of operations research in NATO; on the contrary, it was "particularly valuable . . . in view of the pitifully small number of properly trained operational research specialists in Western Europe."[94] But it did not educate the kind of person Morse believed was essential if the technique was to be used properly. His ideal of the highly educated civilian consultant and university teacher sitting alongside a senior officer, sharing data, thinking freely and out of the box, and making suggestions whose rationale the military man would understand and act on—this ideal made no sense to the British nor, apparently, to the NATO authorities in SHAPE. The distaste for complex theory and advanced postgraduate training in operations research, the conviction that OR was a practical, pretty nonacademic business, coupled with a strong compartmentalization between the civilian and the military, dealt his model a double blow.

9

Concluding Reflections: Hegemony and "Americanization"

Hegemony

This book explores the place of science and scientists in the implementation of American foreign policy in the first two decades after the war. The use of science as an instrument to project state power abroad and to consolidate it at home was not novel to this period. It was also an important component of the process of colonization and the traditional construction of empire.[1] However, the process of integrating Western Europe into an American-led Atlantic alliance resulted in a very different kind of hegemonic regime than the traditional one—an "informal," coproduced empire, built on consensus rather than coercion.

The construction of the American empire was above all a response to the Cold War geopolitical world order, with its division into two rival world systems. Each hegemon built its own empire by its own means, using its domestic economic, political, and ideological systems as a blueprint and point of reference for how others in its sphere of influence should structure their social worlds. As Gaddis put it so eloquently, by 1947, with the collapse of the United States' hope that the United Nations could manage the postwar system along the lines that it sought,

it was clear that cooperation to build a new order among the nations that had vanquished the old one was not going to be possible. There followed the most remarkable *polarization* of politics in modern history. It was as if a gigantic magnet had somehow come into existence, compelling most states, often even movements and individuals within states, to align themselves along fields of force thrown out either from Washington or Moscow. Remaining uncommitted, in a postwar international system that seemed so compulsively to require commitment, would be no easy matter.[2]

Nor, should it be said, would neutrality be easily tolerated. At the very moment when decolonization was getting into its stride, major powers with global ambitions emerged, seeking to build or extend empires and to consolidate their spheres of influence using instruments that ranged from the outright use of force to irresistible seduction.

Not everybody believed that it was in the interests of the United States to help reconstruct Europe. One American scientist interviewed on behalf of the Ford Foundation in 1955 thought that the United States should leave Europe in its relatively backward state and cream off the best talent, to be trained and employed in the United States. If these views did not prevail, it was partly due to sentiment, to a sense of "decency," and to a belief in historical ties and historical debt. The United States, said Eisenhower, "was related by culture and blood to countries in Western Europe and in this sense was a product of Western Europe." The European peoples, said McGeorge Bundy when he was Kennedy's national security adviser, are "our cousins by history and culture, by language and religion. We are cousins too in our current sense of human and social purpose."[3] As a great power, the United States had responsibilities.

It also had interests. A weak, fragile, and fickle Europe, vulnerable to the attraction of Communist ideology or to invasion by "Russian hordes," was a "clear and present danger" to the security of the United States. By 1947 it was evident to senior policymakers in the U.S. administration that they could not leave continental Western Europe to its fate. The devastation of the war, the popularity and legitimacy of some national Communist parties, and the "German question" dominated the horizons of American policymakers in the immediate postwar period. These justified the massive intervention, or "foreign entanglements," required to align the region with Washington's field of force. Marshall aid and the technological, industrial, and cultural baggage that went with it catalyzed the formation of an economic and political regime on the American model, which NATO protected under an American nuclear umbrella; the sense of an Atlantic community reinforced it with shared values. Linking "weak" states together through supranational structures that restricted the sovereignty of individual members in order to strengthen the whole contained Germany and produced something akin to a "United States of Europe"—a third force firmly aligned with the

United States and helping it to contain Communism. Asked why it was in the United States' interests to encourage the postwar reconstruction of a strong "independent" Europe—"since great states do not usually rejoice in the emergence of other great powers"—McGeorge Bundy shot back, "The immediate answer is in the current contest with the Soviet Union."[4] If Europe and European science were going to be players in that contest, they had to be able to stand shoulder to shoulder with their American allies. As Eisenhower pointed out in 1947, weakness could not cooperate; it could only beg.[5]

The force field of empire that emanated from Washington was coproduced. Consensus rather than coercion was its modus operandi, not simply because this was the way things were done at home, though that surely provided a model, a mindset, and a moral code; it happened this way also because American policymakers were keenly aware that if they were to succeed in building a durable alliance, they had to proceed with caution. "Shall we be free 'to work as we please'?" in Europe, asked Rockefeller trustee Henry Moe in 1944. He doubted it. Europeans no longer had much confidence in America's good intentions. Before the war one could intervene to "foster directly or indirectly, certain 'American-born ideas,' without provoking any suspicion of ulterior motive." Now Europeans were united in their view that, henceforth "Americans will try and impose their views and *'weltanschauung'* upon them; and they are equally united in their 'determination to resist it at all costs.' "[6] In this inauspicious context, the durability of the empire and its resistance to Communism depended on persuading the majority of the population, their representatives, and their standard-bearers that Washington's prescriptions were in their best interests. If hegemony it was, it was to be consensual. Building that consensus required not only the active collaboration of national elites who shared the economic, political, and ideological ambitions of the United States—and had sufficient legitimacy and power to impose their conception of the path that Europe should take on those who thought otherwise; it also required a subtle refashioning of European identity, a gradual implantation of American norms and practices, selectively adapted to local conditions—and the withering away of any illusions that Soviet Communism could provide the liberating lifeworld that it promised.

The imbalance in economic, industrial, and military power between United States and its European allies provided Washington with a diverse toolkit for achieving its ends in Europe. If hegemony was "in the cards," as Maier puts it, it could be coproduced because the United States had so much that the Europeans desperately wanted—including science and technology, transformed by the war into resources as important to the power of the modern state as coal and steel had been before it. In 1939 the U.S. federal government's budget for research was about $75 million; by 1950 the federal budget for R&D was about $1 billion, 90 percent of it for the reorganized Department of Defense and for the new Atomic Energy Commission, which was responsible for the country's nuclear weapons program. The Korean War pushed the defense R&D budget even higher, to $3.1 billion in fiscal year 1953.[7] Vannevar Bush, by defining science as a new frontier, played on an historically sensitive chord to persuade the federal government that the patronage of science in peacetime was now its responsibility. The nation's security, health, economic well-being, and cultural progress depended on strengthening basic science in academic and research institutions. And what was true for America was true for Europe. Its long-term economic and industrial strength, its political will to stand firm against Communism and to defend democratic institutions and values, and its ability to share in the defense of the West and to contribute to the Atlantic alliance—all these things depended on its having a strong scientific base.

American assistance for the postwar reconstruction of science in Europe was not something that happened parallel to, and independently of, the construction of empire: it was part and parcel of the same project and cannot be fully understood without it. The Rockefeller Foundation's support for the French CNRS in 1946 is something of an exception in this regard, for it was not until a few years later that the major fault lines of the Cold War and its alliances were in place. Yet Warren Weaver was determined to use the space opened up by the destruction of France's scientific infrastructure during the war to remodel the community along American lines and orient it outward toward the English-speaking world. He did not draw on the rationale of the Marshall planners; he anticipated it.

The case study approach I have followed does not make for an integrated narrative. Each example throws into relief a different facet of the process of empire building. Underlying them all is the struggle to contain Communism. This is most explicit in the debate over Ephrussi's grant, which itself reflected the Rockefeller Foundation's feelings of vulnerability over the indiscriminate attacks being made in the Congress on anything that smacked of "socialism" and "internationalism," an attempt to impose a narrow nationalistic and jingoistic definition of American identity on a diverse and democratic people. But the Communist threat is obviously also a major theme in the efforts made by Bush and Compton to funnel Marshall aid into science, in the attempts by Rabi and Stone in the early and mid-1950s to build a Western-oriented European physics community with CERN and the Niels Bohr Institute as its key poles, and in NATO's, James Killian's, and Philip Morse's attempts to "train manpower for freedom." As the perceived nature of the Communist threat changed, so too did the means used to discredit it. In the period of "peaceful coexistence" what mattered was not opposing Communism head-on, but promoting a positive image of the United States by exporting the best it had to offer and building an Atlantic community that projected an image of unity and purpose to the newly decolonized states. NATO was now more than simply a military alliance. It was there to weld a community of nations together around a core set of values and to confidently proclaim the scientific and technological strength of the Free World against a program of techno-scientific seduction by an ebullient Soviet Union.

Four men stand out as instruments of American foreign policy: Warren Weaver, Shepard Stone, Isidor I. Rabi, and James Killian. (Others, like Morse, played important if more localized roles.) Their prominence is to some extent an artifact of the case studies I have chosen, themselves determined by the usual considerations that shape the writing of contemporary history: intellectual interest and relevance to ongoing debates in the field (and indeed to current political concerns) and the availability of sources. I have no reason to think, however, that if I had chosen another set of actors, matters would have turned out very differently. These men are emblematic of the perception of America's postwar role in Europe and encapsulate within their behavior how consensual hegemony is exercised in practice. Liberal, internationalist, deeply concerned about Europe and respectful of its culture and traditions, they are also quintessentially

American: anti-Communist but not populist, nationalistic but not jingoist, firmly convinced that the United States, whatever its flaws, had a key role in defending the Free World and its values and determined to use science to promote those values abroad. When Weaver tried to reconfigure French science, or cut off grants to Lamarckists and suspect fellow-travelers, when Rabi got to his feet at a UNESCO meeting to promote a supranational physics laboratory that included a new and only partly sovereign Germany, when Stone supported CERN and the Niels Bohr Institute, and when Killian used NATO to build European basic, civilian science, all were doing so as formal or informal representatives of their government. They may not have been employees of the State Department or, dare one say it, of the CIA, but they certainly shared the values of the liberal internationalist wings in these organizations in the 1950s and worked closely with them. Power, policy, and ideology were fused in these actors; they were individuals in their own right but also bearers of a widely shared, though not universal, conception at home of America's role and responsibilities in the postwar world order.

Their main scientific partners in continental Europe were Louis Rapkine in Paris, Niels Bohr in Copenhagen, Cornelis Bakker at CERN, and Hendrik Casimir at Philips, perhaps also Werner Heisenberg in Germany (for Marshall aid). One science administrator stands out in almost all the case studies—the Belgian Jean Willems. Willems was the head of all three main Belgian science foundations in the 1950s and was omnipresent in European scientific affairs. He was a founding father of CERN, the person who first approached the Ford Foundation about fellowships for the laboratory, the fiscal agent for Ford and NATO's study of how to increase the effectiveness of Western science and a key member of the Armand committee, and a cohost of the first summer school that Morse held in Europe in 1958 near Brussels. Building consensual hegemony requires the consolidation of a transnational elite whose members have shared ideological, political, and scientific goals and who have both the power and the legitimacy at home to promote a joint U.S.-European agenda. It sounds rather inflated to call this handful of people on both sides of the Atlantic a "transnational elite," but my documents only allow me to highlight leaders. Each was locked into a far broader network of allies, friends, and advisers with whom they discussed what they were doing and whom they could count on for support. It is these partly invisible colleges that, when linked by the shared vision of a European science reconstructed with American help and along lines that satisfied the demands of U.S. foreign policy, form the backbone of the hegemonic project.

Political support and money—grants for equipment, conferences, fellowships, summer schools, universities—were the instruments used to reconstruct and reshape science in postwar Europe, the tools used to fashion an informal empire. How necessary were they? Economic historians of the Marshall Plan are still divided over whether European economic reconstruction would have occurred anyway, without American help.[8] By the time the aid began to flow in 1948, they argue, the recovery was already well under way. Yet it seems clear that, even if the aid were not imperative from a strictly economic point of view, it provided hope and inspiration, it defined a vision of a future free from the recent horror of war and the dreadful struggle of everyday life in Europe in the late 1940s. The same can probably be said for scientific reconstruction. The tradition was too deep, the communities too durable, and the determination to recover Europe's past scientific grandeur too strong for science not to flourish again. That said, one must not forget that the search for new knowledge (except perhaps in the atomic field) had a low priority in war-ravaged Europe, and that even if governments accepted that it was an essential national resource, its putative benefits were remote. From this perspective the speed with which the Rockefeller Foundation got back into Europe played a crucial role in helping French science back on its feet. The release of DM 41 million of Marshall aid funds in 1949 to encourage German scientific recovery surely had political effects far in excess of its cash value. Rabi's intervention in Florence in June 1950 further accelerated the legitimation of German physics and inspired Europeans to leapfrog a generation in technology and to build the most powerful accelerator in the world, knowing that they could count on U.S. help and encouragement. The funds that flowed from the Ford Foundation and NATO in the late 1950s were invaluable long after Marshall aid had ended and European economic recovery was beginning to turn into the European economic miracle. American intervention did more than just pump-prime science on the Continent; it provided an ongoing source of economic, political, and intellectual support, and played an important role in reinvigorating, restructuring, and redirecting it.

The measures proposed to transform European science also served a number of domestic needs: to replenish the stockpile of American basic science, whose practitioners had become dominated by the quest for gadgetry; to provide a declassified space in which U.S. researchers could keep abreast of new developments and exploit them to practical ends if they were promising; to build outward-looking national scientific communities in touch with the research frontier and capable of contributing to the eventual technological and economic development of the Continent; to consolidate a united Europe that included Germany; to complement American scientific strength so that Europe could share the technological and ideological burden in the defense of the Free World; and to strengthen an Atlantic alliance welded together by a shared set of values, including the scientific ethos—the critical spirit that was the very essence of democracy. Foundations, scientists, scientific statesmen and some politicians had a mission in Europe in the first two decades after the war: to build a scientifically and technologically capable partner, and a politically and ideologically reliable ally, in the Cold War struggle for supremacy with the Soviet Union.

Consensual (?) Hegemony

As I have stressed, the promoters of American interests and values that I have considered in this book were certainly sensitive to the fact that they might not always be welcome in Europe. If they proceeded cautiously, building consensus where possible and retreating when rebuffed, it was because they believed that that was the way a democratic nation should build its empire. It was also because they were afraid of alienating some of their allies, perhaps pushing them toward Communism, or at least "neutralism," and abandoning others, stripping them of the local legitimacy that they had to promote the American model. Notwithstanding historic bonds of "culture and blood," there was one key respect in which Washington regarded Europeans as an alien tribe: they had very different perceptions of Communism and the extent of the Communist threat. The British, even those who were not left-wing, found Weaver's concern about giving grants to some left-leaning and Communist intellectuals exaggerated. He did not rethink his values; he simply considered his interlocutors quaint and naive. When Gerard Pomerat interviewed Frédéric Joliot-Curie, he found it hard to imagine that such a brilliant man could have political views that were orthogonal to his own. When Waldemar Nielsen reported to Shepard Stone on the meeting at Loumarin, he could not get over the neutralism of some of the French intellectuals and their refusal or inability to see that the United States was a force for good.

Related to this was the difference in perception on the two sides of the Atlantic of where the civilian/military boundary lay and how, if at all, it intersected with higher education. There was nothing in Europe comparable to the militarization of science in postwar America to meet the Communist threat. Joliot-Curie even said that he would never put his knowledge at the service of a capitalism that, he believed, was hell-bent on aggression and confrontation with the Soviet Union. Many people in universities defended a concept of academic freedom that had become largely obsolete in the United States after the war, refusing to do classified work or to work for the military if they possibly could. In a country like Germany, struggling to erase the shame of its past, there was mutual mistrust between the armed forces and "university types." The very first time that Rabi stood up at the NATO Science Committee in 1958 he stressed that, by constituting itself on the basis of merit, and by promoting international scientific exchange through activities like summer schools, he hoped the committee could dampen hostility to NATO and its objectives. Morse's hopes of enfolding highly educated civilian operations researchers into the upper echelons of NATO's military command structure were roundly rejected by both the scientific adviser of the Supreme Headquarters Allied Powers in Europe and by Solly Zuckerman. The American hegemonic enterprise was always in danger of being undermined by a deep-seated internal contradiction. It was intended to make the world safe for democracy by containing the Communist threat, but many European intellectuals enrolled in the project felt that the threat was exaggerated and certainly not worth jettisoning some of their most time-honored values.

These different world views and value sets, materialized in different forms of social organization and in the ways in which science was socially embedded in Europe and the United States, demand that we remain aware that consensus was sometimes only partial, perhaps sometimes purely pragmatic, and sometimes not forthcoming at all. Europe was so desperately in need of material, political, and intellectual support to rebuild its scientific capability in the first five or ten years after the war that it actually had very little room to negotiate and certainly found American help hard to refuse. Niels Bohr and Cornelis Bakker accepted Stone's request that the initial grants to Copenhagen and to CERN not be given to scientists from Communist countries. Bohr even went to discuss the terms of the grant with Stone and the CIA. When U.S. policy changed, and interaction was encouraged, first with the Soviet Union, then with China, Bohr dutifully followed suit. Did he sense that his behavior was incompatible with his spirited defense of internationalism? Was he being pragmatic? Or did he share the views of the Ford Foundation and the U.S. administration on the nature of the Communist threat and how to deal with it? Did he allow the files of Ford Foundation fellows at his institute to be passed afterwards to the CIA, as some authors have suggested?[9] Once economic recovery got under way and Europeans were better able to stand on their own feet, it was also easier for them to say no; indeed, sometimes it was necessary, if their incipient indigenous efforts were to succeed. Hence the rejection of Killian's project, not simply because it required grafting an alien shoot onto a very different social stock, but above all because it threatened to drain national universities of their best staff just as reforms were being implemented in Britain and France. It also became entangled with French President de Gaulle's increasing resentment against *les anglo-saxons,* his view that Britain was little more than a Trojan horse for American interests in Europe, and his rejection of Macmillan's application to join the common market. He was willing to have an international institute of science and technology, but not if the United States was part of it. What continental Europe did now, it had to do alone.[10]

The French were the bête noire, the one perpetual thorn in the side, of U.S. foreign policy makers, including my four key actors.[11] The Americans never trusted them and had a morbid, obsessive fear of the French *non*— itself symptomatic of the people's demand to be independent and not "colonized" by an external power again. The initially popular Communist

Party in France (PCF) was only one reason for the anger that the United States felt toward its ally. After all, Italy also had a Communist Party, and its influence was perhaps even stronger than that of its Gallic comrades— so strong, in fact, that in a moment of hysteria Kennan actually proposed that, rather than have the Italian Communist Party emerge victorious from the elections in April 1948, Italy should be split in two, letting the industrial north fall under Soviet control.[12] In addition, the power of the PCF declined during the period discussed here and, at least among French intellectuals, it was thoroughly discredited by the late 1950s. Lysenkoism had already undermined its credibility among the left; Soviet tanks rolling into Budapest in November 1956 dealt Communism a further blow. No, what U.S. policymakers could not tolerate was French resistance to their grand design for Europe. Rabi gave the game away when he reported that the French delegation in NATO (first André Danjon, director of the Paris Observatory, then physicist Louis Néel), "while not enthusiastic" about what the committee was doing, "has not been obstructive either."[13] It was not just the fear of Communism, but that the French would contest and thwart their plans that dominated the thinking of U.S. policymakers. If Weaver moved into France as quickly as he could, it was to catch the country at a transitional moment, when it was vulnerable and open to new ideas—*his* ideas about how the community should be reconfigured. If Rabi promoted a physics laboratory without a pile, it was to spike plans to pool European capabilities in a joint nuclear venture under French leadership that included a reactor. If, immediately after the Sputnik shock, Stone moved to support a summer school at Les Houches and one organized by Aigrain (as well as a linear accelerator for Hans Halban in Paris) it was to secure French allegiance to the Atlantic community and their contribution to strengthening Western science as a whole.[14] If Seitz and Killian blamed de Gaulle for the collapse of the scheme to export an MIT to Europe, it was not only because the French president had delivered the coup de grâce to a project that was already reeling from the opposition it received in academic and government circles in Britain, France, and Germany. It was above all because he did not want the United States to be any part of the international institute. Indeed, only Morse, of those we have studied, had (publicly stated) qualms about American attitudes to the French, but of course it was among the French military and civilian planners that he felt most at home.[15] It was they, and not the British, nor the NATO defense establishment, that shared his academically oriented conception of operations research and who were convinced that a firm grounding in advanced mathematics and probability had to be included in the researcher's toolbox.

The attitude of scientists like Killian, Rabi, and Seitz toward the French was of a piece with that of the U.S. administration. If France was unwilling to follow American policies in broad outline, Washington put it under immense pressure to toe the line. The most notorious occasion occurred in August 1954, when the French Parliament opposed a plan to integrate German troops into a European army. In a famous threat, Secretary of State Dulles remarked that if the French did not collaborate, it "would compel an agonizing reappraisal of basic United States policy," the implication being that the United States might withdraw from Europe.[16] The Assemblée Nationale did not oblige. Friction with de Gaulle was ongoing after he became the French president in 1958, determined, as he was, to give an enormous impetus to an independent nuclear force de frappe. In matters of nuclear diplomacy—obviously touching on issues far more weighty than those that I have dealt with here—the United States simply refused to see France as an equal partner and wanted to concentrate control over the nuclear deterrent in Washington. In a wonderful moment in the midst of the Cuban missile crisis in 1962, de Gaulle met Kennedy's emissary Dean Acheson as he stepped from his plane and asked him outright: "In order to get our roles clear, have you come . . . to inform me of some decision taken by your President—or have you come to consult me about the decision which he should take?" Acheson replied, "I have come to inform you." The Kennedy administration perceived the Europeans, as Frank Costigliola puts it, as "shaky, often impractical and emotional 'others' of the NATO family, unequal and inexperienced partners who were too ready to compromise with the Soviets (or, conversely prone to resist the compromises decided upon by the Americans) and who, therefore, had to be manipulated rather than consulted by the hard-headed responsible officials in Washington."[17] As Maurice Couve de Murville, one of de Gaulle's prime ministers put it, "We do not blindly follow U.S. policy. That is the real crisis."[18]

I am not, of course, suggesting that the stakes in scientific diplomacy were in any sense comparable to those in nuclear diplomacy. What was common was not the stakes, but rather the tone, the resentment, and the frustration with France for blocking American aspirations. More to the point, I want to emphasize the obvious, that when one speaks of consensual hegemony, it does not mean that the Europeans always agreed to do what the representatives of the United States suggested or always accepted the proposals that they made. As in any process of consensus building, agreements had to be brokered, and once brokered successfully, they had to be consolidated. The American empire was not all of a piece. It was built up over time and, like lava flowing down a mountainside, sometimes its advance was thwarted, sometimes it had to mould itself to local contingencies, sometimes it had to flow around obstacles, leaving them intact, and sometimes it was diverted away from entire sections of the mountain face. But its gradual onward movement was inexorable, and no one could remain unaware of its presence.

"Americanization"

The Americans whom I have discussed in this book had a clear idea about what had to be done to improve the state of science in Europe after the war: it needed to be "Americanized." Weaver saw a crying need for a polycentric, internally competitive, outward-looking, English-speaking French scientific community, freed from the monopolistic grip of Paris and the baleful influence of extreme left-wing political views, which could only paralyze creative scientific thought. Rabi believed that an institution modeled on Brookhaven provided the most suitable structure for the organization of European high-energy physics, with governments instead of universities pooling resources. Killian was convinced that the outstanding interdisciplinary program that had been put in place at MIT, fusing engineering with science and triangulating graduate education with the military and industry, was the one and only way to exploit emerging fields or disciplines that had immense social potential. Morse took the social relations between civilian experts and military executives that he had worked out with the U.S. Navy during the war as paradigmatic for how to improve decision making in all spheres of social life. He infused them with ever more sophisticated scientific and technological tools to optimize decisions in increasingly complex social systems, and he set out to proselytize his version of operations research to anyone who would listen. In short, all of these men were driven by a missionary fervor to export American models of science and education to Western Europe, believing that the American way of doing things was the one best way, and that it had to be emulated by anyone, anywhere who wished to make the most of his scientific and technical resources for social well-being, industrial strength, and military preparedness. Nothing else made sense, for these were the very ways of doing things that had made America the leading scientific and technological power after the war. These were the keys to modernization, to successful scientific collaboration, and to burden sharing to meet the Communist threat.

From the European perspective, adopting these proposals meant more than just changing a set of local practices. The ideas that the United States was exporting became charged with multiple meanings as they crossed the Atlantic; they were associated, through a complex set of short-circuited connections, with America tout court—the America promoted by the Marshall Plan and celebrated in Hollywood movies, the America of mass production and mass consumption, the America driven by a market concept of democracy that had no place for class distinctions, an America that prized equality and shunned elitism and difference, and the America of sometimes rabid anti-Communism and political conformism. Several European visitors to laboratories in the United States in the 1930s and 1940s read scientific life there through these polyvalent lenses and suffered severe bouts of culture shock. They were at once fascinated by the energy, organization, and efficiency of the research groups they came into contact with and contemptuous of the mass production and mass consumption of scientific output, the leveling and the suffocating uniformity that came from subsuming the individual to the crowd.

Lawrence's Berkeley laboratory, one of the obligatory points of passage for budding cyclotroneers in the 1930s, was one such locale. Visitors noted the organization of the workplace with its division of labor, collaborative teamwork, camaraderie and crewing, all of which respected no intellectual or academic hierarchy. They commented on the obsession to improve the performance of a capricious and complex machine, the ceaseless tinkering and incremental improvements to the hardware in an endless search for greater efficiency, all at the expense of actually doing physics research. The means had become an end in itself: "I do not even know what substances are being bombarded or exactly what is being done," wrote Lawrence in 1937. Maurice Nahmias, sent by Joliot-Curie to learn the art, was appalled by the "mania for gadgets or a post-infantile fascination for scientific meccano games," as if good physics could flow spontaneously from a high-performance instrument: he preferred reading. Walter Elsasser, a Ph.D. from Göttingen, wrote Joliot-Curie that "Americans are mostly coarse types, very good workers but without many ideas in their heads. . . . Their number is impressive, it is true, but one should not worry too much about their technical facilities. It will be a long time before they get from them what they can."[19] In European eyes, Lawrence's laboratory was populated with boorish, parochial, unimaginative machine builders whose creativity had been sacrificed to the quasi-military dictates of a giant charged-particle production factory.

Ten years later, another European visitor to a very different spot had very similar reactions, explicitly tying American mass culture to American research culture. Jean-Paul Gaudillière has described the deep unease, mingled with admiration, experienced by Jacques Monod during a visit to the United States in 1946. The Pasteurian biochemist attended a symposium at Cold Spring Harbor and toured a number of other laboratories. While on Long Island, he had a picnic one Sunday with some colleagues at Jones Beach, where he was appalled by the vulgarity of the place, its cement roads, its ice cream parlors, its loudspeakers, its comically dressed lifeguards, and above all the "fifty thousand cars and five hundred thousand people sweating in the heat." This teeming mass made such an impression on Monod that when he visited the research group at the marine biological station at Woods Hole a few weeks later he could not but see similarities. As he put it to his wife,

Very big laboratories, huge library, three seminars a week, impressive organization, etc. The idea that 350 biologists are working here, that they accumulate observations; that they complete experiments, measurements, weightings; that they operate Warburg apparatus, centrifuges, and microtomes while piling up articles. All this has a somehow depressing effect on me. I am used to think that my work is something rare, highly personal, something I have almost invented. In my understanding, this is what makes it valuable. Here it is no longer possible to cherish such illusions. I feel the same way I felt on Jame's [sic] Beach, when facing 50,000 cars and 500,000 bathers[20]

—that is, stripped of his identity and his uniqueness, his individual creativity devalued, and his personal contribution but a small and insignificant part of the whole.

It was Shepard Stone who realized that it was not enough to offer the elite of European science political support, equipment grants, money for conferences, summer schools, and fellowships. He shared their values (he was the only one of my key actors who was not directly involved in the world of science and engineering), and he understood that, to ensure their allegiance, one had to promote an alternative definition of America to that which so appalled Monod, an America that was rich in culture, music, and art, an America that respected individual liberty and self-expression. In exporting the American model to the European scientific community, it was not just a form of social organization and a set of practices that was at stake; it was also the very definition of America itself, and with it, the notion of what "Americanization" might mean.

Whence the awe and the excitement that was mingled with fear and contempt? Monod and his colleagues were stunned by the quantity and the quality of the data that poured out of American laboratories: "I must confess that the vast number of new results and their importance baffled the three of us," he wrote home.[21] They were overwhelmed by the resources and the facilities for the research: the library in New Haven was "an enormous, sumptuous edifice packed with treasures for the bibliophile; perfect organization, immense catalogs with twelve classificatory systems."[22] Standing on the very cusp of the research frontier, they were drawn by the appeal of America as a bearer of all that was new and modern, a symbol of freedom.[23] This is why Weaver enthusiastically supported the French CNRS in 1946: he was encouraged by Joliot-Curie's "urgency to break with the immobility of the past." This is why Raymond Latarjet was enthusiastic about the proposed international institute of science and technology. The only way to shake the French educational system out of its sclerosis was to begin something new, different, and exciting, using shock tactics to force the existing institutions to reform themselves more rapidly. This was an open-ended experiment that left one space to adapt and to innovate. It was an attempt to escape the dead weight of the past, to establish "the tradition of the future," as Killian put it. Pierre Guillaumat and General Olie could not hide their sympathy for his views. "European prejudices towards the United States," writes Reinhold Wagnleitner, "characterized by the dialectical tension between total fascination with America and the aggressive-defensive cultural feeling of superiority would always have a decisive impact on the development of European-U.S. relations."[24]

To resolve that tension, an "American model" was neither passively accepted nor totally rejected. As Jonathan Zeitlin and Gary Herrigel have suggested, it is more productive to think about "Americanization" as a contested historical project, in which methods and practices developed in the United States were selectively adapted and modified to suit local circumstances, producing innovative hybrids that in turn called for the reform and reshaping of indigenous practices and institutions.[25] Weaver may have failed to reconfigure the French scientific community along American lines; Killian may have failed to establish an MIT in Europe; and Morse may have failed to create an operations research community in NATO that was a carbon copy of the one that he had built in Cambridge. What they, and Stone and Rabi did, though, was "of the greatest importance in bringing U.S. scientific sophistication to Europe," and of raising "the prestige of the U.S. in intellectual and scientific circles."[26] European science was not "Americanized": this term is too cumbersome and too analytically crude to be of much use. But we can say that America's scientific accomplishments remained an omnipresent point of reference and a constant source of pressure for change in Europe, while U.S. recognition of European achievements was an essential source of scientific credibility and scientific capital.[27]

One reason that Rabi gave for promoting a high-energy physics laboratory in June 1950 was that the United States' scientific community wanted to be able to talk to someone in Europe again. The phrase was thick with meaning. Being part of a community with a shared universe of discourse also meant having access to comparable experimental equipment, both comparable in kind and, perhaps more fundamentally, in performance. In the age of techno-science, the research frontier was honed by instrumentation. Those who lacked it, or who lacked instruments that were deemed to perform adequately, quickly lagged behind the leaders, their results dismissed as partial, obsolete, or irrelevant. To implant American "scientific sophistication" in Europe and to train scientists and produce knowledge that was credible and pertinent in the eyes of their American colleagues, European laboratories needed to have access to the best equipment on the market.

They also needed to use that equipment according to current norms of "best practice." The "universality" of laboratory science, as Robert Kohler has stressed, depends on the ability to produce knowledge that bears no trace of its local origins.[28] Effacing place means factoring out not simply the equipment that was used, but also the conditions of its use as well as experimental methods, techniques, and protocols. Universality requires the standardization of equipment and also the standardization of practices. Summer schools, conferences, fellowships in foreign laboratories, apprenticeships—these were essential complements to shared equipment, as Weaver realized when he made his first French grant, as Stone recognized when he gave thousands of dollars for fellowships to CERN and the Niels Bohr Institute, and as the NATO Science Committee appreciated. This circulation of people, skills, and ideas served to transmit the kind of knowledge, tacit and explicit, required to produce credible information within the community. It served to define, from the multitude of questions that one could ask of nature, those that would be rewarded and that would increase scientific capital. Strengthening Western science and building an Atlantic community required training people who circulated within a common geopolitical scientific space, along with instruments, procedures, interpretative techniques, and data that collectively made of them producers and consumers of reliable knowledge.

The informal empire built by Washington in continental Europe in the first two decades after the war largely defined the boundaries of this space. As Rabi put it in an interview in 1973, "You see it used to be very important to us to talk to the British. When the British declined in influence we lost somebody to talk to. Now we talk to the Russians more readily than to anyone else, in a sense, with similar problems. But if Europe ever became something, we'd have civilized people to talk to."[29] This book has described how a number of key Americans set out to "make something" out of Europe and of European science as part of their civilizing mission to "make the world safe for democracy."

Notes

1   Basic Science and the Coproduction of American Hegemony

1 Charles S. Maier, "Alliance and Autonomy: European Identity and U.S. Foreign Policy Objectives in the Truman Years," in *The Truman Presidency*, ed. Michael J. Lacey (Washington, D.C.: Woodrow Wilson International Center for Scholars and Cambridge University Press, 1989), 275.

2 I have addressed the issue in John Krige, "History of Technology after 9/11: Technology, American Power and 'Anti-Americanism,'" *History and Technology* 19:1 (2003), 32–39.

3 Maier, "Alliance and Autonomy," 275.

4 Ronald E. Doel, "Scientists as Policymakers, Advisors and Intelligence Agents: Linking Contemporary Diplomatic History with the History of Contemporary Science," in *The Historiography of Contemporary Science and Technology*, ed. Thomas Söderqvist (Amsterdam: Harwood Academic Publishers, 1997), 220. This is one of the first systematic attempts by an historian of science to address the issue of science and foreign policy, particularly regarding "science in black." It also has a fine general survey of the literature on science and foreign relations. For recent case studies by historians of science, see, for example, Ronald E. Doel, "Evaluating Soviet Lunar Science in Cold War America," *Osiris*, 2nd ser., 7 (1992), 44–70; Jacob Darwin Hamblin, "Visions of International Scientific Cooperation: The Case of Oceanic Science, 1920–1955," *Minerva* 38:4 (2000), 393–423; Hamblin, "The Navy's 'Sophisticated' Pursuit of Science: Undersea Warfare, the Limits of Internationalism, and the Utility of Basic Research, 1945–1956," *Isis* 93 (2002), 1–27.

5 For the point of view of some insiders of the early period, see Wallace R. Brode, "National and International Science," *Department of State Bulletin* 42 (9 May 1960), 735–739; Arnold W. Frutkin, *International Cooperation in Space* (Englewood Cliffs, N.J.: Prentice-Hall, 1965); George B. Kistiakowsky, "Science and Foreign Affairs," *Department of State Bulletin* (22 February 1960), 276–283; Eugene B. Skolnikoff, *Science, Technology and American Foreign Policy* (Cambridge: MIT Press, 1967).

6 Doel, "Scientists as Policymakers," 220.

7 Joseph Manzione, "'Amusing and Amazing and Practical and Military': The Legacy of Scientific Internationalism in American Foreign Policy, 1945–1963," *Diplomatic History* 24:1 (2000), 49.

8 This point may surprise some historians of the natural or physical sciences, but it will not shock historians of the social sciences. For a recent example, see Mark Solovey, "Project Camelot and the 1960s Epistemological Revolution: Rethinking the Politics-Patronage-Social Science Nexus," *Social Studies of Science* 31:2 (2001), 171–206. See also Allan Needell, "'Truth Is Our Weapon': Project TROY, Political Warfare and Government-Academic Relations in the National Security State," *Diplomatic History* 17:3 (1993), 399–420. On the hegemonic aims of the foundations, two classics are Robert F. Arnove, *Philanthropy and Cultural Imperialism: The Foundations at Home and Abroad* (Bloomington: Indiana University Press, 1982), and Donald Fisher, "The Role of Philanthropic Foundations in the Reproduction and Production of Hegemony: Rockefeller Foundations and the Social Sciences," *Sociology* 17:2 (1983), 206–233. One historian of science, Lily E. Kay, has tackled this directly; see "Rethinking Institutions: Philanthropy as an Historiographic Problem of Knowledge and Power," *Minerva* 35:3 (1997), 283–293.

9 There are of course other notions of hegemony, notably that developed by the Italian Marxist Antonio Gramsci, whose analyses of the role of intellectuals in consolidating hegemony will surely bear fruit if applied to my topic. See in this regard, for example, Giles Scott-Smith, *The Politics of Apolitical Culture: The Congress for Cultural Freedom, the CIA and Post-war American Hegemony* (London: Routledge, 2002). I shall say a little about the congress in chapter 6.

10 Tony Judt, "Dreams of Empire," *New York Review of Books* 51:17 (4 November 2004).

11 John Lewis Gaddis, *We Now Know: Rethinking Cold War History* (Oxford: Oxford University Press, 1997), 27.

12 Melvyn P. Leffler, *A Preponderance of Power: National Security, The Truman Administration, and the Cold War* (Stanford: Stanford University Press, 1992).

13 John Lewis Gaddis, *Surprise, Security, and the American Experience* (Cambridge: Harvard University Press, 2004), 30, 106.

14 This formulation is inspired by Jonathan Zeitlin's, "Introduction" to *Americanization and Its Limits: Reworking US Technology and Management in Postwar Europe and Japan*, ed. Jonathan Zeitlin and Gary Herrigel (Oxford: Oxford University Press, 2000), 1–50.

15 Tony Smith, "Making the World Safe for Democracy in the American Century," *Diplomatic History* 23:2 (1999), 179.

16 Maier, "Alliance and Autonomy," 276.

17 Judt, "Dreams of Empire," 38.

18 The remark that the United States "doesn't do empire" is apparently mistakenly attributed to Vice President Dick Cheney. Tony Judt described its messy ori-

gins in his presentation "Fashioning an American Imperial Identity," Atlanta Seminar in Comparative History and Society, Georgia Institute of Technology, Atlanta, 4 February 2005.

19 Charles S. Maier, "An American Empire? The Problems of Frontiers and Peace in Twenty-First-Century World Politics," *Harvard Magazine* 105:2 (2002), 28–31, www.harvardmagazine.com/on-line/1102193.html.

20 Maier, "Alliance and Autonomy," 276. According to Joseph S. Nye Jr., "Soft co-optive power is just as important as hard command power. If a state can make its power seem legitimate in the eyes of others, it will encounter less resistance to its wishes. If its culture and ideology are attractive, others will more willingly follow. If it can establish international norms consistent with its society, it is less likely to have to change. If it can support institutions that make other states wish to channel their activities in ways the dominant state prefers, it may be spared the costly exercise of coercive or hard power." Nye, "Soft Power," *Foreign Policy* 80 (1990), 167.

21 Gaddis also writes, "American hegemony as FDR conceived it was now to be global, but in contrast to anything John Quincy Adams could ever have imagined, *it was to arise by consent*" (emphasis original). *Surprise, Security, and the American Experience*, 54.

22 Charles Maier, "The Politics of Productivity: Foundations of American International Economic Policy after World War II," in Maier, *In Search of Stability: Explorations in Historical Political Economy* (Cambridge: Cambridge University Press, 1987), 148. In Maier's view, "Perhaps the best term for the postwar Western economy would be that of consensual American hegemony. 'Consensual' can be used because European leaders accepted Washington's leadership in view of their needs for economic and security assistance. Hegemony derives from Washington's ability to establish policy guidelines binding on the West."

23 Geir Lundestad has gone further than either Gaddis or Maier in describing the Europeans as actually *inviting* the United States to construct a hegemonic regime during the early Cold War. See his "Empire by Invitation? The United States and Western Europe, 1945–1952," *Journal of Peace Research* 23:3 (1986), 263–277; Lundestad, "'Empire by Invitation' in the American Century," *Diplomatic History* 23:2 (1999), 189–217; Lundestad, *"Empire" by Integration: The United States and European Integration, 1945–1997* (Oxford: Oxford University Press, 1998). Lundestad has somewhat backed off from this oxymoron, which in my view excessively dilutes the notion of hegemony.

24 For example, in the recent collection edited by Shelia Jasanoff, *States of Knowledge: The Co-Production of Science and the Social Order* (London: Routledge, 2004).

25 Ann Stoler, "Degrees of Imperial Sovereignty," paper presented at the Atlanta Seminar in Comparative History and Society, Georgia Institute of Technology, Atlanta, February 4, 2005.

26 Tony Smith, *America's Mission: The United States and the Worldwide Struggle for Democracy* (Princeton University Press, 1995).

27 Alan Brinkley, "The Concept of an American Century," in *The American Century in Europe,* ed. R. Laurence Moore and Maurizio Vaudagna (Ithaca: Cornell University Press, 2003), 8.

28 Gaddis, *We Now Know,* 34.

29 Ibid., 9.

30 Robert Kagan, *Of Paradise and Power: America and Europe in the New World Order* (New York: Alfred A. Knopf, 2003), 78.

31 Quoted in Brinkley, "The Concept of an American Century," 18–19.

32 Quoted in Kagan, *Of Paradise and Power,* 78.

33 The quotations from Carter and Reagan are from Geir Lundestad, *The American "Empire" and Other Studies of US Foreign Policy in a Comparative Perspective* (Oxford: Oxford University Press, 1990). George W. Bush is cited in Judt, "Dreams of Empire," 38.

34 Federico Romero, "Democracy and Power: The Interactive Nature of the American Century," in Moore and Vaudagna, *The American Century in Europe,* 55.

35 Stanley Hoffmann, "Why Don't They Like Us?" *American Prospect* 12:20, 19 November 2001.

36 Maier quoted by Geir Lundestad, *The American "Empire" and Other Studies,* 39.

37 In the copious literature on this see, particularly, Daniel J. Kevles, "The National Science Foundation and the Debate over Postwar Research Policy, 1942–1945," *Isis* 68 (1977), 5–26; Michael Aaron Dennis, "Reconstructing Sociotechnical Order: Vannevar Bush and US Science Policy," in Jasanoff, *States of Knowledge,* 225–253; Daniel J. Kleinman, *Politics on the Endless Frontier: Postwar Research Policy in the United States* (Durham: Duke University Press, 1995). See also Daniel J. Kevles, "Principles and Politics in Federal R&D policy, 1945–1990: An Appreciation of the Bush Report," in *Vannevar Bush: Science— The Endless Frontier* (Washington D.C.: National Science Foundation, 40th Anniversary, 1950–1990, Report NSF 90-8), ix–xxxiii.

38 David A. Hounshell, "The Evolution of Industrial Research in the United States," in *Engines of Innovation: U.S. Industrial Research at the End of an Era,* ed. Richard S. Rosenbloom and William J. Spencer (Boston: Harvard Business School, 1996), 42.

39 For an example of how the U.S. Navy benefited from having access to relatively freely available British oceanographic data, see Hamblin, "The Navy's 'Sophisticated' Pursuit of Science." Fusion research in the 1950s is another instance of classification coming at a high price. When the American, British, and Soviet researchers all decided to declassify their main results in anticipation of the 1958 Atoms for Peace conference in Geneva, they found that they had all made very little progress, and that each had as much difficulty in confining a plasma as the other. See Joan Lisa Bromberg, *Fusion: Science, Politics and the Invention of a New Energy Source* (Cambridge: MIT Press, 1982).

40 Michael Aaron Dennis, "'Our First Line of Defense': Two University Laboratories in the Postwar American State," *Isis* 85 (1994), 451.

41 Peter Galison, "Removing Knowledge," *Critical Inquiry* 31:1 (2004), 230.

42 On this theme, which will be dealt with in more detail later, see for example, David Hollinger, "The Defense of Democracy and Robert K. Merton's Formulation of the Scientific Ethos," *Knowledge and Society: Studies in the Sociology of Culture Past and Present* 4 (1983), 1–15; Hollinger, "Science as a Weapon: *Kulturkämpfe* in the United States during and after World War II," *Isis* 86 (1995), 440–454.

43 Clay to Oliver P. Echols, 4 October 1946, cited in John Gimbel, *Science, Technology, and Reparations: Exploitation and Plunder in Postwar Germany* (Stanford: Stanford University Press, 1990), 140.

44 Minutes of Meeting No. 95 at Bohemian Grove, 19 August 1947, NARA, AEC Records, RG326, E67A, box 46, folder 3. See also "Foreign Distribution of Radioisotopes," Majority Submission to the State Department, undated, but following on the Bohemian Grove meeting. For the general context see Angela N. H. Creager, "Tracing the Politics of Changing Postwar Research Practices: The Export of 'American' Radioisotopes to European Biologists," *Studies in History and Philosophy of Biology and Biomedical Science* 33C (2002), 367–388. See also Creager, "The Industrialization of Radioisotopes by the U.S. Atomic Energy Commission," in *The Science-Industry Nexus: History, Policy, Implications,* ed. Karl Grandin, Nina Wormbs, and Sven Widmalm (Sagamore Beach, Mass.: Science History Publications, 2004), 141–167; John Krige, "The Politics of Phosphorus-32: A Cold War Fable Based on Fact," *Historical Studies in the Physical and Biological Sciences* 36:1 (2005), 71–91.

45 Hounshell, "The Evolution of Industrial Research," 41.

46 As Oppenheimer put it in 1949, defending the policy of distributing radioisotopes to Europe for research and medicinal purposes, the reasons for doing so lay "in fostering science; in making cordial effective relations with the scientists and technical people in western Europe; in assisting the recovery of western Europe; in doing the decent thing." J. Robert Oppenheimer, "Investigation into the United States Atomic Energy Project," Hearings before the U.S. Congress Joint Committee on Atomic Energy, 81st Congress, Part 7, 13 June 1949, 283. Cited in Creager, "Tracing the Politics of Changing Postwar Research Practices," 384.

47 Geir Lundestad, *The American "Empire" and Other Studies,* 11.

2 Science and the Marshall Plan

1 Bush to Forrestal, 23 December 1947, NARA, RG 330, RDB E341, box 603, folder 3.

2 This section relies heavily on Melvyn Leffler, *A Preponderance of Power: National Security, the Truman Administration, and the Cold War* (Stanford: Stanford University Press, 1992), chaps. 4 and 5.

3 Quoted in Geir Lundestad, *"Empire" by Integration: The United States and European Integration, 1945–1997* (Oxford: Oxford University Press, 1998), 167.

4 The data is from David Ellwood, *Rebuilding Europe: Western Europe, America and Postwar Reconstruction* (London: Longman's, 1992), 63.

5 The remark was made by the editor of *Foreign Affairs*, Hamilton Fish Armstrong, in an article published in the journal in July 1947—quoted by Ellwood, *Rebuilding Europe*, 76.

6 Ibid.

7 Ibid., 69. See also Leffler, *A Preponderance of Power*, 145.

8 Leffler, *A Preponderance of Power*, 149.

9 John Lewis Gaddis, "Was the Truman Doctrine a Real Turning Point?" *Foreign Affairs* 52:2 (1974), 386–402. See also Gaddis, "The Insecurities of Victory: The United States and the Perception of the Soviet Threat after World War II," in *The Truman Presidency*, ed. Michael J. Lacey (Cambridge: Woodrow Wilson International Center for Scholars and Cambridge University Press, 1989), 235–271.

10 Gaddis, "Was the Truman Doctrine a Real Turning Point?" 402.

11 Quoted in Ellwood, *Rebuilding Europe*, 69.

12 For Clayton's memo, see Leffler, *A Preponderance of Power*, 159.

13 For this quotation and the citations from Marshall's speech, see Ellwood, *Rebuilding Europe*, 85–86.

14 Ibid., 85.

15 William I. Hitchcock, *France Restored: Cold War Diplomacy and the Quest for Leadership in Europe, 1944–1954* (Chapel Hill: University of North Carolina Press, 1998), 72–73.

16 Quoted in John Lewis Gaddis, *We Now Know: Rethinking Cold War History* (Oxford: Oxford University Press, 1997), 42.

17 For this paragraph see Ellwood, *Rebuilding Europe*, 83–84.

18 The first two were fused into a "bizone" in December 1946; Berlin, in the Soviet zone, was itself divided into four zones.

19 Leffler, *A Preponderance of Power*, 152.

20 Ibid., 201.

21 Marshall, in a memo of 21 July 1947. Ibid., 188.

22 Ibid., 202.

23 The words are those of the French ambassador to Prague, Maurice Dejean, quoted in Hitchcock, *France Restored*, 93.

24 For a discussion of the elections and their impact, see John Krige, "The Politics of Phosphorus-32: A Cold War Fable Based on Fact," *Historical Studies in the Physical and Biological Sciences* 36:1 (2005), 71–91.

25 David W. Ellwood, "The Propaganda of the Marshall Plan in Italy in a Cold War Context," in *The Cultural Cold War in Western Europe, 1945–1960,* ed. Giles Scott-Smith and Hans Krabbendam (London: Frank Cass, 2003), 225–236; Ellwood, "Italian Modernisation and the Propaganda of the Marshall Plan," in *The Art of Persuasion: Political Communication in Italy from 1945 to the 1990s,* ed. Luciano Cheles and Lucio Sponza (Manchester: Manchester University Press, 2001), 23–48. For the unions, see Federico Romero, *The United States and the European Trade Union Movement, 1944–1951* (Chapel Hill: University of North Carolina Press, 1992).

26 Leffler, *A Preponderance of Power,* 196. See also Ellwood, *Rebuilding Europe,* 115–116.

27 Leffler, *A Preponderance of Power,* 196.

28 Ellwood, *Rebuilding Europe,* 116.

29 James E. Miller, "Taking Off the Gloves: The United States and the Italian Elections of 1948," *Diplomatic History* 7:1 (1983), 35–55. According to Ellwood, the Vatican played a far greater role in mobilizing the people against the Communists than did United States efforts. See his "Italian Modernisation," 26.

30 Quoted in Hitchcock, *France Restored,* 96.

31 Ellwood, *Rebuilding Europe,* 119.

32 Alan S. Milward, *The Reconstruction of Western Europe, 1945–1951* (Berkeley: University of California Press, 1984), argued that the Marshall Plan was not essential to European economic recovery, which would have proceeded apace anyway. For reactions see, for example, Ellwood, *Rebuilding Europe,* chap. 8, and Leffler, *A Preponderance of Power,* 159–161.

33 This integrationist agenda has been stressed by Michael Hogan, *The Marshall Plan: America, Britain, and the Reconstruction of Western Europe, 1947–1952* (Cambridge: Cambridge University Press, 1987).

34 Charles Maier, "The Marshall Plan and the Division of Europe," *Journal of Cold War Studies* 7:1 (2005), 173.

35 See especially David W. Ellwood, "The Propaganda of the Marshall Plan in Italy in a Cold War Context," in Scott-Smith and Krabbendam, *The Cultural Cold War in Western Europe,* 225–236; Richard Kuisel, *Seducing the French: The Dilemma of Americanization* (Berkeley: University of California Press, 1993); Reinhold Wagnleitner, *Coca-Colonization and the Cold War: The Cultural Mission of the United States in Austria after the Second World War* (Chapel Hill: University of North Carolina Press, 1994).

36 Reinhold Wagnleitner, "The Empire of the Fun, or Talkin' Soviet Union Blues: The Sound of Freedom and U.S. Cultural Hegemony in Europe," *Diplomatic History* 23:3 (1999), 506.

37 Ellwood, "The Propaganda of the Marshall Plan in Italy," describes this very well. For the total failure of the Communist parties to relate to this consumption ideology, see Marc Lazar, "The Cold War Culture of the French and Italian

Communist Parties," in Scott-Smith and Krabbendam, *The Cultural Cold War in Western Europe*, 213–224.

38 Lawrence S. Kaplan, *The Long Entanglement: NATO's First Fifty Years* (Westport: Praeger, 1999), 3.

39 The Brussels Pact is reproduced as appendix A in Lawrence S. Kaplan, *The United States and NATO: The Formative Years* (Lexington: University Press of Kentucky, 1984). The signatories were Britain, Belgium, France, Luxembourg, and the Netherlands.

40 Kaplan, *The Long Entanglement*, 16.

41 This was emphasized by Foreign Minister Bidault and the Minister of War immediately after the vote on 17 June. See Leffler, *A Preponderance of Power*, 217.

42 Kaplan, *The Long Entanglement*, 13.

43 Ireland was one of the sixteen countries that participated in the Marshall Plan, namely, Austria, Belgium, Britain, Denmark, France, Greece, Iceland, Ireland, Italy, Luxembourg, the Netherlands, Norway, Portugal, Sweden, Switzerland, Turkey, and the US-UK bizone of Germany, under Allied supervision. It is noteworthy that of these only the two neutrals, Sweden and Switzerland, along with Austria, Luxembourg, and Ireland (and of course the bizone) were not also members of NATO by 1952, when Greece and Turkey joined the nine original signatories of the North Atlantic Treaty.

44 Norway had also recently been approached by the Soviet Union suggesting a nonaggression pact between the two countries.

45 For example, including Italy in the scheme, which the United States felt was essential, was only possible by making major concessions to France (which insisted in having her North African territories covered by the treaty). It also required a huge propaganda campaign in Italy itself, where a petition circulated by the left garnered six million signatures and where even the conservative government was a reluctant partner. See Ellwood, "The Propaganda of the Marshall Plan in Italy"; Ellwood, "Italian Modernisation and the Propaganda of the Marshall Plan," in Cheles and Sponza, *The Art of Persuasion*, 23–48.

46 Kaplan, *The Long Entanglement*, 32–33.

47 The North Atlantic Treaty is reproduced in Kaplan, *The United States and NATO*, appendix C.

48 Kaplan, *The Long Entanglement*, 62.

49 Quoted in Ronald W. Pruessen, "Cold War Threats and America's Commitment to the European Defense Community: One Corner of a Triangle," *Journal of European Integration History* (1966), 59.

50 Greece and Turkey acceded to the treaty in February 1952, thus protecting access to the Mediterranean and Europe's southeastern flank.

51 Leffler, *A Preponderance of Power*, 177–178.

52 Bush to Forrestal, 23 December 1947, NARA, RG 330, RDB E341, box 603, folder 3.

53 Forrestal to Bush, 31 December 1947, NARA, RG 330, RDB E341, box 603, folder 3.

54 Forrestal to Marshall, 31 December 1947, NARA, RG 330, RDB E341, box 603, folder 3.

55 Michael A. Dennis, "Reconstructing Sociotechnical Order: Vannevar Bush and US Science Policy," in *States of Knowledge: The Co-Production of Science and the Social Order,* ed. Sheila Jasanoff (London: Routledge, 2004), 245.

56 Harvey M. Sapolsky, *Science and the Navy: The History of the Office of Naval Research* (Princeton: Princeton University Press, 1990). See also Sapolsky, "Academic Science and the Military: The Years since the Second World War," in *The Sciences in the American Context: New Perspectives,* ed. Nathan Rheingold (Washington D.C.: Smithsonian Institution Press, 1979), 379–399.

57 NARA RG 330, RDB E341, box 27, folder 2. The report was attached to a memo of 31 August 1947 by T. A. Solberg, Rear Admiral USN, to the chairman of the Research and Development Board (RDB 110/2.1, ONR:N-102, serial number 15746).

58 NARA RG 330, RDB E341, box 27, folder 3, contains the appendices.

59 "Rehabilitation of Science in Europe," 1–2, for the quotations in this paragraph.

60 Ibid., 25–26.

61 I thank NARA archivist Lawrence H. McDonald, Modern Military Records (NWCTM), at College Park, for his efforts in finding one page of the appendix in the public domain: NARA, RG 263, entry 22, folder 77 (ORE58-48), box 3, 1 November 1948, a "Notice to Recipients of Appendices to ORE 58-48, published 27 October 1949." It deals with corrections to a table of estimates made in mid-1948 of the output of crude steel in the USSR, in the "European orbit" (apparently satellite countries in Eastern Europe), and in Western Europe, excluding the United Kingdom but including West Germany.

62 Price and Solberg to chairman of the RDB, "Rehabilitation of Science in Europe," undated memo, NARA, RG 330, RDB E341, box 27, folder 2, document RDB 110/3.4; Scott to Compton, "Progress Report on Rehabilitation of Science in Europe," memo of 10 March 1949, NARA, RG 330, RDB E341, box 603, folder 3.

63 Price and Solberg to chairman of the RDB, "Rehabilitation of Science in Europe."

64 Ibid.

65 Compton to Lemnitzer, 21 January 1949, NARA, RG 330, RDB E341, box 27, folder 2, document RDB 110/3.

66 Rabi to Compton, 16 February 1949, NARA, RG 330, RDB E341, box 603, folder 3.

67 Forrestal, circular letter, 6 January 1949, RDB 110/3.1, NARA, RG 330, RDB E341, box 27, folder 2.

68 Compton to Lemnitzer, 21 January 1949, document RDB 110/3.

69 Agenda, 19th Meeting of the RDB, undated but shortly after 1 February 1949, NARA, RG 330, RDB E341, box 603, folder 3. I assume that the board concurred in the recommendations proposed in this agenda.

70 Patterson to Compton, "Progress Report on Rehabilitation of Science in Europe," memo of 24 February 1949, NARA, RG 330, RDB E341, box 603, folder 3.

71 Compton to Bernardini, 16 March 1949, NARA, RG 330, RDB E341, box 603, folder 3.

72 Gross to Compton, memo of 10 March 1949, NARA, RG 330, RDB E341, box 603, folder 3. The president's address is available at http://www.truman library.org/hstpaper/point4.htm.

73 Patterson to Compton, "Progress Report on Rehabilitation of Science in Europe," 24 February 1949.

74 Beckler, "Rehabilitation of European Science," memo for Technical Intelligence Files, 9 March 1949, NARA, RG 330, RDB E341, box 603, folder 3.

75 Gross to Lemnitzer, letter of 22 April 1949, NARA, RG 330, RDB E341, box 603, folder 3.

76 For these developments, and Berkner in particular, see Allan A. Needell, *Science, Cold War and the American State: Lloyd V. Berkner and the Balance of Professional Ideals* (Amsterdam: Harwood Academic Publishers, 2000), chap. 5. I will deal with the May 1950 "Berkner" report again in chapter 3.

77 Compton to Rabi, 16 March 1949, NARA, RG 330, RDB E341, box 603, folder 3 (emphasis added).

78 D. J. Beckler, "Rehabilitation of European Science," memo of 9 March 1949, NARA RG 330 RDB E341, box 603, folder 3.

79 Compton to Rabi, 16 March 1949.

80 This data is from a study by Karin Orth, *Forschungspolitik in der Bundesrepublik: Die Präsidien der deutschen Forschungsgemeinschaft zwischen "Freiheit der Wissenschaft" und "Planungseuphorie,"* chap. 2, kindly made available to me by Orth and Prof. Helmuth Trischler. The money was distributed by the Notgemeinschaft der Deutschen Wissenschaft, which had been reconstituted in January 1949. See Cathryn Carson and Michael Gubser, "Science Advising and Science Policy in Post-War West Germany: The Example of the Deutscher Forschungsrat," *Minerva* 40:2 (2002), 165.

81 Compton to Bernardini and to Rabi, 16 March 1949, NARA, RG 330, RDB E341, box 603, folder 3.

82 Gianni Battimelli and Ivana Gambaro, "Da via Panisperma a Frascati: gli acceleratori mai realizzati," *Quaderno di Storia della Fisica* 1 (1997), 332. I shall say more about the state of physics in Italy below.

83 On the Monnet Plan, see William I. Hitchcock, *France Restored: Cold War Diplomacy and the Quest for Leadership in Europe, 1944–1954* (Chapel Hill: University of North Carolina Press, 1998), 29–40.

84 Michel Pinault, *Frédéric Joliot-Curie* (Paris: Editions Odile Jacob, 2000), 319–323.

85 Ibid., 323.

86 See Dominique Pestre, "Louis Néel, le magnétisme et Grenoble. Récit de la création d'un empire physicien dans la province française, 1940–1965," *Cahiers pour l'Histoire du CNRS* 8 (Paris: CNRS, 1990), 163; Robert Gilpin, *France in the Age of the Scientific State* (Princeton: Princeton University Press, 1968).

87 I make no pretence of providing a comprehensive list of material dealing with Britain's postwar scientific and technological capability. In the nuclear field, which was of the greatest concern to the RDB, the classic text is Margaret Gowing, *Independence and Deterrence: Britain and Atomic Energy, 1945–1952* (New York: St. Martin's Press, 1974). For an overview, see Norman J. Vig, "Policies for Science and Technology in Great Britain: Postwar Development and Reassessment," in *Science Policies of Industrial Nations*, ed. T. Dixon Long and Christopher Wright (New York, 1975), 62–109. For the major commitment made in postwar Britain to invest in R&D by the state and by industry, see David E. H. Edgerton, "British R&D after 1945: A Reinterpretation," in *Science and Technology Policy: An International Perspective*, ed. Carlye Honig (London: British Library, 1995), 126–132. See also Edgerton, "Research, Development and Competitiveness," in *The Future of UK Competitiveness and the Role of Industrial Policy*, ed. Kirsty Hughes (London: Policy Studies Institute, 1994), 40–54. For the underlying strength of British physics before the war, see Jeff Hughes, "Redefining the Context: Oxford and the Wider World of British Physics, 1900–1940," in *Physics in Oxford 1839 –1939: Laboratories, Learning, and College Life*, ed. Robert Fox and Graeme Gooday (Oxford: Oxford University Press, 2005), 267–300.

88 "Rehabilitation of Science in Europe," part 2, "Rehabilitation of Science in England," NARA, RG 330, RDB E341, box 27, folder 2.

89 For all quotations in this paragraph see "Summary of Conclusions and Recommendations," 18 January 1949, NARA, RG 330, RDB E341, box 27, folder 2, document RDB 110/3.2 (emphasis added).

90 Edoardo Amaldi's contribution on Italy is his "Notes from Abroad," *Physics Today* 1:1 (1948), 35–37.

91 E. J. Schremp, "Report on Research Conditions in Italy in the Physical Sciences," 17 June 1948, Technical Report OANAR-17-48, NARA, RG 330 RDB E341, box 603, folder 3.

92 Hans Lewy, "Needs of Italian and French Science and Possible American Aid under ERP," 21 March 1948; "Organization of Italian Scientific Life," 22 March 1948, NARA, RG 330, RDB E341, box 603, folder 3.

93 T. Coor, "Technical Memorandum. University Physics in Italy – X: General," 23 May 1949, NARA RG 330 RDB E341, box 603, folder 3.

94 The standard participant account of the Rome work is Edoardo Amaldi, "Personal Notes on Neutron Work in Rome in the 30s and Post-war European Collaboration in High-Energy Physics," in *Proceedings of the International School of Physics "Enrico Fermi": Course LVII; History of Twentieth-Century Physics,* ed. Charles Weiner (New York: Academic Press, 1977), 294–351.

95 On Lawrence's laboratory, see John L. Heilbron and Robert W. Seidel, *Lawrence and His Laboratory: A History of the Lawrence Berkeley Laboratory,* vol. 1 (Berkeley: University of California Press, 1989). On foreign traffic through the laboratory in the 1930s, see John L. Heilbron, "The First European Cyclotrons," *Rivista di Storia della Scienza* 3 (1986), 1–44.

96 Michelangelo De Maria, "Fermi, un fisico da via Panisperma all'America," *Le Scienze* 2:8 (1999), 51.

97 For the account that follows, see De Maria, "Fermi," 50 passim; Battimelli and Gambaro, "Da via Panisperma a Frascati," 319–333.

98 Battimelli and Gambaro, "Da via Panisperma a Frascati," 328–329 for the quotations (my translation).

99 The laboratory is described by G. Bernardini, C. Longo, and E. Pancini in *Physics Today* 1 (June 1948), 26–27.

100 Edoardo Amaldi, "Notes from Abroad."

101 Quoted in Giuliana Gemelli, "Epsilon Effects: Biomedical Research in Italy between Institutional Backwardness and Islands of Innovation (1920s–1950s)," in *Managing Medical Research in Europe: The Role of the Rockefeller Foundation (1920s – 1950s),* ed. Giuliana Gemelli, Jean-François Picard, and William H. Schneider (Bologna: Clueb, 1999), 188.

102 Bernardini to Compton, 15 February 1949, NARA, RG 330, RDB E341, box 603, folder 3. The ONR officer's figure is in Coor's "Technical Memorandum," 2.

103 As described by Gemelli, "Epsilon Effects," 187.

104 Bernardini to Compton, 15 February 1949.

105 De Gasperi apparently said this to John Gunther, who wrote a famous portrait of Europe's interwar ruling classes. There was probably more than a little truth in his claim. By 1950 there was immense concern about how the ERP was being implemented in Italy. Its aims to remodel the country along U.S. lines had fallen foul of deeply embedded local, personal, political, and industrial structures and traditions. Much of the money donated to the government had disappeared without a trace. Leading industrialists argued that the productivity measures the plan encouraged may have been appropriate for the United States, where salaries were high and technology advanced, but were irrelevant in Italy, where salaries were low and production technology correspondingly unsophisticated (except perhaps in automobile manufacturing). Mass consumption patterns were slow to gain ground among a people that valued the personal preparation of food highly and in which the extended family played such an important supportive function.

Indeed from 1950 onward a totally different approach to "selling" the plan in Italy was put in place, an approach that was far more sensitive than before to local needs and practices. See Ellwood, "Italian Modernisation and the Propaganda of the Marshall Plan."

106 Amaldi, "Notes from Abroad," 37.

107 "Report on Research Conditions in Italy in the Physical Sciences," ONR Technical Report OANAR-17-48, prepared by E. J. Schremp, 17 June 1948, 29, NARA, RG 330, RDB E341, box 603, folder 3.

108 "Rehabilitation of Science in Europe," part 5, "Rehabilitation of Science in Italy," 19, NARA, RG 330, RDB E341, box 27, folder 2.

109 E. M. Powers, Memorandum for the Air Force Secretary, Research and Development Board, 22 December 1948, quoting von Karman, NARA, RG 330, RDB E341, box 603, folder 3.

110 Quoted in *Time* magazine, 26 April 1948.

111 For an extensive discussion of this issue, see David Cassidy, "Controlling German Science, I: U.S. and Allied Forces in Germany, 1945–1947," *Historical Studies in the Physical and Biological Sciences* 24:1 (1994), 197–235; Cassidy, "Controlling German Science, II: Bizonal Occupation and the Struggle over West German Science Policy," *Historical Studies in the Physical and Biological Sciences* 26:2 (1996), 197–239. See also Alan Beyerchen, "German Scientists and Research Institutions in Allied Occupation Policy," *History of Education Quarterly* 22:3 (1982), 289–299; Richard H. Beyler and Morris F. Low, "Science Policy in Post-1945 West Germany and Japan," in *Science and Ideology: A Comparative Perspective*, ed. Mark Walker (London: Routledge, 2003), 197–203.

112 The relevant clauses are quoted in full in Beyerchen, "German Scientists and Research Institutions," 290–291. See also Clarence G. Lasby, *Project Paperclip: German Scientists and the Cold War* (New York: Atheneum, 1971), 54.

113 Quoted in Lasby, *Project Paperclip*, 59.

114 On British influence, see Beyerchen, "German Scientists and Research Institutions." On British policy in the occupied zones, see Carl Glatt, "Reparations and the Transfer of Scientific and Industrial Technology from Germany: A Case Study of the Roots of British Industrial Policy and of Aspects of British Occupation Policy in Germany between Post-World War II Reconstruction and the Korean War," 3 vols. (Ph.D. diss., European University Institute, Florence, September 1994).

115 Mitchell G. Ash, "Denazifying Scientists and Science," in *Technology Transfer out of Germany after 1945,* ed. Matthias Judt and Burghard Ciesla (Amsterdam: Harwood Academic Publishers, 1996), 61–79; Johannes Bähr, Paul Erker, and Geoffrey Giles, "The Politics of Ambiguity: Reparations, Business Relations, Denazification and the Allied Transfer of Technology," in Judt and Ciesla, *Technology Transfer out of Germany*, 131–144; Mark Walker, "The Nazification and Denazification of Physics," in Judt and Ciesla, *Technology Transfer out of Germany*, 49–59.

116 On the pillage see, above all, John Gimbel, *Science, Technology, and Reparations: Exploitation and Plunder in Postwar Germany* (Stanford: Stanford University Press, 1990), and the collection of articles in Judt and Ciesla, eds., *Technology Transfer out of Germany*. On French activities, see Corine Defrance, "La mission du CNRS en Allemagne (1945–1950): Entre exploitation et contrôle du potentiel scientifique allemand," *Revue pour l'Histoire du CNRS* 5 (2001), 54–65; Marie-France Ludmann-Obier, "La mission du CNRS en Allemagne (1945–1950)," *Cahiers pour l'Histoire du CNRS 1939–1989* 3 (1989), 73–84. On the complaints, see Cassidy, "Controlling German Science, II: Bizonal Occupation," 220.

117 On the export of experts see Lasby, *Project Paperclip*. See also James McGovern, *Crossbow and Overcast* (New York: William Morrow, 1964).

118 On resistance to the employment of technicians in factories by both managers and shopfloor workers, see Marie-France Ludmann-Obier, "Un aspect de la chasse aux cerveaux: les transferts de techniciens allemands en France: 1945–1949," *Relations Internationales* 46 (Summer 1986), 195–208. On opposition by the CNRS to having German scientists in France, see Defrance, "La mission du CNRS en Allemagne (1945–1950)." She quotes Prof. L. Cagniard, the CNRS representative who was responsible for the French scientific mission in occupied Germany, as saying, "It does not seem to me that a technician or a savant must be a superman just because he is German. We must not fall prey to the same fad as have our Allies, nor be taken in by the chatter of Germans about themselves or their work. We have in France a large number of young researchers—not to speak of older people— who would surely be just as good as the Germans if we gave them the money and the equipment that we would have to give to these foreigners. Our policy towards the Germans should be to get as much information out of them as possible on their new ideas, on the results they have obtained and the methods they used to get those results. We must treat them like milking cows that we abandon once we have finished with them" (pp. 61–62; my translation).

119 Cited in Lasby, *Project Paperclip,* 189. Chapter 5 is a comprehensive account of the protests.

120 Philip Morrison, "*Alsos:* The Story of German Science," *Bulletin of the Atomic Scientists* 3:12 (1947), 354, 365.

121 Max von Laue, "The Wartime Activities of German Scientists," *Bulletin of the Atomic Scientists* 4:4 (1948), 103, and replies, 104–105.

122 Mitchell Ash, Review of *Surviving the Swastika: Scientific Research in Nazi Germany,* by Kristie Macrakis, *Isis* 85:4 (1994), 727–729, at 728. Macrakis is more indulgent in *Surviving the Swastika* (New York: Oxford University Press, 1993), suggesting that most scientists simply wanted to get on with their research unimpeded by political concerns so that the "middle ground of outward accommodation and withdrawal into one's scientific work was most common" (p. 200). For a recent analysis, see *Politics and Science in Wartime: Comparative International Perspectives on the Kaiser Wilhelm Institute,* ed. Carola Sachse and Mark Walker, *Osiris,* 2nd ser., 20 (2005).

123 He was deeply harmed, telling John von Neumann who visited Göttingen in June 1933 that if he was forced to abandon his mathematical institute "an essential part of his life-work will be lost." Von Neumann to Veblen, 19 June 1933, quoted in Charles Wiener, "A New Site for the Seminar: The Refugees and American Physics in the Thirties," in *The Intellectual Migration,* ed. D. Fleming and B. Bailyn (Cambridge: Harvard University Press, 1966), 206.

124 The following year the ONR sent Hans Bethe from Cornell on a fact-finding mission to Germany. H. A. Bethe, "Trip to Germany of H. A. Bethe," 31 August 1948, Memorandum for the Office of the Assistant Naval Attache for Research, American Embassy, London, Samuel Goudsmit Papers, AIP, box 28, file 49. I thank Cathryn Carson for providing this reference for me.

125 N. Artin and R. Courant, "Report on Impressions of Scientific Work in Goettingen and Hamburg, Germany—July 1947," 5 November 1947; Artin and Courant, "Summary Report on Conditions of Science in Germany," August 1947, NARA, RG 330, RDB E341, box 603, folder 3.

126 Artin and Courant, "Report on Impressions," for this paragraph.

127 Ibid.

128 Artin and Courant, "Summary Report."

129 German scientists' inability to admit that they had accommodated themselves to an odious regime, their unrepentant attitude, and their willingness to blame all their ills on the occupying Allied authorities without accepting that they were at least partly to blame for their own misfortune has been remarked on frequently. See, for example, Ruth Lewin Sime, *Lise Meitner: A Life in Physics* (Berkeley: University of California Press, 1996), 335–364; Ute Deichman, "The Expulsion of German-Jewish Chemists and Biochemists and Their Correspondence with Colleagues in Germany after 1945: The Impossibility of Normalization," in *Science in the Third Reich,* ed. Margit Szöllösi-Janze (Oxford: Berg, 2001), 243–283. On the rationalizations for working temporarily on the bomb in particular, see, notably, Mark Walker, "Legends Surrounding the German Atomic Bomb," in *Science, Medicine and Cultural Imperialism,* ed. Teresa Meade and Mark Walker (New York: St. Martin's Press, 1991), 178–204. On the attitudes of medical faculties and practicing physicians, see Paul Weindling, "'Out of the Ghetto': The Rockefeller Foundation and German Medicine after the Second World War," in *Rockefeller Philanthropy and Modern Biomedicine: International Initiatives from World War I to the Cold War,* ed. William H. Schneider (Bloomington: Indiana University Press, 2002), 208–222.

130 Artin and Courant, "Report on Impressions."

131 The Soviet Union made a huge effort to lure not only scientists but also artists and writers in Germany into the Communist camp after the war. A 1945 decree on the restoration of German cultural life in the Soviet zone called for the "mobilization of art in the struggle against Fascism and the re-education of the German people in the spirit of true internationalism." The Soviets encouraged the fine arts and created "cultural centers for German artists, provided them with

means of sustenance and work, and organized exhibits of German art." They became far less tolerant and open after 1947, as Cold War tensions grew, attacking modern art (which was also heavily criticized by many young Germans, some of whom suggested that abstract artists should be "stuck in concentration camps") and art that was not "Socialist-Realist." The United States, for its part, also "developed a fine arts policy to encourage the emergence of a new West German aesthetic paradigm in the Cold War context." See Cora Sol Goldstein, "The Control of Visual Representation: American Art Policy in Occupied Germany, 1945–1949," in Scott-Smith and Krabbendam, *The Cultural Cold War in Western Europe,* 283, 285, 288.

132 Artin and Courant, "Summary Report," for all the quotations in this paragraph.

133 "Rehabilitation of European Science," 17, NARA, RG 330, RDB E341, box 3, folder 27. Germany is discussed on pp. 11–17.

134 Bethe, "Trip to Germany of H. A. Bethe."

135 The phrase is quoted by Weindling, "'Out of the Ghetto,'" 209.

136 Gregg was struck by the lack of guilt in people he spoke to: on his very first day he wrote, "The Germans still seem to me to be strangers to self-reproach and the responsibilities that attend freedom." The purple prose he used in his diary after leaving Germany and Austria gives one an idea of how disgusted he was by his encounters with the medical community: "It isn't that you can vomit what you have already had to eat," wrote Gregg, "—you can't—but at least you don't have to sit smiling and eat more and more." Quoted in Weindling, "'Out of the Ghetto,'" 216.

137 Officers Conference, Monday, 7 April 1947, RFA RG 1.2, series 717, box 4, folder 32.

138 Richard Courant had been supported by the Rockefeller Foundation during the first years of his appointment at New York University. The bonds established before the war were strengthened after the German scientist obtained U.S. citizenship in 1942; this enabled him to work closely with Weaver's Applied Mathematics Panel in the framework of Bush's OSRD. Courant's fortunes flourished after the war, culminating in the establishment of the Courant Institute of Mathematical Sciences in the early 1960s—its main building was baptized "Warren Weaver Hall." See Reinhard Siegmund-Schultze, "Rockefeller Support for Mathematicians Fleeing from the Nazi Purge," in *The "Unacceptables": American Foundations and Refugee Scholars between the Two Wars and After,* ed. Giuliana Gemelli (Brussels: Peter Lang, 2000), 83–194.

139 Coversheet for a memo from Weaver to Havighurst, 24 August 1947, RFA RG 1.2, series 717, box 4, folder 32.

140 The data is tabulated in Weindling, "Out of the Ghetto," 215, from whom the quotation regarding Russia is also taken.

141 Compton to Adams and Compton to G. H. A. Clowes, Lilly Research Laboratories, Woods Hole, Mass., 13 October 1949, NARA, RG 330, RDB E341,

box 603, folder 1. He also sought advice from George Scatchard, a professor of physical chemistry at MIT who had succeeded Adams as research control officer in the American zone in Germany in summer 1946. Scatchard to Compton, 21 October 1949, NARA, RG 330, RDB E341, box 603, folder 1.

142 On the byzantine negotiations around the development of structures to manage German science policy, see Carson and Gubser, "Science Advising and Science Policy"; Cassidy, "Controlling German Science, II: Bizonal Occupation"; Cathryn Carson, "New Models for Science in Politics: Heisenberg in West Germany," *Historical Studies in the Physical and Biological Sciences* 30 (1999), 115–171; Michael Eckert, "Primacy Doomed to Failure: Heisenberg's Role as Scientific adviser for Nuclear Policy in the FRG," *Historical Studies in the Physical and Biological Sciences* 21:1 (1990), 29–58.

143 Quoted in Eckert, "Primacy Doomed to Failure," 35.

144 The Notgemeinschaft was renamed the Deutsche Forschungsgemeinschaft and the DFR was folded into it: a singular victory for the cultural ministers of the German states, and a singular defeat for Heisenberg.

3 The Place of CERN in U.S. Science and Foreign Policy

1 See above all Armin Hermann, John Krige, Ulrike Mersits, and Dominique Pestre, *History of CERN,* vol. 1, *Launching the European Organisation for Nuclear Research* (Amsterdam: North Holland, 1987); Dominique Pestre and John Krige, "Some Thoughts on the History of CERN in the 50s and 60s," in *Big Science: The Growth of Large Scale Research,* ed. Peter Galison and Bruce Hevly (Stanford: Stanford University Press, 1992), 78–99; John Krige, "What Is 'Military' Technology? Two Cases of U.S.-European Scientific and Technological Collaboration in the 1950s," in *The United States and the Integration of Europe: Legacies of the Postwar Era,* ed. Francis Heller and John Gillingham (New York: St. Martin's Press, 1996), 307–338. For an account by two of the "founding fathers," see Edoardo Amaldi, "Personal Notes on Neutron Work in Rome in the 30s and Post-war European Collaboration in High-Energy Physics," in *Proceedings of the International School of Physics "Enrico Fermi": Course LVII; History of Twentieth-Century Physics,* ed. Charles Wiener (New York: Academic Press, 1977), 294–351; Lew Kowarski, "The Making of CERN—An Experiment in Cooperation," *Bulletin of the Atomic Scientists* 11:10 (1955), 381; Kowarski, "An Account of the Origin and Beginnings of CERN," *CERN Yellow Report CERN 61-10* (10 April 1961).

2 The analysis that follows is an elaborated version of John Krige, "Isidor I. Rabi and CERN," *Physics in Perspective* 7:2 (2005), 150–64. My familiarity with the topic triggered a spate of requests for articles on the birth of CERN, as 2004 was its fiftieth birthday. These requests were satisfied with two: "Isidor I. Rabi and the Birth of CERN" *Physics Today* 57:9 (2004), 44–48; John Krige, "CERN, l'atome piégé par le 'plan Marshall,'" *La Recherche* 379 (October 2004), 64–69.

3 On Rabi and the birth of CERN see, in addition to the sources just cited, John S. Rigden, *Rabi: Scientist and Citizen* (Cambridge: Harvard University Press, 2000), chap. 17.

4 *Records of the General Conference of UNESCO, Fifth Session, Florence, 1950; Resolutions,* Section B, Resolution 2.21, 38 and 364–5. This resolution and the debate on it on 7 June 1950, is available in the CERN Archives, Geneva (hereafter CERN).

5 "American Scientist Expresses Views on UNESCO-Sponsored Science Research Centers," UNESCO Press Release 311, Paris, 9 June 1950, LoC, Rabi Papers, box 26, folder 5.

6 For Rabi's time in Europe, see Rigden, *Rabi: Scientist and Citizen,* chap. 3.

7 Lew Kowarski, Interview with Rabi, 6 November 1973, pp. 2–3, CERN; "American Scientist Expresses Views on UNESCO-Sponsored Science Research" for the quotations.

8 "Council on Foreign Relations. Study Group on Aid to Europe. Statement of Mission," 19 January 1949, document M-1, 2, LoC, Rabi Papers, box 25, folder 3.

9 "Draft Digest of Discussion of First Meeting of the Group on Aid to Europe," 17 January 1949, p. 3, LoC, Rabi Papers, box 25, folder 3.

10 Kowarski, Interview with Rabi, 6 November 1973, p. 13. The section in square brackets reads as follows: "You see it used to be very important to us to talk to the British. When the British declined in influence we lost somebody to talk to. Now we talk to the Russians more readily than to anyone else, in a sense, with similar problems. But if Europe ever became something, we'd have civilized people to talk to."

11 Counterpart funds were those funds obtained from the sale of Marshall Plan goods that the Economic Cooperation Administration, which administered the plan, kept for its own uses.

12 Rabi claims that straight after he spoke in Florence, Pierre Auger, French cosmic-ray physicist and director of UNESCO's Department of Natural Sciences, turned to him and said gratefully, "Now you have given us the possibility, the marching orders. We can do something, before that we were tied." Kowarski, Interview with Rabi, 6 November 1973, pp. 5–6.

13 Dominique Pestre discusses these in Herman et al., *History of CERN,* chap. 2.

14 For a general account of the French CEA see Spencer R. Weart, *Scientists in Power* (Cambridge: Harvard University Press, 1979), especially chaps. 15–19. See also Gabrielle Hecht, *The Radiance of France: Nuclear Power and National Identity after World War II* (Cambridge: MIT Press, 1999).

15 Michel Pinault, *Frédéric Joliot-Curie* (Paris: Odile Jacob, 2000), 432–433.

16 Ibid., 436.

17 "Note sur la coopération atomique européenne," November 1949, CERN, Lew Kowarski interview file (my translation).

18 "Note sur la création d'un organisme coopératif de recherche atomique en Europe occidentale," unsigned, dated Paris April/May 1950, attached to Lew (Kowarski) to Pierre (Auger), 20 June 1950, CERN, CHIP 30. The relationship between this note and that written in November 1949, after Truman's announcement of the successful Soviet A-bomb test, has been discussed by Pestre in Herman et al., *History of CERN.*

19 Pinault, *Frédéric Joliot-Curie,* 433.

20 Robert P. Crease, *Making Physics: A Biography of Brookhaven National Laboratory, 1946–1972* (Chicago: University of Chicago Press, 1999), chaps. 5 and 7. In addition to having a graphite pile and an accelerator complex crowned by the 3 GeV Cosmotron, BNL started with a small medical and life sciences division.

21 Rigden, *Rabi,* 220, writes that between 1947 and 1952, when they served together on the GAC, the two men "grew even closer to each other. They carried many concerns in common and worked together, whether in Rabi's living room or in Oppenheimer's institute office."

22 Kowarski sent him a copy of his "Note" of April/May 1950 on 20 June 1950, ten days after Rabi's speech. He mentioned that "Francis" (Perrin) may have already told Auger about the scheme and added that copies of the note were circulating in high administrative (Dautry–CEA) and diplomatic (de Rose–Quai d'Orsay) circles in France.

23 Kowarski, Interview with Rabi, 6 November 1973, pp. 3–4. Rabi, "The Cultural and Scientific Meaning of CERN," LoC, Rabi Papers, box 26, folder 6. This speech, probably given at the closing ceremony of the ISR in the mid-1980s, also makes the connection with Brookhaven quite explicit. Rigden (*Rabi,* 235) quotes Amaldi's wife, Ginestra, as saying that he suggested Brookhaven as a model while waiting for a tram in Rome shortly after the war. This anecdote, almost certainly apocryphal, originated with Robert Jungk, *The Big Machine* (New York: Scribner, 1968), 29.

24 Kowarski, Interview with Rabi, 6 November 1973, p. 3.

25 Interview with I. I. Rabi for the History of CERN by E. Amaldi, Columbia University, New York, 2 March 1983, CERN, file W411, Italian Scenario.

26 Nor did he mind perhaps working closely with German physicists so soon after the war. See Dominique Pestre, "From Revanche to Competition and Cooperation: Physics Research in Germany and France," paper presented at the conference "Society in the Mirror of Science: The Politics of Knowledge in Modern France," Berkeley, 30 September–1 October 1988.

27 Kowarski, "An Account of the Origin and Beginnings of CERN," 3 (emphasis added).

28 Within a few years Europeans set up a collaborative organization in the field of atomic energy. In March 1957 Euratom was established by the Treaties of Rome, which also established the European Economic Community.

29 For a detailed account of Italy joining CERN, with a special emphasis on Amaldi's role, see the chapter by Lanfranco Belloni in Herman et al., *History of CERN*.

30 CERN, Italian Scenario, Interview with Amaldi; Rabi to Amaldi, 30 May 1983, CERN, Rabi drawer.

31 For a detailed account of Germany joining CERN, with a special emphasis on Heisenberg's role, see the chapter by Armin Herman in Herman et al., *History of CERN*.

32 Interviews with Amaldi and Kowarski.

33 Maria Ozietzki, "The Ideology of Early Particle Accelerators: An Association between Knowledge and Power," in *Science, Technology and National Socialism*, ed. Monika Renneberg and Mark Walker (Cambridge: Cambridge University Press, 1994), 255–270, gives considerable detail on the (essentially nonmilitary) motivations behind accelerator construction in the Third Reich.

34 Walther Bothe, Havighurst Interview, 9 October 1947, p. 197, RFA, RG 1.1, series 717, box 3, folder 19.

35 H. A. Bethe, American Embassy, Office of the Naval Attache, London, Memorandum [on trip to Germany], 31 August 1948, AIP archives, Samuel Goudsmit Papers, box 28, file 49. I thank Cathryn Carson for giving me a copy of this document.

36 The history of this device is told by Burghard Weiss, "The 'Minerva' Project: The Accelerator Laboratory at the Kaiser Wilhelm Institute/Max Planck Institute of Chemistry: Continuity in Fundamental Research," in Renneberg and Walker, *Science, Technology and National Socialism*, 271–290.

37 Havighurst Interview, Prof. Werner Heisenberg, Goettingen, 20 September 1947, p. 79, RFA, RG 1.1, series 717, box 3, folder 19.

38 Bethe, memorandum of 31 August 1948, 6.

39 This project is described by Allan Needell, "'Truth Is Our Weapon': Project TROY, Political Warfare, and Government-Academic Relations in the National Security State," *Diplomatic History* 17 (Summer 1993), 399–420. I shall say more about it in chapter 6.

40 These two quotations are from "Project TROY Report to the Secretary of State," vol. 1 (1951), 61, 58. A sanitized version of this top-secret, four-volume report was released by the State Department to Allan Needell in March 1991. I thank him for letting me see his copy.

41 See Robert E. Kohler, *Landscapes and Labscapes: Exploring the Lab-Field Border in Biology* (Chicago: University of Chicago Press, 2002), chap. 1; Lorraine Daston, "Objectivity and the Escape from Perspective," *Social Studies of Science* 22:4 (1992), 597–618; and of course Steven Shapin and Simon Schaffer, *Leviathan and the Air Pump: Hobbes, Boyle and the Experimental Life* (Princeton: Princeton University Press, 1985).

42 On the 184-inch, see John L. Heilbron, and Robert W. Seidel, *Lawrence and His Laboratory: A History of the Lawrence Berkeley Laboratory*, vol. 1 (Berke-

ley: University of California Press, 1989). More generally, see John L. Heilbron, Robert W. Seidel, and Bruce Wheaton, "Lawrence and His Laboratory: Nuclear Science at Berkeley," *LBL News Magazine* 6:3 (1981).

43 Richard P. Feynman, *Physics Today* 1 (1948), 8–10, at 10.

44 Peter Galison, *Image and Logic: A Material Culture of Microphysics* (Chicago: University of Chicago Press, 1997). See also Kohler, *Landscapes and Labscapes.*

45 International Science Policy Survey Group, *Science and Foreign Relations: International Flow of Scientific and Technological Information* (U.S. Department of State Publication 3860, General Foreign Policy Series 30, released May 1950), 3. For a description of the history of the report, see Allan A. Needell, *Science, Cold War and the American State: Lloyd V. Berkner and the Balance of Professional Ideals* (Amsterdam: Harwood Academic Publishers, 2000), especially 141–144.

46 Alan Waterman, "Some Thoughts on International Communication in Science," *Physics Today* 6:1 (1953), 6–9. The NSF had been invited to speak about the effect of immigration laws on U.S. science.

47 Richard Harrison Shryock, "American Indifference to Basic Science in the Nineteenth Century," in his collection *Medicine in America: Historical Essays* (Baltimore: Johns Hopkins University Press, 1966), 71–89, is a classic statement of the historical roots of this position.

48 For a fine analysis of the changing profile of the basic/applied science distinction in the United States, see Ronald Kline, "Construing 'Technology' as 'Applied Science': Public Rhetoric of Scientists and Engineers in the United States, 1880–1945," *Isis* 86 (1995), 194–221.

49 *Science and Foreign Relations,* 3.

50 Waterman, Statement before the Commission on Immigration and Naturalization. The same view was propounded in a study produced for the Council on Foreign Relations a few years later: "It is chastening but useful to realize that virtually every new weapon of the Second World War was the outgrowth of European, not American thinking. To be sure we have displayed our customary ability in organizing, applying and mass-producing the products of this thinking, but a roster of the personnel connected not only with the Manhattan project but with the subsequent development of thermonuclear weapons will show our dependence on foreign scientific thought and theory." Henry L. Roberts, *Russia and America: Dangers and Prospects* (New York: Harper and Bros. for the Council on Foreign Relations, 1956), 69–70.

51 Edward Teller, "Back to the Laboratories," *Bulletin of the Atomic Scientists* 6:3 (1950), 71–72.

52 Quoted in Heilbron, Seidel, and Wheaton, "Lawrence and His Laboratory," 62.

53 Quoted in Silvan S. Schweber, *In the Shadow of the Bomb: Oppenheimer, Bethe and the Moral Responsibility of the Scientist* (Princeton: Princeton University Press, 2000), 17.

54 It was declassified some years ago and Allan Needell kindly made a copy available to me. For a general discussion of science and foreign policy, see Ronald E. Doel, "Scientists as Policymakers, Advisors and Intelligence Agents: Linking Contemporary Diplomatic History with the History of Contemporary Science," in *The Historiography of Contemporary Science and Technology,* ed. Thomas Söderqvist (Amsterdam: Harwood Academic Publishers, 1997), 215–244.

55 Berkner report, Classified Appendix, "Scientific Intelligence," 10. For more on this dimension see John Krige, "Atoms for Peace, Scientific Internationalism and Scientific Intelligence," *Global Power Knowledge,* ed. John Krige and Kai-Henrik Barth, *Osiris,* 2nd ser., 21 (2006), 161–181.

56 Ibid., 5.

57 Ibid., 10.

58 The Berkner report also recommended that science attaches in major foreign capitals engage in scientific intelligence gathering. For an evaluation of that program, see Wilton Lexow, "The Science Attache Program," *Studies in Intelligence* (1 April 1966), 21–27, available on http://www.foia.cia.gov. I would like to thank Jahnavi Phalkey for drawing this unusual document to my attention.

It should be mentioned that several U.S. physicists have told me that they were regularly "debriefed" informally by agents of both the FBI and the CIA after having made trips to Europe. Doel, "Scientists as Policymakers," confirms this; indeed, he takes it for granted.

59 For a brief description of the plan, see, for example, David W. Ellwood, *Rebuilding Europe: Western Europe, America and Postwar Reconstruction* (London: Longman, 1992), 169–171.

60 The Communists touched a sensitive chord. The Western political base of the member states deeply disturbed the Swiss authorities, who feared that their image of political neutrality was seriously endangered and who at one point even threatened to withdraw from the program. See Bruno J. Strasser and Frédéric Joye, "L'atome, l'espace, les molecules: Les pays neutres dans la coopération scientifique européenne (1949–1969)," unpublished manuscript.

61 Translated clippings, LoC, Rabi Papers, box 25, folder 6.

62 For the British position, and Chadwick in particular, see my detailed account in Herman et al., *History of CERN.* A brief version is John Krige, "Scientists as Policymakers: British Physicists' 'Advice' to their Government on membership of CERN (1951/52)," in *Solomon's House Revisited: The Organization and Institutionalization of Science; Nobel Symposium 75,* ed. T. Frängsmyr (Canton, Mass.: Science History Publications, 1990), 270–291.

63 Bohr's fears on this score are described in John Krige, "Felix Bloch and the Creation of a 'Scientific Spirit' at CERN," *Historical Studies in the Physical and Biological Sciences* 32:1 (2002), 57–69.

64 Rigden, *Rabi: Scientist and Citizen,* 237, gives the text.

65 Ibid.

4 The Rockefeller Foundation in Postwar France

1 Fosdick to Weaver, 12 May 1948, RFA, RG 3, series 900, box 25, folder 199.

2 Interview by Charles T. Morrissey and Ronald J. Grele with Shepard Stone, 12 December 1972, pp. 12–13, FFA.

3 Volker R. Berghahn, "Philanthropy and Diplomacy in the 'American Century,'" *Diplomatic History* 23:3 (1999), 399–402.

4 Waldemar Nielsen, cited by Berghahn, "Philanthropy and Diplomacy," 399.

5 Robert E. Kohler, "Warren Weaver and the Rockefeller Foundation Program in Molecular Biology: A Case Study in the Management of Science," in *The Sciences in the American Context: New Perspectives,* ed. Nathan Rheingold (Washington, D.C.: Smithsonian Institution Press, 1979), 252.

6 For a description of how Weaver put his program in place and pruned it to the requirements of the foundation's trustees, see Kohler, "Warren Weaver"; Robert E. Kohler, "The Management of Science: The Experience of Warren Weaver and the Rockefeller Foundation Program in Molecular Biology," *Minerva* 14:3 (1976), 276–306. See also Robert E. Kohler, *Partners in Science: Foundations and Natural Scientists, 1900–1945* (Chicago: University of Chicago Press, 1991).

7 Mina Rees, "Warren Weaver, July 17, 1894, November 24, 1978," *Biographical Memoirs, National Academy of Sciences* 57 (1987), 514–518, describes his extensive wartime achievements.

8 Warren Weaver, *Scene of Change: A Lifetime in American Science* (New York: Scribner, 1970), 72.

9 Kohler, "Warren Weaver," 258.

10 Pnina G. Abir-Am, "The Discourse of Physical Power and Biological Knowledge in the 1930s: A Reappraisal of the Rockefeller Foundation's 'Policy' in Molecular Biology," *Social Studies of Science* 12 (1982), 341–382.

11 Barbara Land, Interview with Warren Weaver, vol. 3, p. 430, RFA, RG 900.

12 Ibid., 438.

13 Kohler, *Partners in Science,* 396–404. For the intricacies of this transformation, see Toby Appel, *Shaping Biology: The National Science Foundation and American Biological Research* (Baltimore: Johns Hopkins University Press, 2000).

14 Jean-François Picard, *La Fondation Rockefeller et la recherche médicale* (Paris: PUF, 1999); Jean-François Picard and William H. Schneider, "From the Art of Medicine to Biomedical Science in France: Modernization or Americanization," in *American Foundations and Large-Scale Research: Construction and Transfer of Knowledge,* ed. Giuliana Gemelli (Bologna: Clueb, 2001), 91–114; William H. Schneider, "War, Philanthropy and the National Institute of Hygiene in France," *Minerva,* 41:1 (2003), 1–23.

15 On these French programs in the natural sciences see Kohler, *Partners in Science,* 175–179, 309–310, 376–378.

16 Detailed references to the work of Dosso and Zallen (for the CNRS grant) and to Burian and Gayon (for the Ephrussi grant) will be made at appropriate points in the text.

17 Rapkine to Miller, 4 August 1945; Rapkine to Weaver, 4 August 1945, RFA, RG 1.1, series 500D, box 11, folder 126. Unless otherwise stated, all citations in this section are from these two long letters.

18 The brief biography of Rapkine that follows is based on the account given by Doris Zallen, "Louis Rapkine and the Restoration of French Science after the War," *French Historical Studies* 17:1 (1991), 6–37. For a more complete account of his life, see Diane Dosso, *Louis Rapkine et la mobilisation scientifique de la France libre*, Thèse pour l'obtention du Diplôme de Docteur de l'Université Paris 7, Spécialité: Épistémologie et histoire des sciences et des institutions scientifiques (Paris, 1998).

19 On this period in his life see Diane Dosso, "The Rescue of French Scientists. Respective Roles of the Rockefeller Foundation and the Biochemist Louis Rapkine (1904–1948)," in *The "Unacceptables": American Foundations and Refugee Scholars between the Two Wars and After,* ed. Giuliana Gemelli (Brussels: Peter Lang, 2000), 195–215.

20 Doris Zallen, in her otherwise fine studies of Rapkine's role in persuading the foundation to support French science, completely overlooks this dimension. She sees the letters as describing "the condition of the basic sciences in France" or as informing the foundation "about the state of French science." She makes no mention of the central issue in Rapkine's letters, the divisions inside the French scientific community, their pride, and their sensitivity to receiving outside help. This is perhaps because her analysis is biased toward the outcome of the negotiations, which was a major equipment and conference grant. Zallen, "Louis Rapkine"; Zallen, "The Rockefeller Foundation and French Research," *Cahiers pour l'Histoire du CNRS* (1989), 35–58. The issue is partly addressed, but still underplayed by Diane Dosso in her Ph.D. thesis, *Louis Rapkine et la mobilisation scientifique de la France libre*, section 5.2.

21 The quotes about Weil are from Weaver to Fosdick, "N.S. Aid to France—Post-War Transition Period (1–3 years)," memo of 19 November 1945, appendix A, RFA, RG 1.2, series 500D, box 3, folder 30.

22 Rapkine to Miller, 4 August 1945.

23 Also listed were Magat, Marszak, Francis Perrain (sic) [Perrin], Rosenblum, Winter, and Wurmser. I say a good deal more about Ephrussi in chapter 5.

24 "GRP Diary, 5 June 1947 (Paris), Fromageot Interview, Desnuelle," RFA, RG 1.2, series 500, box 6, folder 54.

25 "GRP Diary, 3 June 1947 (Paris), Rapkine Interview," RFA, RG 1.2, series 500D, box 6, folder 54.

26 On collaboration, notably among intellectuals, see Philippe Burrin, *France under the Germans: Collaboration and Compromise* (New York: New Press, 1996), chap. 20.

27 Weaver to Fosdick, "N.S. Aid to France—Post-War Transition Period (1–3 years)," appendix A. In addition to Moricard, the list includes two mathematicians, Gaston Julia and Charles Pisot; a statistician, Georges Darmois; three physicists, Eugène Darmois, Dunoyer, and Thibaud; and a biophysicist, J. Chevallier. Rapkine claimed later that Thibaud "badly maltreated Langevin during the occupation and that while he never actually collaborated with the Germans he did accept and take advantage of posts offered by the Vichy government" ("GRP Diary, 3 June 1947 [Paris], Rapkine Interview, Desnuelle"). It should be noted that Pomerat now felt that Rapkine was not a reliable source. The officer noted in his diary that Rapkine had "overemphasized the Desnuelle case," and he provisionally "discounted" the information given here about Thibaud.

28 Rapkine to Miller, 4 August 1945.

29 Rapkine to Weaver, 1 February 1946, RFA, RG 1.2, series 500D, box 31, folder 3. In fact, the foundation was loath to make such an invitation, for the Joliots' Communism was well known in the United States. Irène did eventually visit the United States in 1948 but was detained at Ellis Island. The hostile reactions among the French Communist Party and press, as well of the Joliots themselves, can be well imagined.

30 "Meniere's Disease is a very disturbing illness, presenting patients with hearing loss, pressure in the ear, tinnitus, severe imbalance and vertigo," www.ear surgery.com.

31 Rapkine to Jean Brachet, 30 January 1946, cited in Zallen, "Restoration," 23; Weaver to Barnard, memo of 20 May 1949, RFA, RG 1.2, series 500D, box 3, folder 25.

32 I use the term "intimacy" deliberately, as it is has been a site of some contestation in the literature. Robert Kohler's use of the term in his *Partners in Science* has been severely criticized in a review essay by Pnina Abir-Am, "'New' Trends in the History of Molecular Biology," *Historical Studies in the Physical and Biological Sciences* 26:1 (1995), 183. I shall return to this issue and to Kohler's notion of "partnership" to describe the relationship between men like Weaver and their grantees later in this chapter.

33 Miller to Weaver, memo of 13 August 1945, RFA, RG 1.1, series 500D, box 11, folder 126.

34 Weaver to Miller, handwritten memo of 17 August 1945, RFA, RG 1.1, series 500D, box 11, folder 126.

35 Weaver to Fosdick, "N.S. Aid to France—Post-War Transition Period (1–3 years)."

36 Miller to Rapkine, 23 and 30 August 1945, RFA, RG 1.1, series 500D, box 11, folder 126.

37 Rapkine to Jean Brachet, 30 January 1946, quoted in Zallen, "Rapkine and French Science," 23 (my translation).

38 Rapkine to Sarah Rapkine, 13 November 1945, quoted in Zallen, "Rapkine and French Science," 23.

39 Weaver to Fosdick, "N.S. Aid to France—Post-War Transition Period (1–3 years)."

40 Weaver claimed that this was his idea, although Teissier attributed it to Rapkine at his funeral oration in spring 1949. Weaver took this as showing how successful he had been in getting the French to adopt proposals coming from the foundation as their own. Weaver to Barnard, memo of 20 May 1949, RFA, RG 1.2, series 500D, box 3, folder 25.

41 For the conditions in the laboratory, see John L. Heilbron, "The First European Cyclotrons," *Rivista di Storia della Scienza* 3 (1986), 25–26, 31–33; Michel Pinault, *Frédéric Joliot-Curie* (Paris: Odile Jacob, 2000), 170–172, 245–246.

42 On the political ambiguities of Joliot's "technological collaboration," see Burrin, *France under the Germans,* 310–317.

43 For life in the laboratory in wartime see several contributions to Monique Bordry and Pierre Radvanyi, eds., *Oeuvre et engagement de Frédéric Joliot-Curie* (Les Ulis: EDP Sciences, 2001), 52–60. See also Spencer Weart, *Scientists in Power* (Cambridge: Harvard University Press, 1979), 158, 164–166.

44 On Lépine, see Jean Paul Gaudillière, "Paris-New York Roundtrip: Transatlantic Crossings and the Reconstruction of the Biological Sciences in Post-War France," *Studies in History and Philosophy of Biology and Biomedical Sciences* 33 (2002), 398–400. See also Jean-Paul Gaudillière, *Inventer la biomédecine: La France, l'Amérique et la production des savoirs du vivant (1945–1965)* (Paris: Éditions la Découverte, 2002), chap. 3.

45 Dominique Pestre, "Louis Néel, le magnétisme et Grenoble: Récit de la création d'un empire physicien dans la province française, 1940–1965," *Cahiers pour l'Histoire du CNRS* 8 (1990), 49 (my translation).

46 Rapkine to Miller, 4 August 1945, RFA, RG 1.1, series 500D, box 11, folder 126.

47 Weaver, memo of 6 December 1945, RFA, RG 1.2, series 500D, box 3, folder 30.

48 "Compte rendu sur la mission du CNRS en Allemagne, 25 novembre 1946," Archives Nationales, Paris, F800284/115, published in abbreviated form in Corine Defrance, "La mission du CNRS en Allemagne (1945–1950): Entre exploitation et contrôle du potentiel scientifique allemand," *Revue pour l'Histoire du CNRS* 5 (2001), 54–65 (my translation). This paragraph is based on the contents of this document and comments in the article. See also Marie-France Ludmann-Obier, "La mission du CNRS en Allemagne (1945–1950)," *Cahiers pour l'Histoire du CNRS 1939–1989* 3 (1989), 73–84.

49 Taking FF 119 = $1, the exchange rate on 1 October 1946, the date for which the French figure is given in the original document, reproduced in Defrance, "La Mission du CNRS en Allemagne."

50 Weaver to Fosdick, "N.S. Aid to France—Post-War Transition Period (1–3 years)."

51 "Interviews. L. Rapkine. Thursday, December 6, 1945," RFA, RG 1.2, series 500D, box 3, folder 30.

52 Rapkine to Weaver, 1 February 1946; Frédéric Joliot-Curie and Georges Teissier to Weaver, 28 February 1946, RFA, RG 1.2, series 500D, box 31, folder 3.

53 Joliot-Curie and Teissier to Weaver, 28 February 1946.

54 Docket, Meeting of the Board of Trustees, 3 April 1946, "Centre National de la Recherche Scientifique," RFA, RG 1.2, series 500D, box 3, folder 30.

55 Weaver, "Proposed French Items," memo of 3 January 1946, RFA, RG 1.2, series 500D, box 3, folder 31.

56 Rapkine to Weaver, 1 February 1946, RFA, RG 1.2, series 500D, box 31, folder 3.

57 Weaver to Teissier, 9 April 1946, RFA, RG 1.2, series 500D, box 3, folder 30.

58 Docket, Meeting of the Board of Trustees, 3 April 1946, "Centre National de la Recherche Scientifique."

59 Joliot-Curie and Teissier to Weaver, 28 February 1946.

60 Weaver to Fosdick, "N.S. Aid to France—Post-War Transition Period (1–3 years)," for all quotations in this paragraph.

61 For a history of the CNRS, see Jean-François Picard, *La république des savants: La recherche française et le CNRS* (Paris: Flammarion, 1990).

62 For this account see Jean-Francois Picard, "La création du CNRS," *Revue pour l'Histoire du CNRS* 1 (1999), 50–66.

63 Pinault, *Frédéric Joliot-Curie*, 261. Picard, "La création du CNRS," agrees.

64 These are described and some of the founding documents presented in Giro-lamo Ramunni, "La réorganisation du Centre de la Recherche Scientifique, 7 septembre 1944," *Revue pour l'Histoire du CNRS* 3 (2000), 60–70.

65 William I. Hitchcock, *France Restored: Cold War Diplomacy and the Quest for Leadership in Europe, 1944–1954* (Chapel Hill: University of North Carolina Press, 1998), 12–13.

66 Not even his closest friends realized that from May 1941 Joliot-Curie was the president of the most powerful resistance movement in the northern sector, the Front National (FN) de Lutte pour la Libération et l'Indépendance de la France. He was also president of the directing committee of the FN Universitaire and a founding member of the group l'Université Libre. See Pinault, *Frédéric Joliot-Curie*, 235–242. Joliot's views of the social function of science originated in the humanist, rationalist context of the so-called Arcouest group in the 1930s, many of them left-wing physicists who believed that scientific rationality transcended partisan political views and was a force for liberation in a planned, humane society.

67 The quote is my translation from Pinault, *Frédéric Joliot-Curie*, 252.

68 For example, Louis Néel, in a memorandum drafted probably at the end of 1944 or early in 1945, argued that it was essential to create specialized, well-equipped laboratories in France which, in physics and chemistry at least, needed to be transdisciplinary (meaning bringing together researchers of all the requisite qualifications to tackle a particular problem), and to combine fundamental, applied, and industrial research under one roof. Pestre, "Louis Néel," 36.

69 Weart, *Scientists in Power*, 214. Joliot attended the lavish celebration of the 220th anniversary of the USSR Academy of Sciences in June 1945, which Stalin himself decreed should be opened to foreign visitors. He came away impressed by how the Soviet Academy of Sciences was able to coordinate the work of a number of specialized institutes, having them collectively attack an important agricultural problem. See David Caute, *Communism and the French Intellectuals, 1914–1969* (New York: Macmillan, 1964), 305. See also Pinault, *Frédéric Joliot-Curie*, 374–379.

70 Dominque Pestre has described the many valences that surrounded this divide in interwar France. It referred, he tells us, not merely to different forms of material practice. There as elsewhere it was embedded in institutional and cultural but also political divides. "Pure" research was the domain of men of culture, savants, generally left-wing and working in universities, free of any ties to industry and the state. Applied research or innovation was the business of engineers and polytechnicians, who put established knowledge at the service of industry, the military, and the state authorities. Pestre, "Le renouveau de la recherche à l'École polytechnique et le laboratoire de Louis Leprince-Ringuet, 1936–1965," in *La formation polytechnicienne, 1794–1994,* ed. Bruno Belhoste, Amy Dahan Dalmedico, and Antoine Picon (Paris: Dunod, 1994), 345. See also Caute, *Communism and the French Intellectuals*, 304.

71 Joliot-Curie was deeply hostile to the "abstract" learning promoted by the École Polytechnique to the point of criticizing it heavily for France's defeat in an interview published in a Parisian pro-Vichy newspaper on 15 February 1941. See Burrin, *France under the Germans*, 315.

72 Ramunni, "La réorganisation du Centre de la Recherche Scientifique"; Pinault, *Frédéric Joliot-Curie*, 300–314 (my translation).

73 In 1941 the Ministry of Health set up the INH (Institut National d'Hygiène). In 1942 the colonies established the Office de Recherches Scientifiques Coloniales. Institutes for research into welding and "pétroles" were set up at the instigation of industry in 1944, as was the Centre National d'Études des Télécommunications (CNET), an interministerial body. Ignoring all consultation with the CNRS, the air ministry set up ONERA (Office National d'Études et de Recherches Aéronautiques) in 1945. Agriculture also established an Institut National de la Recherche Agronomique (INRA). The Commissariat à l'Énergie Atomique (CEA), which Joliot-Curie moved to in October 1945, was thus part of a more general movement in which the state and industry were taking an increasing interest in the role of science and technology in a modern society.

74 Picard, "La création du CNRS," quotes Teissier as saying that "French scientists are often accused of having resolutely ignored applied research, but one must not, by going from one extreme to the other, sacrifice pure science for the benefit of applied science, for only it can prepare the future" (p. 64; my translation).

75 Caute, *Communism and the French Intellectuals,* 306. This is Caute's summary of an article published by Teissier in 1949 entitled "Choisir entre deux Civilisations."

76 Picard, "La creation du CNRS," 66 (my translation).

77 "Budget Primitif de 1946," CNRS document, dated by hand 8 October 1945 (presumably by RF staff), RFA, RG 1.2, series 500D, box 3, folder 30.

78 Weaver to Fosdick, "N.S. Aid to France—Post-War Transition Period (1–3 years)."

79 Ibid. Unless otherwise stated, all the quotations in this paragraph are from this memo and from Docket, Meeting of the Board of Trustees, 3 April 1946, "Centre National de la Recherche Scientifique."

80 "From AG's Diary, Paris—October 18, 1945," RFA, RG 1.2, series 500D, box 3, folder 30; Weaver, "Proposed French Items," memo of 3 January 1946.

81 JM to WW, memo of 14 December 1945, RFA, RG 1.2, series 500D, box 3, folder 30.

82 Weaver to Fosdick, "N.S. Aid to France—Post-War Transition Period (1–3 years)."

83 Weaver, "Proposed French Items," 3 January 1946 (emphasis original). The interesting case of André Chevallier, the first director of the Institut National d'Hygiène set up by the Vichy government in November 1941, gives one some idea of how carefully the foundation had to tread. Chevallier, who was appointed director more for his expertise than for his political allegiance, did important work on food shortages in Marseille with Rockefeller support in the early period of the war. As director of the INH, he provided a major stimulus to research into medicine and public health during the occupation. Chevallier quickly established contact with the foundation after liberation, and having survived one charge of collaborating with the Germans, he was promptly accused by the new Communist minister of health of being a collaborator—with the Americans. See William H. Schneider, "War, Philanthropy and the National Institute of Hygiene in France," *Minerva* 41:1 (2003), 1–23; Jean-François Picard, "The Institut National d'Hygiène and Public Health in France, 1940–1946," http://picardp1.ivry.cnrs.fr/INH.html.

84 Weaver to Fosdick, "N.S. Aid to France—Post-War Transition Period (1–3 years)."

85 Docket, Meeting of the Board of Trustees, 3 April 1946, "Centre National de la Recherche Scientifique."

86 For the evolution of the CNRS budget in this period (normalized to 1982 francs), see Jean-François Picard, *La république des savants,* 112–113. Interpret-

ing the data and the financial significance of the grant is further complicated by the repeated devaluations of the franc against the dollar. In October 1946 the official exchange rate was FF 119/$. The franc was devalued on 26 January 1948, when the rate became about FF 215/$. It was devalued again on 19 September 1949, to FF 350/$, closer to the "free franc" which was about FF 333/$ in summer 1949.

87 There was some knee-jerk criticism of U.S. support from the intellectuals of the PCF. See, for example, Gérard Vassails, "Les monopoles financiers américains contre la science française," *La Pensée* 41 (March 1952), 121–136.

88 Teissier to the secretary of the Rockefeller Foundation and to Weaver, undated but received in the office on or about 8 May 1946, RFA, RG 1.2, series 500D, box 3, folder 31.

89 "Rockefeller Gift to be of Great Help to our Laboratories," translation of article from *Le Figaro,* 6 September 1946, RFA, RG 1.2, series 500, box 3, folder 32.

90 Weaver to Teissier, 9 April 1946, RFA, RG 1.2, series 500D, box 3, folder 31.

91 Ibid.

92 "Advancing the Sciences through Discussion," Excerpt from Trustees' Confidential Monthly Report, February 1952, RFA, RG 1.2, series 500D, box 2, folder 20.

93 Zallen gives the team members rather than the research topics; interested readers are referred to her "Rockefeller Foundation" or "Louis Rapkine."

94 Around the time the grant was being carved up by the CNRS, Weaver suggested that Natural Sciences Division funds could be allocated to "preclinical medical subjects without worrying over Divisional programs" in the foundation. Weaver to Rockfound, cable of 16 May 1946, RFA, RG 1.2, series 500D, box 3, folder 31.

95 Rapkine to Weaver, 8 December 1946, RFA, RG 1.2, series 500, box 3, folder 32.

96 "GRP Diary. 19 May 1946 (Paris)," RFA, RG 1.2, series 500D, box 3, folder 31; "Excerpt from letter May 23, 1946, GRP to WW," RFA, RG 1.2, series 500D, box 31, folder 3.

97 "GRP Diary, October 26, 1949, Professor Oleg Yadoff," RFA, RG 1.2, series 500D, box 3, folder 25. Yadoff claimed that just four of the beneficiaries of the CNRS were not Communists or at least "communizers." As an applied physicist he was hoping that the foundation could give 133 books to engineering faculties in Paris and Grenoble, especially the latter, "to encourage good Franco-American relationships in an area where there is already great sympathy for the United States." I return to Yadoff in the next chapter.

98 Friedmann to Heins, 14 September 1946 (two letters), 16 September 1946, 17 December 1946, 18 December 1946 (for the cards), RFA, RG 1.2, series 500, box 3, folder 32.

99 Doris Zallen, "Louis Rapkine and the Restoration of French Science after the War," *French Historical Studies* 17:1 (1991), 6–37, has a list that is organized chronologically within topics and covers the full ten years in which the grant was operative.

100 "Symposium #12. Liste des personnes invitées," RFA, RG 1.2, series 500D, box 2, folder 23.

101 The list of Nobel Prize winners is taken from the Trustees' Confidential Report 2/52, "Advancing the Sciences through Discussion," RFA, RG 1.2, series 500D, box 2, folder 20. For the list of invitees see Bernheim to Gillette, "Symposium #32," 6 March 1950, RFA, RG 1.2, series 500D, box 3, folder 26.

102 Pestre, "Louis Néel," 64.

103 Bernheim to Pomerat, "Symposium: Ferromagnetism and Antiferromagnetism," 5 December 1949, RFA, RG 1.2, series 500D, box 3, folder 25; Bernheim to Pomerat, "Symposia #36 and #38," 24 March 1952, RFA, RG 1.2, series 500D, box 3, folder 27.

104 Resolution RF52058, 4 April 1952, RFA, RG 1.2, series 500D, box 2, folder 20.

105 "Special Emergency Grant in Aid Fund—Scientific Equipment—the Netherlands," 16 November 1945, RFA, RG 1.2, series 650, box 1, folder 6.

106 Docket, Meeting of the Board of Trustees, 3 April 1946, "Centre National de la Recherche Scientifique." This concept of opportunity, which we find so often in Weaver's writings, is thick with meaning. As David Hounshell pointed out recently, destructive and "shocking" events, like wars or acts of terrorism, "provide opportunities for individuals and institutions to shape the future in ways which would not be possible under normal conditions. Often, these opportunists actually shape how the events or developments that have shocked the nation are to be understood both in the halls of government and on main street USA. . . . The opportunities and the opportunists feed off each other, such that profound change—in policies, institutions, and programs—occurs, thereby altering the course of history." Hounshell, "After September 11, 2001: An Essay on Opportunities and Opportunism, Institutions and Institutional Innovation, Technologies and Technological Change," *History and Technology* 19:1 (2003), 40.

107 Weaver to Fosdick, "N.S. Aid to France—Post-War Transition Period (1–3 years)" (emphasis added).

108 "Advancing the Sciences through Discussion," excerpt from Trustees' Confidential Monthly Report, February 1952, RFA, RG 1.2, series 500D, box 2, folder 20, for both quotations in this paragraph. I suspect that Weaver wrote this report.

109 Weaver to Barnard, memo of 20 May 1949, RFA, RG 1.2, series 500D, box 3, folder 25.

110 Docket, Meeting of the Board of Trustees, 3 April 1946, "Centre National de la Recherche Scientifique." See also Weaver to Fosdick, "N.S. Aid to France—Post-War Transition Period (1–3 years)."

111 I am grateful to Dan Kevles for drawing my attention to this point.

112 RFA, RG 1.2, series 500D, box 3, folder 30, Memo Weaver to Fosdick, 19 November 1945. Joliot-Curie and Teissier, in their letter of 28 February 1946 accepting this grant, changed the list slightly to read "USA, England and the Dominions, Scandinavia, etc." It still had an Anglo-Saxon bias, Canada (presumably) had been added—and the "etc."? Did they have Eastern Europe and the USSR in mind?

113 C. J. Gorter, *Physics Today* 1:1 (May 1948), 9, 35.

114 "Advancing the Sciences through Discussion," excerpt from Trustees' Confidential Report, February 1952.

115 Ibid.

116 Abir-Am, " 'New' Trends"; Kohler, *Partners in Science*. See also Abir-Am, "The Discourse of Physical Power."

117 Donald Fisher, "The Role of the Philanthropic Foundations in the Reproduction and Production of Hegemony: Rockefeller Foundations and the Social Sciences," *Sociology* 17:2 (1983), 206–233; Robert F. Arnove, ed., *Philanthropy and Cultural Imperialism: The Foundations at Home and Abroad* (Boston: G. K. Hall, 1980). For a more recent analysis from a Gramscian point of view, see Inderjeet Parmar, "To Relate Knowledge and Action: The Impact of the Rockefeller Foundation on Foreign Policy Thinking during America's Rise to Globalism 1939–1945," *Minerva* 40:3 (2002), 235–263.

118 Fisher, "The Role of the Philanthropic Foundations," 223; Arnove, *Philanthropy and Cultural Imperialism*, 18.

119 Abir-Am, "Discourse of Physical Power"; Lily E. Kay, "Rethinking Institutions: Philanthropy as an Historiographic Problem of Knowledge and Power," *Minerva* 35 (1997), 283–293. See also Kay, *The Molecular Vision of Life: Caltech, the Rockefeller Foundation and the Rise of the New Biology* (New York: Oxford University Press, 1993).

120 As Kohler puts it, the discourse of power and hegemony seems too blunt an instrument to dissect the subtle and varied social experiences of collaboration and patronage; *Partners in Science*, 391.

121 Murad and Zylberman claim that in the area of public health in France too the foundation in the 1920s "intended no less than to remake civilization along American lines" and strove to export an American model across the Atlantic, "to teach French doctors American methods." When the Rockefeller Commission moved into the country to deal with the massive outbreak of tuberculosis during World War I, it knew that "a delicate job of diplomacy was awaiting it," for it had to "direct the French in the right way" while not arousing the feeling that "foreigners are coming to assume functions which belong to the French" (the words are those of the officers). The homology between this case and mine is noteworthy; consensual hegemony captures the essentials of the relationship between the foundation and the French in both cases. See Lion Murard and Patrick Zylberman, "Seeds for French Health Care: Did the Rockefeller Founda-

tion Plant the Seeds between the Two World Wars?" *Studies in History and Philosophy of Biology and Biomedical Science* 31:3 (2000), 463–475.

5 The Rockefeller Foundation Confronts Communism in Europe and Anticommunism at Home

1 Department of State, *France: Policy and Information Statement,* memo of 15 September 1946, NARA 711.51/9-1546, 1. I thank Tim Stoneman for procuring this document for me.

2 Ibid., 1.

3 Ibid., 2.

4 Quoted in Kristin Ross, *Fast Cars, Clean Bodies: Decolonization and the Reordering of French Culture* (Cambridge: MIT Press, 1998), 186. Ross is quoting from B. Mazon, "Fondations américaines et sciences sociales en France: 1920–1960" (thesis, Écoles des Hautes Études en Sciences Sociales, 1985), 103.

5 Fosdick to Weaver, 12 May 1948, RFA, RG 3, series 900, box 25, folder 199.

6 For anti-Americanism in France, see Richard Kuisel, *Seducing the French: The Dilemma of Americanization* (Berkeley: University of California Press, 1993). For Austria, see Reinhold Wagnleitner, *Coca-Colonization and the Cold War: The Cultural Mission of the United States in Austria after the Second World War* (Chapel Hill: University of North Carolina Press, 1994).

7 Rapkine to Weaver, 1 February 1946, RFA, RG 1.2, series 500D, box 31, folder 3.

8 Michel Pinault, *Frédéric Joliot-Curie* (Paris: Odile Jacob, 2000), 276–285; Spencer Weart, *Scientists in Power* (Cambridge: Harvard University Press, 1979), 204.

9 Samuel A. Goudsmit, *Alsos* (Los Angeles: Tomash, 1983), 96–99; Weart, *Scientists in Power,* 204.

10 Pinault, *Frédéric Joliot-Curie,* 374–375.

11 Record of discussion between Weaver and Rapkine, Monday, 12 November 1945, RFA, RG 1.1, series 500D, box 11, folder 126.

12 "GRP Diary, Monday, February 16, 1948, Miss G. Friedmann (telephone)," RFA, RG 1.2, series 500D, box 2, folder 23. Friedmann was the secretary of the CNRS's office in New York. See also Kamen to Friedmann, 22 February 1948, Pomerat to Friedmann, 26 February 1948.

13 Friedmann to Pomerat, 4 March 1948, RFA, RG 1.2, series 500D, box 2, folder 23.

14 "GRP Diary, Monday, February 16, 1948, Miss G. Friedmann (telephone)."

15 Kamen to Friedmann, 17 February 1948, RFA, RG 1.2, series 500D, box 2, folder 23.

16 This is described in some detail in Jessica Wang, *American Science in an Age of Anxiety: Scientists, Anticommunism and the Cold War* (Chapel Hill: University of North Carolina Press, 1999).

17 Lawrence Badash, "Science and McCarthyism," *Minerva* 38:1 (2000), 63, notes that Kamen fought and a won a libel suit against the *Chicago Times,* and eventually got a passport in 1955. Jessica Wang (private communication) says that she has seen no credible evidence to suggest that Kamen was a security risk.

18 Wang, *American Science in an Age of Anxiety,* 85–86.

19 "Officers Conference—March 18, 1948 at 2.30pm," RFA, RG 3, series 900, box 25, folder 199.

20 Trustee's meeting, "Policy re Program in Europe, 4/6–7/48," RFA, RG 3, series 900, box 25, folder 199.

21 "Officers Conference—March 18, 1948 at 2.30pm."

22 Fosdick to Weaver, 6 May 1948, RFA, RG 3, series 900, box 25, folder 199. This view was confirmed by a Romanian refugee, Dr. Nicholai Georgescu, who told foundation officers that in Romania "immediately you were known to have had any connection with America, particularly an old connection and one in which you had been in America for some length of time, you were immediately black-listed." He suggested that the officers no longer write to people there better to protect them from the Romanian authorities. "Dr Nicholai Georgescu," memo of 6 July 1948, RFA, RG 3, series 900, box 25, folder 199.

23 Trustee's meeting, "Policy re Program in Europe, 4/6–7/48."

24 Weaver to Fosdick, 13 April 1948, RFA, RG 3, series 900, box 25, folder 199.

25 Trustee's meeting, "Policy re Program in Europe, 4/6–7/48."

26 Strode to Grant, 14 April 1948, RFA, RG 3, series 900, box 25, folder 199.

27 Weaver to Fosdick, 13 April 1948. A few years later Weaver adopted a more moderate position, saying that biology and agriculture, unlike the physical sciences, were "largely innocuous from a security point of view" in terms of their content. This shows how much anxiety the events in Europe in spring 1948 produced. Weaver to Barnard and Kimball, memo of 6 February 1952, RFA, RG 3, series 900, box 25, folder 200.

28 "Officers Conference—March 18, 1948 at 2.30pm."

29 Weaver to Fosdick, memo of 2 May 1948, RFA, RG 3, series 900, box 25, folder 199.

30 Weaver to Fruton, 2 May 1948, RFA, RG 3, series 900, box 25, folder 199.

31 Weaver to Fosdick, memo of 2 May 1948.

32 Weaver to Fosdick, 13 April 1948. The names were added by hand in the margin.

33 "Excerpts from WW's diary notes on visits in London," RFA, RG 3, series 900, box 25, folder 199. All material in this paragraph and the next is from this source.

34 16 April 1948, 109, RFA, RG 905, series Pom 1948.

35 For a history of the institute, see Michel Morange, "L'institut de biologie physico-chimique, de sa fondation à l'entrée dans l'ère moléculaire," *Revue pour l'Histoire du CNRS* 7 (2002), 32–41.

36 Robert E. Kohler, *Lords of the Fly: Drosophila Genetics and the Experimental Life* (Chicago: University of Chicago Press, 1994), 208–232, describes the collaboration between Beadle and Ephrussi. See also Richard M. Burian and Jean Gayon, "The French School of Genetics: From Physiological and Population Genetics to Regulatory Molecular Genetics," *Annual Review of Genetics* 33 (1999), 321–322.

37 "Diary WW, Paris—Thursday, May 11, 1950," RFA, RG 1.2, series 500, box 5, folder 42. See also an earlier, similarly enthusiastic description by Pomerat, "GRP Diary, Paris,—May 22, 1946," RFA, RG1.2, series 500D, box 31, folder 3.

38 On the highly specific path that genetics took in France, see Richard M. Burian and Jean Gayon, "Genetics after World War II: The Laboratories at Gif," *Cahiers pour l'Histoire du CNRS* 7 (1990), 25–48; Richard M. Burian, Jean Gayon, and Doris Zallen, "The Singular Fate of Genetics in the History of French Biology, 1900–1940," *Journal of the History of Biology* 21:3 (1998), 357–402; Burian and Gayon, "The French School of Genetics"; Jean-Paul Gaudillière, "Molecular Biology in the French Tradition? Redefining Local Traditions and Disciplinary Patterns," *Journal of the History of Biology* 26:3 (1993), 473–498; Jean-François Picard, "Un demi-siècle de génétique de la levure au CNRS: De la biologie moléculaire à la génomique," *Revue pour l'Histoire du CNRS* 7 (2002), 42–49. See also Jean-Paul Gaudillière, *Inventer la biomédecine: La France, l'Amérique et la production des savoirs du vivant (1945–1965)* (Paris: Éditions la Découverte, 2002); Michel Morange, *A History of Molecular Biology* (Cambridge: Harvard University Press, 2000). "Entretien avec Philippe L'Héritier," http://picardp1.ivry.cnrs.fr/~jfpicard/LHeritier.html is useful for an insider view.

39 "GRP Diary, 11 May 1950 (Paris)," RFA, RG 1.2, series 500, box 5, folder 42; "Diary WW, Paris—Thursday, May 11, 1950."

40 Burian and Gayon list "at least four" causes for the resistance to genetics in France from about 1910 to 1930: a variety of intellectual traditions that together led to a general skepticism regarding how genes act to transmit traits across generations, the connection of genetics with eugenics, the failure of academic biology to interact meaningfully with agricultural research, and the loss of a generation of young men in World War I. Burian and Gayon, "The French School of Genetics," 317–318. See also Burian and Gayon, "Genetics after World War II," 29; Burian, Gayon, and Zallen, "The Singular Fate of Genetics," 401–402.

41 Burian, Gayon and Zallen, "The Singular Fate of Genetics," 378–379.

42 Ibid.

43 Weaver to Ephrussi, 15 February 1950, RFA, RG1.2, series 500, box 5, folder 42.

44 Burian and Gayon, "Genetics after World War II," have described some elements of the story I tell here and have reproduced in full the important letter from Weaver to Ephrussi of 15 February 1950. However, these authors did not contextualize the attitudes of the foundations' officers, relating them to American anti-Communism at the time. Reacting to their paper, Dominique Pestre wondered what the attitudes of the foundation in the affair "showed of the perception—or the misperception—that the Americans then had of Europe, and of France in particular" (Dominique Pestre, Commentary on *Colloque sur l'Histoire du CNRS des 23 et 24 octobre 1989*, www.ivry.cnrs.fr/politiques_de_la_science/sciences_ex_shs_4). What follows will answer his question.

45 Barnard to Watson, 18 August 1949, Barnard to Division Directors, memo of 18 August 1949; "Excerpt from Memorandum to principal officers by CIB—March 9, 1950," RFA, RG3, series 900, box 25, folder 199.

46 Indeed, a few weeks before Ephrussi applied for his grant in summer 1949, Pomerat had suggested that as long as Teissier remained the director of the CNRS, a renewal of the 1946 award to the organization was "perhaps a potentially questionable undertaking likely to lead to difficulties." "Excerpt from diary item of May 13, 1949, G. Teissier," RFA, RG1.2, series 500D, box 3, folder 25.

47 The earlier classics on the Lysenko affair were Zhores A. Medvedev, *The Rise and Fall of T. D. Lysenko* (New York: Columbia University Press, 1969), and David Joravsky, *The Lysenko Affair* (Cambridge: Harvard University Press, 1970). Loren R. Graham has a useful summary in *Science in Russia and the Soviet Union: A Short History* (Cambridge: Cambridge University Press, 1994). A little-known and stimulating "left-wing" interpretation of Lysenko is Richard Lewontin and Richard Levins, "The Problem of Lysenkoism," in *The Radicalisation of Science: Ideology of/in the Natural Sciences*, ed. Hilary Rose and Steven Rose (London: Macmillan, 1976), 32–64. I have benefited most from the analysis based on newly opened Soviet archives by Nikolai Krementsov, *Stalinist Science* (Princeton: Princeton University Press, 1997).

48 Kohler, *Lords of the Fly*, has a picture of Muller with the first generation of Russian drosophilists in Moscow in 1922 on 119.

49 The most widely known was his technique of "vernalization." In this procedure, farmers were encouraged not to plant winter varieties of crops in the autumn. Instead, they were advised to wet and chill the seeds in an environment protected from the harsh frosts that could destroy them if they were in the ground (and which had in fact destroyed thirty-two million acres of winter wheat in 1927–28 and 1928–29).

50 Krementsov, *Stalinist Science,* chap. 6.

51 Graham, *Science in Russia,* 133.

52 Krementsov, *Stalinist Science,* 183.

53 Julian Huxley, "Soviet Genetics: The Real Issue," *Nature,* 18 June 1949, 935. This is the first of two articles.

54 What follows relies heavily on Dominque Lecourt, *Proletarian Science? The Case of Lysenko* (London: New Left Books, 1977). This information is at 18, 26.

55 Anon., "The Birth of Zoë," *The Economist,* 25 December 1948, 1057.

56 Pinault, *Frédéric Joliot-Curie,* 424 (my translation; emphasis added).

57 Ibid., 425 (my translation).

58 The Salle Wagram is situated close to the Arc de Triomphe in the heart of Paris and has had a long and glorious history that stretches back to the early nineteenth century. Today it is marketed thus: "The Salle Wagram has been a remarkable space for cultural expression since the tastes, the desires, the flights of enthusiasm and of hope, the anxieties, the sufferings and the dreams of Parisians are there incarnated, since the walls of this hall speak softly in the wind of their history, and resonate with its steps, with its dances, with its smiles, with its creativity and with its *art de vivre*," www.sallewagram.com (my translation). That the PCF should call people together in a space known for having "prestigious conferences and political meetings," there to lay down the party line on Lysenko, is at once indicative of its social power at the time and of the immense historical importance its leadership attached to the occasion.

59 Ibid., 439.

60 Louis Althusser, "Unfinished History," introduction to Lecourt, *Proletarian Science? The Case of Lysenko,* 13, 8.

61 "Interview: GRP, Boris Ephrussi, February 28, March 1 and 2, 1950; March 3, 1950"; "Diary: WW, February 28, 1950. Professor Boris Ephrussi, CNRS, Paris, France," RFA, RG1.2, series 500, box 5, folder 42.

62 "Diary: WW, February 28, 1950. Professor Boris Ephrussi, CNRS, Paris, France."

63 Ibid.

64 "Interview: GRP, Boris Ephrussi, February 28, March 1 and 2, 1950; March 3, 1950."

65 Ephrussi to Weaver, 16 March 1950, RFA, RG1.2, box 5, folder 42.

66 For the circumstances, see Burian and Gayon, "Genetics after World War II," 34. See also Weart, *Scientists in Power,* 258.

67 "Interview: GRP, Boris Ephrussi, February 28, March 1 and 2, 1950; March 3, 1950."

68 For the circumstances, see Weart, *Scientists in Power,* 259–261.

69 "Interview: GRP, Boris Ephrussi, February 28, March 1 and 2, 1950; March 3, 1950."

70 Ibid.

71 "Diary: WW, February 28, 1950. Professor Boris Ephrussi, CNRS, Paris, France."

72 "Interview: GRP, Boris Ephrussi, February 28, March 1 and 2, 1950; March 3, 1950."

73 "Diary: WW, February 28, 1950. Professor Boris Ephrussi, CNRS, Paris, France" for this paragraph.

74 Pinault, *Frédéric Joliot-Curie,* 461.

75 Philippe L'Héritier, "Entretien avec Philippe L'Héritier," http://picardp1.ivry .cnrs.fr/~jfpicard/LHeritier.html.

76 Burian and Gayon, "The French School of Genetics," 319–320; L'Héritier, "Entretien avec Philippe L'Héritier."

77 Burian and Gayon, "The French School of Genetics," 330–331, describe this aspect of Ephrussi's research. For his admission of guilt, see "Interview: GRP, Boris Ephrussi, February 28, March 1 and 2, 1950; March 3, 1950."

78 L'Héritier, "Entretien avec Philippe L'Héritier." See also the various papers by Burian and Gayon for the place of Lamarckism in France.

79 "Interview: GRP, Boris Ephrussi, February 28, March 1 and 2, 1950; March 3, 1950."

80 Weaver to Terroine, 6 March 1950, RFA, RG1.2, series 500, box 5, folder 42.

81 "Interview: GRP, Boris Ephrussi, February 28, March 1 and 2, 1950; March 3, 1950"; "Diary: WW, February 28, 1950. Professor Boris Ephrussi, CNRS, Paris, France."

82 Appropriation "RF50034, Centre National de la Recherche Scientifique," 5 April 1950, RFA, RG 1.2, series 500, box 5, folder 42.

83 For this affair and the eventual establishment of a research group at Gif, see also Burian and Gayon, "Genetics after World War II," 37–40.

84 "GRP Diary, 9 September, 1950, Boris Ephrussi," RFA, RG1.2, series 500, box 5, folder 42.

85 "GRP Diary, June 6, 1951, Dupouy," RFA, RG1.2, series 500, box 5, folder 42.

86 Weaver diary, "Monday, June 18, 1951, Professor Boris Ephrussi," RFA, RG1.2, series 500, box 5, folder 42. On L'Héritier hunting, see his comments in "Entretien avec Philippe L'Héritier."

87 "WW's Diary, November 26, 1951, Dr. B. Ephrussi," RFA, RG1.2, series 500, box 5, folder 42.

88 "GRP Diary, 3 June 1952 (Paris)," RFA, RG1.2, series 500, box 5, folder 43.

89 Pomerat to Weaver, 27 March 1953, RFA, RG1.2, series 500, box 5, folder 43.

90 In fact, the domestic situation in the United States that I will describe made it politically impossible for Pomerat to suggest that the foundation make an award to the CNRS to support a known Communist like Teissier.

91 "GRP Diary 9 June 1953 (Paris)," RFA, RG1.2, series 500, box 5, folder 43. The grant was officially rescinded by the trustees' meeting on 25 September 1953. RFA, RG1.2, series 500, box 5, folder 42.

92 Wang, *American Science in an Age of Anxiety,* 274.

93 Ibid., 274, 275.

94 Ibid., 278. See also the special number of the *Bulletin of Atomic Scientists,* October 1952, and John Krige, "La science et la sécurité civile de l'Occident," in *Les sciences pour la guerre, 1940–1960,* ed. Amy Dahan and Dominique Pestre (Paris: Éditions de l'École des Hautes Études en Sciences Sociales, 2004), 373–401.

95 "Investigation of Certain Educational and Philanthropic Foundations. Speech of Hon. E. E. Cox of Georgia in the House of Representatives, Wednesday, August 1, 1951," RFA, RG 3.2, series 900, box 14, folder 85.

96 Rhind to Barnard, "Cox Resolution re Investigation of Foundations," memo of 18 October 1951, RFA, RG3.2, series 900, box 14, folder 85.

97 This was done by House Resolution 561 of the Second Session of the 82nd Congress.

98 E. E. Cox to the Rockefeller Foundation, 2 October 1952, RFA, RG3.2, series 900, box 14, folder 85. See also "Excerpt from Minutes of the Board of Trustees, 4/2/52," RFA, RG3, series 900, box 25, folder 199.

99 "Answers of the Rockefeller Foundation to Questionnaire Submitted by the Select Committee of the House of Representatives of the Congress of the United States," RFA, RG3.2, series 900, box 14, folder 89.

100 This was to be distinguished from the names of people or organizations merely *mentioned* in hearings or reports of the Committees or the names of persons or organizations *criticized* by witnesses giving evidence before the committees. The foundation restricted its list of suspects to people or organizations specifically identified as such by the committees themselves.

101 RFA, RG3.2, series 900, box 15, folder 90, contains the list, which is sometimes removed from copies of the entire report.

102 These conclusions distressed those who were still convinced that behind the activities of the Carnegie, Ford, and Rockefeller foundations "lies the story of how Communism and Socialism are financed in the United States." A new resolution in the House created yet another committee chaired by Carroll Reece from Tennessee to investigate tax-exempt foundations. The Reece Committee filed its report with the Clerk of the House on 16 December 1954. It made no attempt at objectivity. With one exception, all the witnesses it interrogated were anti-foundation. It concluded that the foundations were promoting socialistic doctrine by selectively supporting certain kinds of research in the social sciences and by fostering socialist teaching in schools to the detriment of "true Americanism." Those thus accused were not permitted a public hearing but only the chance to reply to the charges brought against them by sworn statements or briefs. The two

Democratic representatives on the five-person committee appended a spirited minority report stating, "The foundations have been indicted and convicted under procedures that can only be characterized as barbaric," adding that the Reece report "should never have been published." (This last remark is from "Statement by Angier L. Goodwin, member of the Cox Committee," attached to J. M. Ripley to W. C. Cobb, 3 January 1955, RFA, RG3.2, series 900, box 14, folder 86. For more general information on the Reece report, see RFA, RG3.2, series 900, box 14, folder 86, "Reece Committee Report," *New York Times,* 21 December 1954; "Representative Comment on the Reece Report, Office of Publications, The Rockefeller Foundation, January 17, 1955.")

While the Reece committee deliberated, Senator McCarran made an assault on foundations from another direction. He proposed that the Internal Revenue Code should be amended to punish severely tax-exempt foundations, as well as charitable, educational, and religious institutions that supported subversive organizations or individuals. This amendment was killed in committee, but only on condition that in the next session of Congress a bill would be passed placing responsibility on the foundations, if they were to retain their tax-exempt status, to certify to the absence of "left-wing bias" in the recipients of their funds (Dean Rusk to Senator Williams, 7 July 1954; Excerpt from memorandum of S. D. Cornell, National Academy of Sciences, to Detlev Bronk, 20 December 1954, RFA, RG3, series 900, box 25, folder 200).

103 Lindsley F. Kimball, "The Rockefeller Foundation Vis a Vis National Security," memo dated 19 November 1951, RFA, RG3, series 900, box 25, folder 20.

104 From "Statement of the Rockefeller Foundation and the General Education Board before the Special Committee to Investigate Tax Exempt Foundations, August 3, 1954," RFA, RG3, series 900, box 25, folder 200.

105 "Answers of the Rockefeller Foundation to Questionnaire Submitted by the Select Committee of the House of Representatives of the Congress of the United States," 46 (emphasis added).

106 Willits to Barnard, memo of 19 September 1951, RFA, RG3, series 900, box 25, folder 200.

107 From "Statement of The Rockefeller Foundation and The General Education Board before The Special Committee to Investigate Tax Exempt Foundations, August 3, 1954."

108 RFA, RG 905, series Pom, 1949, 273–274.

109 "GRP Diary, October 26, 1949, Professor Oleg Yadoff," RFA, RG1.2, series 500D, box 3, folder 25.

110 RFA, RG 905, series Pom 1951, 129–130. Monod also explained his position in a special number of the *Bulletin of the Atomic Scientists* (October 1952), devoted to the visa issue.

111 Quoted in Lecourt, *Proletarian Science? The Case of Lysenko,* 21. This is from an article written on 15 September 1948.

112 RFA, RG 905, series Pom 1951, 130.

113 RFA, RG 905 Pom 1951, 153–154.

114 Ibid. (emphasis added).

115 "Answers of the Rockefeller Foundation to Questionnaire Submitted by the Select Committee of the House of Representatives of the Congress of the United States," 46.

116 Julian Huxley, "Soviet Genetics: The Real Issue," *Nature,* 18 June 1949, 935–942.

117 Weaver to Ephrussi, 15 February 1950.

118 Weaver to Barnard and Kimball, memo of 6 February 1952, RFA, RG3, series 900, box 25, folder 200.

119 HAD to JHW, "Some Suggestions for Approaches to Questions about RF's Willingness to Make Grants in Support of 'Socialist' or other 'Un-American' or 'Subversive' Ideas," memo of 1 December 1952, RFA, RG 3, series 900, box 25, folder 200. See also JHW to Dean Rusk, "Suggestions concerning policy for dealing with subversives or potential subversives in RF recommendations," memo of 27 January 1953, RFA, RG 3.2, series 900, box 14, folder 85.

120 Weaver to Ephrussi, 15 February 1950.

121 The paragraph that follows is inspired by David Hollinger, "The Defense of Democracy and Robert K. Merton's Formulation of the Scientific Ethos," *Knowledge and Society: Studies in the Sociology of Culture Past and Present* 4 (1983), 1–15; Hollinger, "Science as a Weapon in *Kulturkämpfe* in the United States during and after World War II," *Isis* 86 (1995), 440–454; Everett Mendelsohn, "Robert K. Merton: The Celebration and Defense of Science," *Science in Context* 3:1 (1989), 269–289. See also Yaron Ezrahi, "Science and the Problem of Authority in Democracy," in *Science and Social Structure: A Festschrift for Robert K. Merton,* ed. Thomas F. Gieryn (New York: New York Academy of Sciences, 1980), 43–60. For the historical evolution of the argument see Roy Macleod, "Science and Democracy: Historical Reflections on Present Discontents," *Minerva* 35 (1997), 369–384; Jessica Wang, "Merton's Shadow: Perspectives on Science and Democracy since 1940," *Historical Studies in the Physical and Biological Sciences* 30:2 (1999), 279–306. See also Philip Mirowski, "The Scientific Dimensions of Social Knowledge and Their Distant Echoes in 20th-Century American Philosophy of Science," *Studies in the History and Philosophy of Science* 35 (2004), 283–326.

122 Thomas F. Gieryn, *Cultural Boundaries of Science: Credibility on the Line* (Chicago: University of Chicago Press, 1998). See also Gieryn, "Boundaries of Science," in *Handbook of Science and Technology Studies,* ed. Shelia Jasanoff, Gerald E. Markle, James C. Petersen and Trevor Pinch (London: SAGE, 1995), 393–443.

123 "GRP Diary, 30 March 1955 (Paris)," RFA, RG1.2, series 500D, box 3, folder 29.

124 "GRP Diary, 25 May 1956 (Paris)," RFA, RG1.2, series 500, box 5, folder 43.

125 Trustees' meeting, 10 October 1956, Appropriation 56162, Centre National de la Recherche Scientifique, RFA, RG12, series 500, box 5, folder 42. A further award of $47,000 for three years, Appropriation RF60187, was made on 21 October 1960.

126 "Official Indices Check: Professor Boris Ephrussi," RFA, RG1.2, series 500, box 5, folder 43.

127 DR (Rusk) to LFK (Kimball), "Security Check," memo of 14 March 1957, RFA, RG3, series 900, box 25, folder 200.

128 Nicolas Rasmussen, "Instruments, Scientists, Industrialists and the Specificity of 'Influence': The Case of RCA and Biological Electron Microscopy," in *The Invisible Industrialist: Manufactures and the Production of Scientific Knowledge*, ed. Jean-Paul Gaudillière and Ilana Löwy (London: Macmillan, 1998), 173–208. See also Nicolas Rasmussen, *Picture Control: The Electron Microscope and the Transformation of Biology in America, 1940–1960* (Stanford: Stanford University Press, 1997).

129 Pnina G. Abir-Am, "The Discourse of Physical Power and Biological Knowledge in the 1930s: A Reappraisal of the Rockefeller Foundation's 'Policy' in Molecular Biology," *Social Studies of Science* 12 (1982), 341–382; Abir-Am, "The Rockefeller Foundation and the Rise of Molecular Biology," *Nature Reviews Molecular Biology* 3:1 (2002), 65–70. See also Abir-Am, "The Rockefeller Foundation and Refugee Biologists: European and American Careers of Leading RF Grantees from England, France, Germany and Italy," in *The "Unacceptables": American Foundations and Refugee Scholars between the Two Wars and After*, ed. Giuliana Gemelli (Brussels: Peter Lang, 2000), 217–240.

6 The Ford Foundation, Physics, and the Intellectual Cold War in Europe

1 "Policy Planning Staff Memorandum," Washington, D.C., 4 May 1948. The document is available online at www.state.gov, Foreign Relations of the United States, Truman Administration, 1945–1950, Emergence of the Intelligence Establishment, Psychological and Political Warfare, Doc. 269.

2 "Report by the National Security Council on Coordination of Foreign Information Measures," Washington, D.C., 17 December 1947, NSC 4, NSC 4/A. The document is available online at www.state.gov, Foreign Relations of the United States, Truman Administration, 1945–1950, Emergence of the Intelligence Establishment, Psychological and Political Warfare, Doc. 252, 253.

3 NSC10/2, "National Security Council Directive on Office of Special Projects," Washington, D.C., June 18, 1948. The document is available online at www.state.gov, Foreign Relations of the United States, Truman Administration, 1945–1950, Emergence of the Intelligence Establishment, Psychological and Political Warfare, Doc. 292. Covert actions were those actions "so planned and executed that any US Government responsibility for them is not evident to unau-

thorized persons and that if uncovered the US government can plausibly disclaim any responsibility for them."

4 Kennan's memo suggested that the State Department should be responsible for covert activities, which the CIA contested: NSC10/2 "placed the responsibility" for covert actions "within the structure of" the CIA. For the bureaucratic infighting on this, see John Prados, *Keepers of the Keys: A History of the National Security Council from Truman to Bush* (New York: William Morrow, 1991), 50–56. See also Scott Lucas, "Campaigns of Truth: The Psychological Strategy Board and American Ideology, 1951–1953," *International History Review* 18:2 (1996), 279–302.

5 Marc Lazar, "The Cold War Culture of the French and Italian Communist Parties," in *The Cultural Cold War in Western Europe, 1945–1960,* ed. Giles Scott-Smith and Hans Krabbendam (London: Frank Cass, 2003), 214–224, for this paragraph.

6 Pierre Grémion, "The Partnership between the Ford Foundation and the Congress for Cultural Freedom in Europe," in *The Ford Foundation and Europe (1950s–1970s): Cross-Fertilization of Learning in Social Science and Management,* ed. Giuliana Gemelli (Brussels: European Interuniversity Press, 1998), 137.

7 This is described in Richard Pells, *Not Like Us: How Europeans Have Loved, Hated, and Transformed American Culture since World War II* (New York: Basic Books, 1997), 64–68.

8 This editorial was written by the novelist A. A. Fadyeev and is approvingly quoted by the American author Paul Robeson in his "Thoughts on Winning the Stalin Peace Prize," also entitled "Here's My Story," *Freedom* (January 1953).

9 For more detail on this conference and its repercussions, see Frances Stonor Saunders, *Who Paid the Piper? The CIA and the Cultural Cold War* (London, Granta Books, 1999), chap. 3; Giles Scott-Smith, *The Politics of Apolitical Culture: The Congress for Cultural Freedom, the CIA and Post-war American Hegemony* (London: Routledge, 2002), 94. The National Council of the Arts, Sciences and Professions, which included many activists that had been involved in the Popular Front in the late 1930s, organized the meeting. The general line of the conference was not wholly one-sided, and some speakers criticized the aggressive policies of both the USSR and the United States.

10 For the Italian figure see David W. Ellwood, "The Propaganda of the Marshall Plan in Italy in a Cold War Context," in Scott-Smith and Krabbendam, *The Cultural Cold War in Western Europe,* 234.

11 Michel Pinault, *Frédéric Joliot-Curie* (Paris: Odile Jacob, 2000), 441–461.

12 Letter from the President to the Speaker on the Need for an Expanded Truth Campaign to Combat Communism, July 13, 1950, Public Papers of the Presidents, Harry S. Truman, at www.trumanlibrary.org, Doc. 190.

13 Richard Kuisel, *Seducing the French: The Dilemma of Americanization* (Berkeley: University of California Press, 1993), 69. In particular, the agreements allowed for a massive penetration of American films into France. The number of

days reserved for the performance of French films was reduced from 50 to 33 percent, and there was a 50 percent decline in French film production during the next year. See Reinhold Wagnleitner, *Coca-Colonization and the Cold War: The Cultural Mission of the United States in Austria after the Second World War* (Chapel Hill: University of North Carolina Press, 1994), 241. As Victoria de Grazia points out, the defeat of fascism paved the way for European states to accede to an American peace premised on the free trade of goods *and* ideas, thereby relinquishing a conception of nationhood that presumed sovereignty over culture. De Grazia, "Mass Culture and Sovereignty: The American Challenge to European Cinemas, 1920–1960," *Journal of Modern History* 61:1 (1989), 53–87.

14 Pells, *Not Like Us,* 56.

15 Ellwood, "The Propaganda of the Marshall Plan in Italy," 230.

16 Kuisel, *Seducing the French,* chap. 4. David W. Ellwood, "Italian Modernisation and the Propaganda of the Marshall Plan," in *The Art of Persuasion: Political Communication in Italy from 1945 to the 1990s,* ed. Luciano Cheles and Lucio Sponza (Manchester: Manchester University Press, 2001), 23–48; Ellwood, "The Propaganda of the Marshall Plan in Italy," 225–236.

17 Tony Judt, *Past Imperfect: French Intellectuals, 1944–1956* (Berkeley: University of California Press, 1992), 195 et seq., discusses the impact of the Blum-Byrnes agreements. See also p. 262. On the difference between European and U.S. cinema, and the appeal of the latter, notably to women, see de Grazia, "Mass Culture and Sovereignty."

18 On "Coca-Colonization," see Kuisel, *Seducing the French,* chap. 3; Kuisel, "Coca-Cola and the Cold War: The French Face Americanisation, 1948–1953," *French Historical Studies* 17:1 (1991), 96–116. The remark by Wenders is cited in Wagnleitner, *Coca-Colonization and the Cold War,* xii.

19 For these examples, see Pells, *Not Like Us,* 79, 81.

20 Eric Johnston, quoted in Lary May, "Movie Star Politics: The Screen Actors' Guild, Cultural Conversion, and the Hollywood Red Scare," in *Recasting America: Culture and Politics in the Age of Cold War,* ed. Lary May (Chicago: University of Chicago Press, 1989), 125–151, at 145.

21 See especially Lawrence Badash, "Science and McCarthyism," *Minerva* 38:1 (2000), 53–80.

22 Jacques Monod, Letter to the American Embassy in Paris published in *Bulletin of the Atomic Scientists* 8:7 (October 1952), 236. This entire issue of the *Bulletin* was devoted to the visa question, and the other information provided here is from this source. See also John Krige, "La science et la sécurité civile de l'Occident," in *Les sciences pour la guerre, 1940–1960,* ed. Amy Dahan and Dominique Pestre (Paris: Éditions de l'École des Hautes Études en Sciences Sociales, 2004), 373–401.

23 Volker R. Berghahn, *America and the Intellectual Cold Wars in Europe: Shepard Stone between Philanthropy, Academy, and Diplomacy* (Princeton: Princeton University Press, 2001), xiii.

24 Harry S. Truman, "Going Forward with a Campaign of Truth," *Department of State Bulletin,* 1 May 1950, 669–672.

25 Massachusetts Institute of Technology, *Project Troy: Report to the Secretary of State, February 1, 1951,* 4 vols. Allan A. Needell kindly made his copy of this report available to me. He has described its genesis, content, and implications in "'Truth Is Our Weapon': Project Troy, Political Warfare, and Government-Academic Relations in the National Security State," *Diplomatic History* 17:3 (1993), 399–420. See also Needell, *Science, Cold War and the American State: Lloyd V. Berkner and the Balance of Professional Ideals* (Amsterdam: Harwood Academic Publishers, 2000), chap. 6.

26 Needell, "'Truth Is Our Weapon,'" 406.

27 *Project Troy: Report to the Secretary of State,* vol. 1, 10.

28 "Excerpt from the Docket, July 15–16, 1952, for the Board of Trustees Meeting, Research Program on International Communication," p. 2, FFA, Grant 52-150. This useful summary is from the MIT application for $875,000 from the Ford Foundation to support the research program of its Center for International Studies (CENIS) on international communication for four years. This was one of the main programs developed from the findings of Project Troy.

29 Clyde Kluckhohn, "Political Warfare—United States vs. Russia," *Project Troy: Report to the Secretary of State,* vol. 2, annex 8, 3.

30 *Project Troy: Report to the Secretary of State,* vol. 1, 51.

31 Robert S. Morrison, "Personnel for Southeast Asia and other Backward Areas," *Project Troy: Report to the Secretary of State,* vol. 3, annex 9, 1 (emphasis added).

32 Scott-Smith, *The Politics of Apolitical Culture.*

33 Laughlin and Miller are quoted in Kathleen D. McCarthy, "From Cold War to Cultural Development: The International Cultural Activities of the Ford Foundation, 1950–1980," *Daedelus* 116:1 (1987), 96, 97. The context was a request for $500,000 from the Ford Foundation to support an intercultural publications program that made works of contemporary American writers available in English, French, German, and Italian.

34 Using this approach, Fulbright fellows were committed to building "democracies of trust" wherever they could. Richard T. Arndt and David Lee Rubin, *The Fulbright Difference, 1948–1992* (New Brunswick, N.J.: Transaction Publishers, 1993). By 1975 this program had offered about 78,000 scientists, academics, teachers, and students from 110 countries the opportunity to travel to the United States, and about 39,000 American professors and students the opportunity to travel abroad. See Wagnleitner, *Coca-Colonization and the Cold War,* 157.

35 John Foster Dulles, "Our Cause Will Prevail," *Department of State Bulletin,* 6 January 1958, 22.

36 Ellwood, "The Propaganda of the Marshall Plan in Italy," 234.

37 Ellwood, "Italian Modernisation," 39.

38 Three major histories of the CCF are Saunders, *Who Paid the Piper?;* Scott-Smith, *The Politics of Apolitical Culture;* and Pierre Grémion, *Intelligence de l'anticommunisme: Le Congrès pour la Liberté de la Culture à Paris (1950–1975)* (Paris: Fayard, 1995). See also Berghahn, *America and the Intellectual Cold Wars in Europe,* chap. 5; Grémion, "The Partnership between the Ford Foundation and the Congress for Cultural Freedom."

39 Berghahn, *America and the Intellectual Cold Wars in Europe,* 132–142.

40 Ibid., 135.

41 On these conferences in Hamburg and Milan, see Scott-Smith, *The Politics of Apolitical Culture,* 143–159; Grémion, "The Partnership between the Ford Foundation and the Congress for Cultural Freedom."

42 Scott-Smith, *The Politics of Apolitical Culture,* 123–124. Scott-Smith tells us that the Fairfield Foundation's purpose was to fund "organizations, groups, and individuals which are engaged in . . . revealing to the nations and peoples of the free world the inherent dangers which totalitarianism poses." Its president until 1962 was James Fleischmann, a Cincinnati millionaire and important patron of the arts in New York and Boston. Klaus Hoblitzelle was the chairman of Republic National Bank in Dallas and a major figure in the Texas business world. While Ford, Rockefeller, and Carnegie did fund individual CCF activities, Scott-Smith claims that, of the foundations, they were far less important than these two. He seems to have underestimated the significance of the Ford Foundation which, according to Berghahn, contributed about half the CCF's budget in 1960.

43 On this issue, see W. Scott-Lucas's interview with Frances Stonor Saunders in Scott-Smith and Krabbendam, *The Cultural Cold War in Western Europe,* 15–40; Jason Epstein, "The CIA and the Intellectuals," *New York Review of Books,* 20 April 1967, 16–21.

44 Quoted by Ellwood, "The Propaganda of the Marshall Plan in Italy," 225.

45 Wagnleitner, *Coca-Colonization and the Cold War,* xii, 29–30 (emphasis original).

46 Quoted by Berghahn, "Philanthropy and Diplomacy," 415–416. See also Berghahn, *America and the Intellectual Cold Wars in Europe,* xi. Nielsen had served as the deputy director of the Information and Film Sections of the Marshall Plan. It disbursed 5 percent of counterpart funds for "propaganda" in Europe for the plan.

47 For an introduction to the vast literature from a radical perspective, see Noam Chomsky et al., *The Cold War and the University: Toward an Intellectual History of the Postwar Years* (New York: New Press, 1997), and Christopher Simpson, ed., *Universities and Empire: Money and Politics in the Social Sciences during the Cold War* (New York: New Press, 1998). For the foundations see, for example, Robert F. Arnove, *Philanthropy and Cultural Imperialism: The Foundations at Home and Abroad* (Boston: G. K. Hall, 1980). More recently, see, for example, Mark Solovey, "Project Camelot and the 1960s Epistemological Revo-

lution: Rethinking the Politics-Patronage-Social Science Nexus," *Social Studies of Science* 31:2 (2001), 171–206.

48 Rabi was a particularly enthusiastic proponent of this view, which was widely shared by many of his colleagues in the 1950s, of course. See particularly Isidor I. Rabi, *My Life and Times as a Physicist* (Claremont, Calif.: Claremont College, 1960), notably the first lecture. Also Isidor I. Rabi, *Science: The Center of Culture* (New York: World Publishing, 1970). For a recent account of Rabi as public intellectual promoting the significance of science as a cultural and moral force, see Michael. A. Day, "In Appreciation: I. I. Rabi; The Two Cultures and the Universal Culture of Science," *Physics in Perspective* 6:4 (2004), 428–476. Of the voluminous literature on international scientific exchange, I particularly like Paul Forman, "Scientific Internationalism and the Weimar Physicists: The Ideology and Its Manipulation in Germany after World War I," *Isis* 64 (1973), 151–180, and Jean-Jacques Salomon, "The *Internationale* of Science," *Science Studies* 1:1 (1971), 23–42. Both strip away the ideology of internationalism to reveal its close link to the pursuit of national interest. See also Elisabeth Crawford, Terry Shinn, and Sverker Sörlin, *Denationalizing Science: The Contexts of International Scientific Practice* (Berlin: Springer, 1992).

49 David Hollinger, "Science as a Weapon in *Kulturkämpfe* in the United States during and after World War II," *Isis* 86 (1995), 440–454. I have discussed this debate at greater length in chapter 4; further pertinent references may be found there.

50 The bones of the argument that follows are in John Krige, "The Ford Foundation, European Physics and the Cold War," *Historical Studies in the Physical and Biological Sciences* 29:2 (1999), 333–361. I thank John Heilbron, the editor of *HSPS*, for removing any copyright restrictions and for allowing me to use the original version freely.

51 In 1950 the principal assets of the foundation consisted of stocks, bonds, cash, and real estate donated by Henry and Edsel Ford during their lifetimes, and by their wills, as well as by Mrs. Clara J. Ford and by the Ford Motor Company.

52 For general accounts of the foundation in this period see, for a view from within, Francis X. Sutton, "The Ford Foundation: The Early Years," *Daedelus* 116:1 (1987), 41–91; for a perceptively critical view from without, see Dwight Macdonald, *The Ford Foundation: The Men and the Millions* (New Brunswick, N.J.: Transaction Publishers, 1989).

53 On Gaither, see David L. Snead, *The Gaither Committee: Eisenhower and the Cold War* (Columbus: Ohio State University Press, 1999), especially 49–51.

54 Martin J. Collins, *Cold War Laboratory: RAND, the Air Force, and the American State, 1945–1950* (Washington: Smithsonian Institution Press, 2002). David Hounshell, "The Cold War, RAND, and the Generation of Knowledge," *Historical Studies in the Physical and Biological Sciences* 27:2 (1997), 237–267, describes the kind of projects Rand was engaged in.

55 Sutton, "The Ford Foundation: The Early Years," 48. Gaither's report is available as a "Report of the Trustees of the Ford Foundation," 27 September 1950 (FFA).

56 H. Rowan Gaither, "The Ford Foundation and Foreign Affairs," address delivered at the Twenty-Five-Year Service Dinner of the Dunwoody Industrial Institute, Minneapolis, Minnesota, 3 May 1956 (New York: Ford Foundation, 1956).

57 Roger L. Geiger, "American Foundations and Academic Social Science, 1945–1960," *Minerva* 28:3 (1988), 315–341.

58 "President's Memorandum to the Board of Trustees," 29 January 1951, 1; "Report of the President to the Board of Trustees," 10 April 1951, 1, 4, both included with the bound copy of the docket excerpts for that year in the FFA.

59 Cited by McCarthy, "From Cold War to Cultural Development," 95.

60 Ibid., 97.

61 Quoted by Berghahn, *America and the Intellectual Cold Wars in Europe,* 146. Bissell called for caution in implementing this program: the foundations were under attack for being soft on Communism and this nonconfrontational approach to the Soviet Union was likely to add fuel to the fire.

62 Sutton, "The Ford Foundation," 84. See also Macdonald, *The Ford Foundation,* chap. 2.

63 Sutton, "The Ford Foundation," 69.

64 Courant to Gaither, 23 August 1951, FFA, Grant 56-154.

65 Niels Bohr, *Open Letter to the United Nations, June 9th, 1950* (Copenhagen: J. H. Schultz, 1950), 8, 13.

66 John Archibald Wheeler, *At Home in the Universe* (New York, 1994), 141.

67 Gaither to Chester C. Davis, "Exchange of Persons," memo of 7 June 1951, FFA, Grant 56-154.

68 Gaither to Joseph M. McDaniel, "Niels Bohr," memo of 30 November 1951, FFA, Grant 56-154.

69 Gaither to Davis, "Exchange of Persons," memo of 7 June 1951.

70 Gaither to McDaniel, "Niels Bohr," 30 November 1951.

71 "President's Report to the Trustees," June 1953, p. 9, FFA.

72 Howard to Gaither, 25 October 1951, FFA, Grant 56-154. Howard was Hoffman's representative in New York when the head office was still in Pasadena.

73 On Bohr's principle, see Finn Aaserud, *Redirecting Science: Niels Bohr, Philanthropy and the Rise of Nuclear Physics* (Cambridge: Cambridge University Press, 1990), 41–42.

74 Howard to Gaither, 25 October 1951; Courant to Bohr, 19 October 1951, NBA.

75 Courant to Howard, 16 May 1952, FFA, Grant 56-154, quoting a letter from Bohr to Courant.

76 "Niels Bohr, International Intellectual Center, Copenhagen, A-762," Speier to Howard, memo of 8 July1952, FFA, Grant 56-154, copied to Gaither and McDaniel.

77 Gaither to Howard, 12 November 1952, FFA, Grant 56-154.

78 Sutton, "The Ford Foundation," 57, 61, 69.

79 Howard to Gaither, 8 December 1952; Katz to Dacy, memo of 15 April 1953, FFA, Grant 56-154.

80 Gaither to Bohr, 19 August 1954, NBA, Ford Foundation, file S-2,1.

81 Berghahn, *America and the Intellectual Cold Wars in Europe*, 182.

82 Sutton, "Ford Foundation," 39.

83 Berghahn, *America and the Intellectual Cold Wars in Europe*, 186. The sum was to cover the globe, of course, including Africa and Asia.

84 Ibid., chaps. 1–3; Volker R. Berghahn, "Shepard Stone and the Ford Foundation," in Gemelli, *The Ford Foundation and Europe (1950s–1970s)*, 69–95.

85 Berghahn, *America and the Intellectual Cold Wars in Europe*, 59–60.

86 Sutton, "Ford Foundation," 28.

87 Charles T. Morrissey and Ronald J. Grele, Interview with Shepard Stone, 12 December 1972, p. 10, FFA.

88 Berghahn, "Shephard Stone," 90.

89 Berghahn, *America and the Intellectual Cold Wars in Europe*, 171.

90 Ibid., 224. This was about 50 percent of the budget for that year.

91 Stone to Bohr, 31 December 1954, 26 May 1955, 8 June 1955, and 28 July 1955, NBA, The Ford Foundation, file S-2,1. The quotation is from the letter of 28 July.

92 Stone to Wheeler, 1 June 1967, and to Aage Bohr, 20 September 1963, FFA, Grant 56-154.

93 Willems to Gaither, 9 May 1955, and Gaither to Stone, 17 May 1955; Stone to Gaither, "Dr. C. J. Bakker," memo of 27 October 1955, FFA, Grant 56-241.

94 Stone to Gaither, 16 May 1955, FFA, Grant 56-241.

95 For more on this, see John Krige, "Atoms for Peace, Scientific Internationalism and Scientific Intelligence Gathering," in John Krige and Kai-Henrik Barth, *Global Power Knowledge, Osiris*, 2nd ser., 21 (2006), 161–181, and the references therein.

96 On Bloch's brief and unhappy tenure as CERN's first director general, see John Krige, "Felix Bloch and the Creation of a Scientific Spirit at CERN," *Historical Studies in the Physical and Biological Sciences* 32:1 (2002), 57–69.

97 "Report on Impressions Received while Attending the International Conference on the Peaceful Uses of Atomic Energy, Geneva, 8–20 August, 1955," FFA, Grant 56-241, R. G. Gustavson.

98 The objector argued that the United States should ensure its leadership by never encouraging the reestablishment of the European scientific community and should do all it could to get the best European students to study in America, with a view to having them stay there. This would help ensure the United States' keeping its leadership in the field. Gustavson, "Report," 13–14.

99 Ernest O. Lawrence, "High Current Accelerators," in *Proceedings of the International Conference on the Peaceful Uses of Atomic Energy: Held in Geneva, 8 August–20 August 1955,* vol. 16, *Record of the Conference* (New York: United Nations, 1956), 64–68, at 64.

100 Gustavson, "Report," 12.

101 Ibid., 8.

102 Bohr to Stone, 16 September 1955, FFA, Grant 56-154.

103 Stone to Price, "CERN (L5-38, European Organization . . .)," memo of 8 December 1958, FFA, Grant 56-241.

104 Stone to the files, "CERN," memo of 1 December 1955, on his dinner with Rabi, FFA, Grant 56-241.

105 Stone to Gaither, "Dr. C. J. Bakker," memo of 27 October 1955, FFA, Grant 56-241, Strauss to Price, 5 December 1955.

106 Stone to the files, memo of 8 December 1955, "CERN (L5-388, European Organization . . .)," FFA, Grant 56-241. Stone spoke with Robert Bowie, assistant secretary of state in the State Department, John Bross, "an expert on European problems," and William Clark, European regional director of the U.S. Information Agency.

107 Price to the files, "Proposed Grant to CERN: Views of . . . Whitman and . . . Lawrence," memo of 30 November 1955, FFA, Grant 56-241; Stone to Bohr, 23 November 1955, FFA, Grant 56-154.

108 Stone to Price, "CERN (L5-388)," memo of 8 December 1958.

109 Stone phoned Bohr a few days before Christmas with the good news (Stone to Price, memo of 23 December 1955, FFA, Grant 56-154). Bohr replied (17 January 1956) and tried to reorient the priorities within the envelope. See also Stone to Bohr, 26 March 1956; Bohr to Stone, 9 April 1956; Bakker to Stone, 20 December 1955, FFA, Grant 56-241.

110 Program action form, release date 14 May 1956, FFA, Grant 56-154.

111 Program action form, release date 3 July 1956, FFA, Grant 56-241.

112 Stone to Gaither, memo of 2 July 1956, FFA, Grant 56-241.

113 Bohr to the Ford Foundation, 3 May 1958; Borgmann to Stone, memo of 10 June 1958; Stone to Nielsen, "Niels Bohr's Institute for Theoretical Physics," memo of 8 August 1958; Stone, memo of 25 June 1958, FFA, Grant 56-154.

114 Stone to Nielsen, 8 August 1958; Bohr to Stone and to the Ford Foundation (attached), 5 September 1958; Paul B. Pearson to Stone, memo of 3 October 1958; Stone to Bohr, 12 March 1959, FFA, Grant 56-154.

115 "Atomic Development," n.d., FFA, Grant 56-241. Stone claimed that "Support for these institutions is in line with the Foundation's over-all program objectives and with such Foundation grants as the one made to the Fund for Peaceful Atomic Development." The fund was a private, nonprofit organization established in 1954 with a board of directors composed mostly of prominent businessmen. Ford awarded it $150,000 in its first year of operation to support its efforts in "diffusing knowledge of a new field, nuclear energy, among the free nations of the world for peaceful purposes." The grant was approved on 17 December 1954 (FFA Reel 495).

116 "Atomic Development"; Stone to Gaither, "Dr. C. J. Bakker," memo of 27 October 1955.

117 Program action form, release date 3 July 1956, FFA, Grant 56-241 (emphasis added).

118 Bohr to Stone, 16 September 1955, FFA, Grant 56-154.

119 Randers to Bloch, 6 January 1955 CERN, file DG-20512.

120 Bakker to Gaither, 29 May 1956, FFA, Grant 56-241.

121 Iverson to Stone, "CERN and Bohr Institute," memo of 9 November 1955, FFA, Grant 56-241.

122 Price to the files, "Proposed Grant to CERN: Views of . . . Whitman and . . . Lawrence," memo of 30 November 1955.

123 Stone to Gaither, 2 July 1956.

124 FFA, Grant 56-241. The list of appointments to CERN paid for by the Ford Foundation grant was appended to Bakker to Stone, 11 May 1959; FFA, Grant 56-154. Bohr's figures are from his "Report of Activities under Grant from the Ford Foundation," 14 May 1956, and cover support from all sources—including grants other than that from Ford.

125 "Request for Grant Action," annotated "Hold for Final Draft in June," n.d., FFA, Grant 56-241 (emphasis added).

126 Stone, on a meeting with Bakker of 9 April 1956, memo, FFA, Grant 56-241.

127 Stone to Bohr, 30 April 1956; Stone to Heald and Price, "Elimination of Restrictions on grant to Bohr's Institute in Copenhagen," memo of 3 March 1958, FFA, Grant 56-154.

128 Stone to Heald and Price, "Elimination of Restrictions on grant to Bohr's Institute in Copenhagen." Just a few months after agreeing to this constraint, on 9 November 1956, Bohr reaffirmed his position taken in June 1950 in an open letter to UN Secretary General Dag Hammarskjöld. He wrote again that "free intercourse and exchange of opinion across all boundaries must form the

foundation of that co-operation in confidence between nations, which in our time is so vital for the future of mankind" (FFA, Grant 56-154).

129 Stone to Bakker, 26 April 1957, FFA, Grant 56-241, written on the day that the announcement of the Polish grant was made.

130 Price to Heald, memo of 25 September 1957, cited by Berghahn, *America and the Intellectual Cold Wars in Europe*, 190.

131 Stone to Heald and Price, "Elimination of Restrictions on Grant to Bohr's Institute in Copenhagen."

132 Ibid.

133 "Report of Activities under Grant from the Ford Foundation," 14 May, 1956; "Report of Activities Sponsored by the Grant from the Ford Foundation, 17 December, 1959," the latter attached to Aage Bohr to J. McDaniel, 14 July 1967, FFA, Grant 56-154.

134 Charles T. Morrissey and Ronald J. Grele, Interview with Shepard Stone, 12 December 1992, pp. 23–24.

135 Kai Bird, *The Chairman: John J. McCloy: The Making of the American Establishment* (New York: Simon and Schuster, 1942), 426–429.

136 Ibid., 429.

137 I have raised this possibility in public on two occasions, at presentations at the Niels Bohr Institute in Copenhagen and at the University of Minnesota, Minneapolis. On both occasions American physicists spontaneously stepped forward to tell me that, indeed, they were regularly approached by CIA (and FBI) agents after trips abroad and questioned about what they saw and learned. See also Ronald E. Doel, "Scientists as Policymakers, Advisors and Intelligence Agents: Linking Contemporary Diplomatic History with the History of Contemporary Science," in *The Historiography of Contemporary Science and Technology*, ed. Thomas Söderqvist (Amsterdam: Harwood Academic Publishers, 1997), 215–244.

138 Saunders, *Who Paid the Piper?* It was this dissimulation, rather than that intellectuals like Spender were engaged in the cultural Cold War, that Saunders deplored. Interview with W. Scott Lucas published in Scott-Smith and Krabbendam, *The Cultural Cold War in Western Europe*, 40.

139 Volker Berghahn, "Philanthropy and Diplomacy in the American Century," 403. Stone had something of a reputation for name-dropping, and this was no exception.

140 Stone to Bohr, 17 June 1958, NBA, Ford Foundation, file S-2,3.

141 Secretary, Royal Danish Academy of Sciences and Letters, to Ford Foundation, 9 March and 8 November 1957, NBA, The Ford Foundation, file S-2,2. Stefan Rozental from the NBI was also active in this program, along with representatives from the Danish Ministry of Education, with support from the Danish Ministry of Foreign Affairs.

142 Jason Epstein, "The CIA and the Intellectuals," 20. This fine article was written at the time when the links between the CIA and the Ford Foundation were exposed.

143 Program action forms associated with Grants 56-241A, 56-241B, and 56-154A, FFA. In Weisskopf to Stone, 26 March 1962, the CERN director general asked the foundation for $750,000. Stone to Weisskopf, 5 February 1963, explains why the grant would be just one-third of that sum. Weisskopf to Stone, 4 March 1963, argues that this cut would hurt Polish scientists particularly badly and asks for additional support just for them. Cullen to Stone, "CERN," memo of 19 April 1963, states that this was obviously "out of the question." For the terminal grant to Copenhagen, see A. Bohr to Stone, 19 February 1965 and the program action form for Grant 56-154B, showing the approval by the trustees on 8–9 December 1966.

144 "Docket Excerpt, Meeting of the Executive Committee, 12/10/59, International Affairs, European Nuclear Research Center (Geneva)," FFA, Grant 56-241.

145 Hill to Heald, "Request for Grant out of Appropriation," memo of 17 January 1963, FFA, Grant 56-241; Stone to Weisskopf, 5 February 1963.

146 Stone to Records Center, "CERN," memo of 20 February 1959, FFA, Grant 56-241, describing a phone call with Weisskopf; Weisskopf to Stone, 4 November 1966.

147 Wheeler to Stone, 11 May 1967, FFA, Grant 56-154.

148 "Docket Excerpt, Meeting of the Executive Committee, 12/10/59, International Affairs, European Nuclear Research Center (Geneva)"; David E. Bell to McGeorge Bundy, "Board of Trustees Docket Item—International Affairs," memo of 26 October 1966 FFA, Grant 56-154.

149 Hill to Heald, "Request for Grant out of Appropriation," memo of 17 January 1963, FFA, Grant 56-241.

150 Wheeler to Stone, 11 May 1967, FFA, Grant 56-154.

151 Weisskopf to Stone, 4 November 1966, FFA, Grant 56-241.

152 "Docket Excerpt, Meeting of the Executive Committee, 12/10/59, International Affairs, European Nuclear Research Center (Geneva)," FFA, Grant 56-241.

153 "Board of Trustees Docket Item—International Affairs," memo, FFA, Grant 56-154.

154 "Docket Excerpt, Meeting of the Executive Committee, 12/10/59, International Affairs, European Nuclear Research Center (Geneva)"; "International Affairs. Institute for Theoretical Physics, Copenhagen," 21 October 1959, FFA, Grant 56-154.

155 "Docket Excerpt, Meeting of the Executive Committee, 12/12/57," FFA Grant 58-35. The Treaties of Rome established the European Common Market and Euratom. Stone admired Monnet, as he did Bohr, and described him as "in

some ways the greatest philosopher I have ever met" (FFA, Interview by Morrissey and Grele with Stone, 25). Monnet, like Bohr, counted "Shep" Stone as one of his "friends." See Jean Monnet, *Mémoires* (Paris, 1976), 546.

156 "Docket Excerpt, Meeting of the Executive Committee, 12/10/59, International Affairs, European Nuclear Research Center (Geneva)."

157 Wheeler to Stone, 11 May 1967, FFA, Grant 56-154.

7 Providing "Trained Manpower for Freedom"

1 "Docket Excerpt, Meeting of the Executive Committee, 12/10/59, International Affairs, European Nuclear Research Center (Geneva)," FFA, Grant 56-241.

2 David Holloway, *Stalin and the Bomb: The Soviet Union and Atomic Energy, 1939—1956* (New Haven: Yale University Press, 1994), 222–223.

3 Richard G. Hewlett and Jack M. Holl, *Atoms for Peace and War, 1953–1961: Eisenhower and the Atomic Energy Commission* (Berkeley: University of California Press, 1989), 250.

4 Quoted in C. Lasby, *Project Paperclip: German Scientists and the Cold War* (New York: Atheneum, 1971), 6.

5 John Krige, "Philanthropy and the National Security State: The Ford Foundation's Support for Physics in Europe in the 1950s," in *American Foundations and Large-Scale Research: Construction and Transfer of Knowledge,* ed. Giuliana Gemelli (Bologna: Clueb, 2001), 3–24.

6 "International Affairs: Scientific Activities in the Atlantic Community," Docket, Board of Trustees Meeting, 13–14 December 1957, FFA.

7 William A. Nierenberg, "The NATO Science Program," *Bulletin of the Atomic Scientists* 21:5 (1965), 45–48.

8 As quoted by David Kaiser, "Cold War Requisitions, Scientific Manpower, and the Production of American Physicists after World War II," *Historical Studies in the Physical and Biological Sciences* 33:1 (2002), 138.

9 Ibid.

10 This has been charted in detail in Paul Forman, "Behind Quantum Electronics: National Security as Basis for Physical Research in the United States, 1940–1960," *Historical Studies in Physical and Biological Sciences* 18:1 (1987), 149–229.

11 Kaiser, "Cold War Requisitions," 153 (emphasis original).

12 David Kaiser, "The Postwar Suburbanization of American Physics," *American Quarterly* 56:4 (2004), 851–888.

13 John L. Rudolph, *Scientists in the Classroom: The Cold War Reconstruction of American Science Education* (New York: Palgrave, 2002), 127.

14 Kaiser, "The Postwar Suburbanization of American Physics," 866 et seq.

15 http://www.gibill.va.gov/education/GI_Bill.htm.

16 Kaiser, "Cold War Requisitions," 153.

17 Ibid., 134 –137.

18 M. H. Trytten, "Our Technical Manpower Supply," Military Industrial Conference on Technical Manpower, Conrad Hilton Hotel, Chicago, 10 February 1955, AmPhilSoc, J. A. Wheeler Papers (hereafter JAW), folder NATO/Jackson Committee, series 6.

19 Ibid., 4, 6, 7, 11, resp.

20 Ibid., 10.

21 Ibid., 9.

22 Ibid., 13–17, for the analysis of the Soviet situation.

23 Ibid., 17.

24 Lewis Strauss, "Freedom's Need for Trained Men," included in U.S. Congress, Joint Committee on Atomic Energy, *Shortage of Scientific and Engineering Manpower,* 84th Cong., 2nd sess., 17 April 1956–1 May 1956, 465, as cited by Benjamin C. Zulueta, "People of Science: American-Educated Chinese and the Cold War of the Classrooms," paper given at History of Science Society Meeting, Austin, Texas, 2004.

25 Melvyn Price to Henry Barton, 29 March 1956, American Institute of Physics, Center for History of Physics Archives, Barton files, box 58, folder 8.

26 *Soviet Scientific and Technical Manpower,* special article attached to Minutes of the Cabinet Meeting, 30 April 1954, Dwight D. Eisenhower Library, Abilene, Texas, Papers of the President, Cabinet Papers, box 3 (Ann Whitman File).

27 "Russia is Overtaking U.S. in Training of Technicians," *New York Times,* 7 November 1954.

28 Nicholas De Witt, *Soviet Professional Manpower: Its Education, Training and Supply* (Washington, D.C.: National Science Foundation, 1955).

29 "Recruitment and Training of Scientists, Engineers and Technicians in NATO Countries and the Soviet Union," Report by Robert Major, Consultant to the Committee of Three, NATO, Document C-M(56)128, 26 November 1956.

30 Marc Trachtenberg, *A Constructed Peace: The Making of the European Settlement, 1945–1963* (Princeton: Princeton University Press, 1999), 150.

31 The archives of the Manpower Committee of the OEEC can be consulted at the Historical Archives of the European Community, European University Institute, Florence, Italy.

32 "Report of the Committee of Three on Non-Military Co-Operation," NATO, Document C-M(56)127 (revised), 10 January 1957, http:www.nato.int/docu/basictxt/b561213a.htm. Unless otherwise stated, all quotations in the next three paragraphs are from this document.

33 NATO Secretary General Lord Robertson, "The Future of the Transatlantic Link," Lisbon, Portugal, 24 October 2001, http://www.nato.int/docu/speech/

2001. Robertson was expressing the solidarity of all the members of the alliance with the United States after the attacks on the World Trade Center six weeks before. See also John Krige, "History of Technology after 9/11: Technology, American Power and 'Anti-Americanism,'" *History and Technology* 19:1 (2003), 32–39.

34 "Members American Advisory Group on NATO Scientific and Technical Personnel," AmPhilSoc, JAW, folder NATO/Jackson Committee.

35 Wheeler to Jackson, 12 February 1957, AmPhilSoc, JAW, folder NATO/Jackson Committee.

36 John Archibald Wheeler, *Geons, Black Holes, and Quantum Foam* (New York: W. W. Norton, 1998), 271.

37 Typewritten notes, each heading on a separate sheet of paper, undated, unsigned, AmPhilSoc, JAW, folder NATO/Jackson Committee.

38 Henry M. Jackson, "Trained Manpower for Freedom: Report by the NATO Parliamentarians' Committee on the Provision of Scientific and Technical Personnel in the NATO Countries . . . as Unanimously Approved by the Third NATO Parliamentarians' Conference, 11–16 November, 1957," NATO, Document RDC/57/408.

39 Henry M. Jackson, "Science and Freedom: Report of the Committee on the Provision of Scientific and Technical Personnel," Fourth Annual NATO Parliamentarians' Conference, 17–21 November 1958, FFA, Grant 58-214.

40 "Task Force on Action by NATO in the Field of Scientific and Technical Cooperation. Report to the Council," NATO, Document C-M(57)130, 4 November 1957.

41 "Verbatim Record of the Eighty-First Meeting of the Council, Public Session," 16 December 1957, NATO Verbatim Record C-VR(57)81(Final), 24 February 1958.

42 From the "Declaration" at the end of the meeting published in *NATO Basic Texts*, 108–116, at 114, NATO.

43 For more details about the people and policies of the NATO Science Committee, see John Krige, "NATO and the Strengthening of Western Science in the Post-Sputnik Era," *Minerva* 38:1 (2000), 81–108.

44 "Science Committee. NATO and Science," NATO, Document AC/137-WP/22, 19 October 1965.

45 The school at Les Houches, which was first held in 1951, was the brainchild of Cécile DeWitt, a French physicist living in the United States. It was modeled on the summer schools that became so popular in America after the war, and it was intended, like them, to inject "new courses, new teaching methods, new people" into the existing educational system which DeWitt judged to be "ossified" in France, according to John Wheeler in an interview with Finn Aaserud (May 1988), Center for History of Physics, American Institute of Physics, College Park, Maryland. The school at Varenna was supported by the Italian Physical Society.

46 Pierre Aigrain was a world authority on quantum electronics. He had a doctorate from the Carnegie Institute of Technology in Pittsburgh and another from the University of Paris. Stone gave him $50,000 for a summer school at the École Normale Supérieure that was attended by about 150 Ph.D.'s and postdoctoral students from all over the world. The success of the school in France led Nevill Mott of Cambridge University to have Stone and the Ford Foundation provide $23,000 for an international summer school at the Cavendish laboratory in 1959, "similar to that organized in Paris except that the emphasis will be on mechanical rather than electrical properties of solids, and there will be much greater emphasis on experimental (as opposed to theoretical) physics." Mott to Stone, 20 September 1958, FFA, from which the quotation is taken.

47 "Science Committee. Financing the Programme of the Science Fellowships and of the Advanced Study Institutes," NATO, Document AC/137-D/174, 20 May 1963.

48 "Science Committee. Summary Record of Meetings . . ." 9–11 July 1958, NATO, Document AC/137-R/2, 22 July 1958, 11. See also "Projet de Fondation Occidentale pour la Recherche Scientifique/Proposals by the French Government Regarding a Western Community Foundation for Scientific Research," document AC/137-D/2, 11 February 1958.

49 "Science Committee. Summary Records of Meetings . . ." 5–8 January 1959, NATO, Document AC/137-R3, 19 January 1959, 7–8, 41. Annex B to these minutes identifies some of the problems the French deemed suitable for a collaborative approach. They included long-range location of submerged submarines, location and identification of ballistic missiles, long-term weather forecasting and artificial modification of meteorological conditions, and the study of lightweight materials with outstanding mechanical and heat-resisting properties.

50 This is the conclusion sadly drawn by Eugene Skolnikoff, who was on the White House staff and assisted the President's Science Advisory Committee at the time. See Eugene B. Skolnikoff, *Science, Technology and American Foreign Policy* (Cambridge: MIT Press, 1967), 183.

51 "Science Committee. Establishment and Operation of Its Subsidiary Bodies," NATO, Document AC/137-D/189, annex 2, 4 October 1963, surveys the work of Mosby's group to date.

52 Aigrain, Interview with "D.C." in *International Science and Technology*, July 1964, 69-74.

53 "Letter to the Defence Research Directors from the Exploratory Group," followed by their report, NATO, Document AC/137(DR)D/4 (Final), AC/74(DR) D/4(Final), attached to Leconte to the Defence Research Directors, 9 December 1963.

54 Ibid., 18–19.

55 Ibid., 32–33.

56 "Provision of Scientific Advice to the NATO Military Authorities—Views of the Science Committee on the Report of the Exploratory Group," NATO, Document AC/137-D/207, 20 January 1964.

57 See "Science Committee. NATO and Science," for the information in this paragraph.

58 "Science and NATO," LoC, Rabi Papers, box 39, folder 7. The document is unsigned and undated but obviously by Rabi; he gives the 1964 figures for the budgets of the fellowship and summer school programs. Unless otherwise stated, all quotations by Rabi in this paragraph are from this paper.

59 "Science Committee. Summary Minutes of the First Meeting . . . ," 26–28 March 1958, NATO, Document AC/137-R/1, 25 April 1958.

60 The topic has been considered from a different angle by Giuliana Gemelli in "Western Alliance and Scientific Diplomacy in the Early 1960s: The Rise and Failure of the Project to Create a European M.I.T.," in *The American Century in Europe*, ed. R. Laurence Moore and Maurizio Vaudagna (Ithaca: Cornell University Press, 2003), 171–192.

61 James R. Killian, "An International Institute of Science and Technology," *NATO Letter* 10:4 (April 1962).

62 Eisenhower to Killian, 23 November 1962, MIT, MC 423, box 14, folder NATO Comm. 1963–64.

63 Francis X. Sutton, "The Ford Foundation's Transatlantic Role and Purposes, 1951–1981," *Fernand Braudel Center Review* 24:1 (2001), 89. Stone also made a contribution of $1 million for the Maison des Sciences de l'Homme in Paris at the time, as described by Sutton. It was actively involved in many other initiatives to export American-style social sciences abroad, of course. See Giuliana Gemelli, ed., *The Ford Foundation and Europe (1950s–1970s): Cross-Fertilization of Learning in Social Science and Management* (Brussels: European Interuniversity Press, 1998).

64 "International Affairs. Churchill College, Cambridge University, General Support," Docket Excerpt, Board of Trustees Meeting, 25 September 1959, FFA, Grant 59-518, for all quotations.

65 As described by Stuart W. Leslie and Robert Kargon, "Exporting MIT: Science, Technology and Nation-Building in India and Iran," in *Global Power Knowledge*, ed. John Krige and Kai-Henrik Barth, *Osiris*, 2nd ser., 21 (2006), 110–130.

66 Brown is writing to Julius Stratton in October 1960, soon after Stratton took over from Killian as president of MIT, Killian having moved on to becoming chairman of the MIT Corporation after a two-year stint as Eisenhower's special assistant for science and technology. The quotation is in Leslie and Kargon, "Exporting MIT."

67 *Increasing the Effectiveness of Western Science* (Brussels: Fondation Universitaire, 1960), FFA, Grant 59-505.

68 The other members were Prof. Dr. med. Wolfgang Bargmann (Kiel), Prof. Dr. Gerhard Hess (Bad Godesberg/Heidelberg), and Prof. G. Puppi (director of physics, University of Bologna).

69 Stone to Willems, 13 July 1959, and reply, 18 July 1959, FFA Grant 59-505.

70 Anon., "International Cooperation in Research," *Nature* 4754, 10 December 1960, 879–881, at 880.

71 There was a second major recommendation also taken up by NATO: to ask the OEEC to establish an "international centre in auxiliary support of science," a "clearing house to expedite the transfer of research equipment."

72 Meeting of the Study Group on the Means of Increasing the Effectiveness of Western Science, 11 September 1959, NATO, Collection SCOM 5-3-03.

73 Casimir, "An Atlantic University," paper presented at the meeting of the Armand group in Paris, 22 October 1959, FFA, Grant 59-505. See also Stone to Nielsen et.al., memo of 27 October 1959.

74 "Summary Record of Joint Meeting of the Science Committee and Study Group . . . on 11 January 1960," NATO, Document Summary Record AC/137-R/6, Part 3, 8 February 1960, for all quotations in this paragraph (emphasis added).

75 Casimir, "An Atlantic University."

76 Akira Iriye, *Cultural Internationalism and World Order* (Baltimore: Johns Hopkins University Press, 1997), 161.

77 *Increasing the Effectiveness,* 12.

78 "NATO Science Program: Forward Planning," Report by Senator Henry M. Jackson to the Scientific and Technical Committee of the NATO Parliamentarians Conference, 21–26 November 1960, 1, 5, MIT, MC 423, box 15, folder NATO Commission 1959–1960. Jackson's speech was reported in the *New York Times,* 2 November 1960, under the title "Europe Institute of Science Urged."

79 *Increasing the Effectiveness,* 12, 10.

80 Stone to Nielsen, "Study of Western Science," memo of 27 October 1959, FFA, Grant 59-505, thus quotes him.

81 *Increasing the Effectiveness,* 12.

82 "The Establishment of an International Institute of Science and Technology," NATO, C-M(60)85, annex 1, 30 September 1960.

83 "Atlantic Institute of Science," *Times,* 31 October 1960; Anon., "International Co-operation in Research," *Nature* 4754, 10 December 1960, 879–881.

84 Nierenberg to Stone, 9 January 1961, FFA, Grant 59-505.

85 Pierre Piganiol, "Comité Consultatif de la Recherche Scientifique et Technique, Réunion du 10 mars 1961: Compte-rendu analytique," CACF, 810401, Art. 123 (my translation here and in all further French documents from this archive). The CCRST was a high-level committee set up in 1958 to advise the government on science policy.

86 "First Notes on I.I.S.T," MIT, MC 423, box 15, folder NATO Comm. 1961 (1), with annotation "handed to JRK by Nierenberg on 1/28/61?"

87 "Note relatif aux problèmes généraux de la coopération internationale en matière de recherche scientifique et technique," 9 February 1961, CACF, 770321, Art. 320, DGRST. The DGRST was established by de Gaulle when he came to power in 1958 to organize, coordinate, and stimulate national science policy; it defined the national research budget.

88 "Proposal for an International Institute of Science and Technology: Report of the Working Group Appointed by the Secretary-General of the North Atlantic Organization," NATO, Document NATO C-M(61)85, 10 October 1961. Also in FFA, Grant 59-505.

89 "Report by the Chairman of the Working Group . . . ," 28 June 1961, 2, MIT, MC 423, box 15, folder NATO Comm. 61.

90 "Proposal for an International Institute of Science and Technology," 6, 8.

91 Ibid., 5, 6.

92 Ibid., 3.

93 Eugene B. Skolnikoff, "The International Institute of Science and Technology," draft memo dated 31 May 1961, MIT, MC 423, box 15, folder NATO Commission 1961 (1). Skolnikoff assisted the Killian panel.

94 "An International Institute of Science and Technology," draft revision of April 1963, MIT, MC 423, box 14, folder NATO Commission 1963–1964, annotated "JRK used some of this material in his lecture at Tech. Univ. of Athens, May 1963."

95 "Proposal for an International Institute of Science and Technology," 2, 8, 38. Also, "International Institute of Science and Technology," "Memorandum for Files," dated 25 September 1961, kindly provided by Eugene B. Skolnikoff.

96 Ibid., 47–49.

97 Eugene B. Skolnikoff, "Memorandum for the President: International Institute of Science and Technology," 13 July 1961, MIT, MC 423, box 15, folder NATO Comm. 61.

98 For the Dutch delegation's anger with the British over not supporting the Killian project, see NATO, Council Meeting, 21 March 1962, document C-R (62)12, 26 March 1962. For the offers from Italy and Luxembourg, see NATO, Science Committee, Summary Minutes of Meeting on 14–15 February 1963, document AC-137-R/15, and annexes 2 and 3, 25 March 1963. For Scotland, see Coleby (NATO Science Office) to Killian, 1 May 1963; for Casimir, see, for example, Coleby to Allis, "Conversation with Professor Casimir, 13th June 1963," and Allis to Killian, 20 June 1963, MIT, MC 423, box 14, folder NATO Commission 1963–1964.

99 Science Committee, Summary Minutes of Meeting on 14–15 February 1963, NATO, Document AC-137-R/15, annex 1, 25 March 1963, for the offer, and

extensive correspondence in MIT, MC 423, box 14, folder NATO Commission 1963–1964.

100 Coleby to Anastassiadès, 5 April 1963, MIT, MC 423, box 14, folder NATO Commission 1963–1964.

101 The OECD was set up in 1961 to replace the OEEC; its membership was expanded to include non-European states, notably the United States.

102 Department of State to Embassy Paris, telegram of 30 January 1962, MIT, MC 423, box 14, folder NATO Comm. 62.

103 Allis to Secretary General, "Visit to Germany, 10th–12th June 1963," memo of 20 June 1963; Allis to Killian, 20 June 1963, MIT, MC 423, box 14, folder NATO Commission 1963–1964.

104 On the birth of ELDO, see John Krige and Arturo Russo, *A History of the European Space Agency, 1958–1987,* vol. 1., *The Story of ESRO and ELDO, 1958–1973* (Noordwijk: ESA SP-1235, April 2000).

105 Todd to Maréchal, 9 August 1962, and attached memo, "An International Institute of Science and Technology: An Alternative Proposal Arising from a Meeting between United States, French and British Representatives on July 16th and 17th, 1962," CACF, 920548, Art. 20, Liasse 3. As a matter of fact, Nierenberg had suggested this kind of compromise to Killian before the working group even got started. His idea was that the core be made up of about five departments and that ten more be scattered. "First Notes on I.I.S.T.," MIT, MC 423, box 15, folder NATO Comm. 1961 (1); the document is undated but was probably written early in 1961.

106 Todd to Killian, 11 August 1962, MIT, MC 423, box 14, folder NATO Commission 1962.

107 Tobin (Office of Science and Technology, Washington) to Killian, 8 April 1963; Coleby to Killian, 26 April 1963; Coleby to Deputy Secretary General, "Killian Report," memo of 9 May 1963, MIT, MC 423, box 14, NATO Commission 1963–1964.

108 His full function, from February 1960 to April 1962, was "ministre délégué auprès du premier ministre, chargé de l'énergie atomique, de la recherche et de la fonction publique."

109 Frederick Seitz, *On the Frontier: My Life in Science* (New York: American Institute of Physics, 1994), 255.

110 For example, Killian to Baranson, 8 May 1964, MIT, MC 423, box 14, folder NATO Commission 1963–64.

111 This in turn was surely part of his renewed hostility to *les anglo-saxons* that was triggered by Kennedy's offer to provide Macmillan with Polaris missiles at a famous meeting in Nassau in December 1962, itself the proximate cause for de Gaulle's veto of the British application to join the EEC on 14 January 1963. On that memorable occasion, de Gaulle announced that if Britain were admitted into Europe, its whole character would be fundamentally changed, and the continental

countries would eventually be absorbed into a "colossal Atlantic community dependent on America and under American control."

112 The following is based on John Krige, "The Birth of EMBO and the Difficult Road to EMBL," *Studies in the History and Philosophy of Biology and Biomedical Sciences* 33 (2002), 547–564.

113 Ibid., 560.

114 It is striking that, while proposing a center for life sciences, the Killian group did not even consult with many of the leading European researchers in the field—not Kendrew, Perutz, or Waddington in the UK; not Jacob, Lwoff, or Monod at the Pasteur Institute in France; not Buzzatti-Traverso in Italy, who headed an important new international marine biology station in Naples.

115 Jean-Marie Palayret, *A University for Europe: Prehistory of the European University Institute in Florence (1948–1976)* (Rome: Istituto Poligrafico e Zecca dello Stato, 1996), 79.

116 Ibid., 51, 77.

117 Ibid., 82. The previous quotation is on 85.

118 "Summary Record of Joint Meeting of the Science Committee and Study Group . . . on 11 January 1960." One of the few noticeable effects of these conflicts on the study groups was the use of the label "institute" rather than "university": the term had raised a host of problems over the equivalence in the status and in the quality of the degrees granted by universities in the different member states of the EEC.

119 "Comité Consultatif de la Recherche Scientifique et Technique, Réunion du 10 mars 1961: Compte-rendu analytique," 13.

120 Stikker included this phrase in a short insert accompanying Killian, "An International Institute." He put the phrase "underdeveloped in science" in quotation marks, which is why I remark that the reference to the newly developing countries was deliberate.

121 Leslie and Kargon, "Exporting MIT."

122 General Olie to several ministers, "Projet O.T.A.N. de création d'un Institut International des Sciences et de la Technologie: Proposition Killian," memo of 31 October 1961, CACF, 920548, Art. 20, Liasse 3.

123 "Comité Consultatif de la Recherche Scientifique et Technique, Réunion du 24 novembre 1961: Compte-rendu analytique," CACF, 810401, Art. 123.

8   "Carrying American Ideas to the Unconverted"

1 Mike Fortun and Sylvan S. Schweber, "Scientists and the Legacy of World War II: The Case of Operations Research (OR)," *Social Studies of Science* 23:4 (1993), 630, and Erik P. Rau, "Morse, Philip McCord," in *American National*

Biography, ed. John A. Garraty and Mark C. Carnes (New York: Oxford University Press, 1999), 15:939. Both imply that Morse was the first to introduce OR to NATO: this is not so.

2 Jan Van der Bliek, ed., *AGARD: The History, 1952–1997* (Neuilly-sur-Seine: NATO Research and Technology Organization, 1999), 2–3.

3 Philip M. Morse, "Background of the NATO O.R. Panel," MIT, MC 75, box 20, folder NATO APOR 10th Anniversary. See also NATO, "Policy for Mutual Support in Operational Research between the United States and Its Allies," AC/137-D/23 attached to circular note from Ramsey, 21 August 1958.

4 Thus Blackett wrote during the war, "In general, one might conclude that relatively too much scientific effort has been expended hitherto in the production of new devices and too little in the proper use of what we have got." Quoted by Erik P. Rau, "Combat Scientists: The Emergence of Operations Research in the United States during World War II" (Ph.D. diss., University of Pennsylvania, 1999), 59.

5 Van der Bliek, *AGARD,* 3–13.

6 Henry M. Jackson, "Trained Manpower for Freedom: Report by the NATO Parliamentarians' Committee on the Provision of Scientific and Technical Personnel in the NATO Countries . . . as Unanimously Approved by the Third NATO Parliamentarians' Conference, 11–16 November, 1957," Part 2, Recommendation 2, NATO, Document RDC/57/408.

7 This brief historical introduction is meant simply to provide the contextual elements needed to make sense of Philip Morse's attempts to promote OR in the framework of NATO. For recent general accounts of the emergence of OR in World War II, see Daniel J. Kevles, *The Physicists: The History of a Scientific Community in Modern America* (New York: Alfred A. Knopf, 1978), chaps. 19 and 20, which is the standard point of departure. See also Fortun and Schweber, "Scientists and the Legacy of World War II"; Rau, *Combat Scientists;* Rau, "The Adoption of Operations Research in the United States during World War II," in *Systems, Experts, and Computers: The Systems Approach in Management and Engineering; World War II and After,* ed. Agatha C. Hughes and Thomas P. Hughes (Cambridge: MIT Press, 2000), 57–92; Andy Pickering, "Cyborg History and the World War II Regime," *Perspectives on Science* 3:1 (1995), 1–48.

The emergence of OR is associated with the development of new forms of radar and their integration into antiaircraft guns, ships, and airplanes. For a general history of radar, see Robert Buderi, *The Invention that Changed the World: How a Small Group of Radar Pioneers Won the Second World War and Launched a Technological Revolution* (New York: Simon and Schuster, 1996). The redefinition of the human/machine interface that came about through integrating men and radar into a complete weapons system spawned cybernetics and eventually the cyborg sciences. For the technical basis for this development in World War II, see the fine accounts by Peter Galison, "The Ontology of the Enemy: Norbert Wiener and the Cybernetic Vision," *Critical Inquiry* 21:1 (1994), 228–266; David A. Mindell, "Automation's Finest Hour: Radar and System Integration in World War II," in Hughes and Hughes, *Systems, Experts and Computers,* 27–56.

8 For the British effort as described here see Fortun and Schweber, "Scientists and the Legacy of World War II," and Rau, *Combat Scientists,* chap. 1.

9 This entire paragraph is based on Kevles, *The Physicists,* 287–308. See also Buderi, *The Invention that Changed the World.*

10 Kevles, *The Physicists,* 300.

11 Bush's position is explained in depth by Rau, "The Adoption of Operations Research," and in Rau, *Combat Scientists.*

12 Kevles, *The Physicists,* 311.

13 Dönitz is quoted in Buderi, *The Invention that Changed the World,* 166; King, in Kevles, *The Physicists,* 313.

14 Kevles, *The Physicists,* 320.

15 For an autobiography, see Philip M. Morse, *In at the Beginnings: A Physicist's Life* (Cambridge: MIT Press, 1977); for a brief biography, see Rau, "Morse, Philip McCord."

16 Morse, *In at the Beginnings,* 170. See also Philip M. Morse, "Must We Always Be Gadgeteers?" *Physics Today* 3 (December 1950), 4–5.

17 For a study of ASWORG, see Rau, *Combat Scientists,* chap. 4.

18 "We wanted technical data to be collected by technical men," Morse wrote, "civilians who knew what to look for, and who were not tempted to distort information to please superior officers." Morse, *In at the Beginnings,* 180.

19 "Operations Research," pamphlet issued by the National Research Council (1950), cited by Fortun and Schweber, "Scientists and the Legacy of World War II," 605.

20 Andrew Pickering, *The Mangle of Practice: Time, Agency, and Science* (Chicago: University of Chicago Press, 1995), 238. See also Pickering, "Cyborg History."

21 Philip M. Morse, "The Growth of Operations Research in the Free World," lecture given at the Operations Evaluation Group 20th Anniversary Conference on Operations Research, 14–16 May 1962, Washington D.C., MIT, MC 75, box 20, folder OEG. All quotations in this paragraph are from this source. Morse remarked that the doctor-patient model had become obsolete in the United States by 1962, when he gave this lecture, and that the relationship had become more equilibrated, more like that between legal counsel and client. The doctor-patient model was, however, the one he took over to Europe with him, believing that the military there were not yet accustomed to having civilian scientists tell them what to do.

22 Erik Rau has stressed this matter of trust in his *Combat Scientists,* 200–205.

23 MIT was joined by other prestigious universities, including the California Institute of Technology and the Case Institute of Technology. Rau, *Combat Scientists,* 339.

24 The task of Rand was to "engage in scientific evaluation and analytical studies to aid the National Military Establishment in the formulation of military plans intended to provide maximum security for the nation at minimum cost in manpower and resources." Cf. Martin J. Collins, *Cold War Laboratory: RAND, the Air Force, and the American State, 1945–1950* (Washington, D.C.: Smithsonian Institution Press, 2002), 161.

25 Morse included a photograph of himself playing a war game at Rand in his autobiography. For Rand's activities in the 1950s, see David Hounshell, "The Cold War, RAND and the Generation of Knowledge," *Historical Studies in the Physical and Biological Sciences* 27:2 (1997), 237–267; Sharon Ghamari-Tabrizi, "Simulating the Unthinkable: Gaming Future War in the 1950s and 1960s," *Social Studies of Science* 30:2 (2000), 163–223.

26 Philip Morse and George Kimball, *Methods of Operations Research* (Cambridge: Technology Press of MIT, 1951).

27 Philip M. Morse, "Background of the NATO O.R. Panel," presented at the APOR Tenth Anniversary Symposium, held 18–20 March 1970 at NATO Headquarters, Brussels, Belgium, MIT, MC 75, box 20, folder NATO APOR 10th Anniversary.

28 Fortun and Schweber, "Scientists and the Legacy of World War II," 610–611.

29 Philip Mirowski writes that "the practitioners [of OR] themselves have suffered recurrent identity crises verging on neuroses, struggling again and again to define themselves to themselves," in "Cyborg Agonistes: Economics Meets Operations Research in Mid-Century," *Social Studies of Science* 29:5 (1999), 611. See also Mirowski, *Machine Dreams: Economics Becomes a Cyborg Science* (Cambridge: Cambridge University Press, 2002).

30 Michel Armatte, "Les sciences économiques reconfigurées par la *pax americana*," in *Les sciences pour la guerre, 1940–1960,* ed. Amy Dahan and Dominique Pestre (Paris: Éditions de l'École des Hautes Études en Sciences Sociales, 2004), 130–173. See also Mark Solovey, "Project Camelot and the 1960s Epistemological Revolution: Rethinking the Politics-Patronage–Social Science Nexus," *Social Studies of Science* 31:2 (2001), 173–179.

31 Robert Solow, *Fortune 53* (February 1956), 148–152, quoted by Fortun and Schweber, "Scientists and the Legacy of World War II," 611.

32 Mirowski, "Cyborg Agonistes," 690, 692. See also Mirowski, *Machine Dreams;* Amy Dahan, "Axiomatiser, modéliser, calculer: Les mathématiques, instrument universel et polymorphe d'action," in Dahan and Pestre, *Les sciences pour la guerre,* 50–71.

33 Morse, *In at the Beginnings,* 263.

34 Quoted in Stuart W. Leslie, "Profit and Loss: The Military and MIT in the Postwar Era," *Historical Studies in the Physical and Biological Sciences* 21:1 (1990), 61. For the growth of MIT's interdepartmental laboratories under predominantly military patronage, see Stuart W. Leslie, *The Cold War and Ameri-*

can Science: The Military-Industrial-Academic Complex at MIT and Stanford (New York: Columbia University Press, 1993).

35 "President Emeritus Julius Stratton dies at 93," http//:web.mit.edu/news office/1994/Stratton-0629.html.

36 Morse, *In at the Beginnings*, 263. More generally, on Killian's support for the social sciences, see Giuliana Gemelli, "Western Alliance and Scientific Diplomacy in the Early 1960s: The Rise and Failure of the Project to Create a European M.I.T.," in *The American Century in Europe*, ed. R. Laurence Moore and Maurizio Vaudagna (Ithaca: Cornell University Press, 2003), 171–192.

37 Quoted in Leslie, "Profit and Loss," 59. Rebecca S. Lowen has stressed the crucial role played by the university administration in encouraging federal, notably military, patronage in the early postwar years, often against the wishes of the faculty. See her "'Exploiting a Wonderful Opportunity': The Patronage of Scientific Research at Stanford University, 1937–1965," *Minerva* 30:3 (1992), 391–421. See also Lowen, *Creating the Cold War University: The Transformation of Stanford* (Berkeley: University of California Press, 1997).

38 The membership is given in Philip M. Morse, "Report on the First Summer Program on Operations Research at the Massachusetts Institute of Technology, June 16–July 2, 1953," *Journal of the Operations Research Society of America* 1:5 (1953), 303–305. The description that follows is derived from this paper.

39 "Training in Operations Research at the Massachusetts Institute of Technology," 3 pp., MIT, MC 75, box 20, folder NATO/APOR Reference File. The document is undated and unsigned but probably no later than 1960.

40 All the quotes here are from his May 1962 lecture, "The Growth of Operations Research in the Free World."

41 Morse had appointed Ramsey the head of physics soon after he was nominated as the first director of the Brookhaven National Laboratory in 1946. See Robert P. Crease, *Making Physics: A Biography of Brookhaven National Laboratory, 1946–1972* (Chicago: University of Chicago Press, 1999), chap. 2. Morse served with Seitz on the governing board of the American Institute of Physics in the mid-1950s.

42 "Note by the United States Delegation Concerning Introductory Courses in Operational Research," NATO, AC/137-D/22, 21 August 1958, lists "scientists and military persons" from Germany, France, Italy, Belgium, Netherlands, Turkey, and Norway as being paid for by the United States. Documentary evidence gives the figures of both thirty and forty Europeans at this summer school, depending on the source. This way of making sense of the numbers seems to be the most plausible one.

43 "Policy for Mutual Support in Operational Research between the United States and Its Allies."

44 Morse, *In at the Beginnings*, 319.

45 Ibid., 320, quoting General Larkin, head of the MWDP in 1959.

46 "Final Report of Project AF19(604)5442: Consultation in Aid of Operations Research in NATO Countries," appendix A, MIT, MC 75, box 18, folder NATO Report on Est. Adv. Panel on O/R, 1959. The next two paragraphs are based on this proposal.

47 "Informal Notes of the First Meeting of APOR," MIT, MC 75, box 20, folder NATO APOR 10th Anniversary. Larkin and the MWDP as well as the indefatigable Jean Willems were involved in the arrangements for the meeting in Brussels.

48 "Final Report of Project AF19(604)5442," appendix D, "Comments of Attendees"; appendix E, "Excerpts from Comments on Brussels Course by Team Members."

49 "Final Report of Project AF19(604)5442: Consultation in Aid of Operation Research in NATO Countries," "Conclusions."

50 In addition to Morse, Kimball, and Koopman, the team comprised George Wadsworth, Ronald Howard, a new young assistant professor in electrical engineering, and Herbert Galliher, Morse's deputy director and administrative assistant in the MIT Operations Research Center.

51 "Final Report of Project AF19(604)5442," appendix F, "Excerpts from Team Reports on Visits to NATO Countries," F-14.

52 For France, see "Final Report of Project AF19(604)5442," appendix F, "Excerpts from Team Reports on Visits to NATO Countries," F-1–F-5.

53 For Germany, see ibid., F-5–F-6. The last quotation is from Morse's autobiography, In at the Beginnings, 325.

54 For Norway, see "Final Report of Project AF19(604)5442," appendix F, "Excerpts from Team Reports on Visits to NATO Countries," F-10–F-14.

55 Morse, In at the Beginnings, 325.

56 Mirowski, Machine Dreams, 182.

57 "Report of the NATO Consultant in Operations Research in the Federal Republic of Germany, 20 September 1961 to 1 July 1963," 1 July 1963, for the Baumgarten report, MIT, MC 75, box 18, folder NATO/APOR 1963.

58 Joseph Weizenbaum, a pioneer in artificial intelligence, cited in Giuliana Gemelli, "Western Alliance and Scientific Diplomacy in the Early 1960s," 175.

59 Olav Wicken, "Cold War in Space Research: Ionospheric Research and Military Communication in Norwegian Politics," in Making Sense of Space: The History of Norwegian Space Activities, ed. John Peter Collett (Oslo: Scandinavian University Press, 1995), 41–73.

60 Jan Rune Holmevik, Educating the Machine: A Study in the History of Computing and the Construction of the SIMULA Programming Languages, STS Report 22/94 (Trondheim, 1994).

61 Bruno Belhoste, Amy Dahan Dalmedico, and Antoine Picon, eds., La formation polytechnicienne, 1794–1994 (Paris: Dunod, 1994).

62 Malavard to Allis, 7 November 1963, MIT, MC 75, box 18, folder NATO/ APOR 1963–4.

63 "Final Report of Project AF19(604)5442," appendix E, "Excerpts from Comments on Brussels Course by Team Members," E-1–E-2. Koopman wrote that "practical experience" could no longer serve as the "sole guide" when weapons and tactics were changing quickly. "War of the future is a battle of brains," he insisted, and to conduct it successfully civilian scientists had an important role to play in military groups. However, it was the "patient," not the "doctor," who took final responsibility for undergoing the recommended treatment, and he had to "answer all the doctor's questions," however reluctant he may be to do so. The discussion of the ideal set of social relationships that should prevail between the operations researcher and the executive decision-maker that he worked with, were spelt out at length in the first chapter of Morse and Kimball's *Methods of Operations Research,* though they did not use the patronizing doctor-patient metaphor there. This was, though, a very important matter for them.

64 "Department of Defense Statement with Respect to Operational Research in the Atlantic Alliance," 17 October 1971, appendix C of Morse's "Review of Operations Research Activities in the Atlantic Alliance," January 1963, MIT, MC 75, box 19, folder NATO Corres. Allis/Aparo.

65 Seitz to Morse, 10 September 1959; Morse to Seitz, 1 October 1959, MIT, MC 75, box 20, folder NATO APOR Panel 1960, First Meeting. See also Morse, *In at the Beginnings,* 324.

66 Koopman to Seitz, 27 November 1959, "Recommendations" to "Final Report of Project AF19(604)5442: Consultation in Aid of Operations Research in NATO Countries," MIT, MC 75, box 20, folder NATO APOR Panel 1960, First Meeting; "Review of Operations Research in NATO," January 1960, appendix B of Morse, "Review of Operations Research Activities in the Atlantic Alliance."

67 Informal minutes of the meeting "On Monday, 28 March . . . ," MIT, MC 75, box 20, folder NATO APOR 10th Anniversary.

68 On Goodeve, see Maurice W. Kirby and R. Capey, "The Origins and Diffusion of Operational Research in the UK," *Journal of the Operational Research Society* 49 (1998), 307–326; Maurice W. Kirby, "Operations Research Trajectories: The Anglo-American Experience from the 1940s to the 1990s," *Operations Research* 48:5 (2000), 661–670.

69 Morse, *In at the Beginnings,* 326.

70 Philip M. Morse, "Background of the NATO O.R. Panel," presented at the APOR Tenth Anniversary Symposium, MIT, MC 75, box 20, folder NATO APOR 10th Anniversary. In this talk, Morse claimed that his original proposals "were reduced to NATO-ese and then were modified by the Science Committee of NATO," but went on to say that most had been carried out. This was only true in the most general sense of the term.

71 "Recommendations of the Advisory Panel on Operational Research of the Office of the Science Adviser of NATO," 1 April 1960, Document NATO SCOM.6-2-01, MIT, MC 75, box 20, folder NATO APOR Panel 1960, First Meeting.

72 Ibid. (emphasis added).

73 Morse, *In at the Beginnings,* 328.

74 "Summary Record of Meetings held . . . on 6th and 7th April, 1960," NATO, Document AC/137-R/7, 12–13, for the quotations in this paragraph.

75 Seitz to Salzmann, 17 February 1960, MIT, MC 75, box 20, folder NATO APOR Panel 1960, First Meeting (emphasis original).

76 "Summary Record of Meetings held . . . on 6th and 7th April, 1960," 13.

77 "NATO Division of Scientific Affairs: Activities in Operational Research," 4 April 1961, appendix A of Morse's "Review of Operations Research Activities in the Atlantic Alliance," suggests this meaning of "indoctrination."

78 "Memorandum for the Record. Subject: IDA Contract SD-72," 13 July1962, signed D. H. Gould, MIT, MC 75, box 19, folder NATO/APOR, IDA.

79 "The Activities of the Advisory Panel on Operational Research," document NATO SA/6-2-04/APOR, 19 August 1963, MIT, MC 75, box 18, folder NATO/ APOR 1963.

80 Memos to Morse and Rothenberg of 4 April 1962 (MIT, MC 75 box 20, folder NATO/APOR Panel, 1963) give revised versions of Baumgarten's, Sasieni's, and Stoller's reports to the APOR meeting held in March 1962. For a far more substantial report by Baumgarten, see "Report of the NATO Consultant . . . in the Federal Republic of Germany," 1 July 1963.

81 "Operational Research Training Programme: Note by the United Kingdom," NATO, AC/137-D/125, 2 February 1962. The next few paragraphs are based on this document.

82 A. L. Oliver, "The Activities of the Advisory Panel on Operational Research," document NATO SA/6-2-04/APOR, 19 August 1963, 5, MIT, MC 75, box 18, folder NATO/APOR, 1963. In 1962 the NATO Science Committee's special budget for OR was $70,000, of which $25,000 was for summer schools and $45,000 for graduate apprenticeships. This was augmented to $115,000 in 1963, distributed $25,000/$90,000 between the two activities. Not all of the money was spent.

83 "Operational Research Training Programme: Note by the United Kingdom."

84 The British did not (in fact could not) stop individual member countries using their NATO fellowships to send someone to do postgraduate work in OR. They were just not going to use that part of the NATO Science Committee's budget dedicated to OR for this purpose and, of course, were unlikely to support any British student's application to London for a NATO fellowship if it was to be used for academic training abroad.

85 "Record of Meeting . . ." of APOR on 19 April 1963, NATO, Document AC/137-D172, 9 May 1963.

86 "The Activities Sponsored by the Advisory Panel on Operational Research," NATO, Document AC/137-D192, 8 October 1963; Morse to Oliver, 3 January 1964, Oliver to Morse, 15 January 1964 (MIT, MC 75, box 19, folder NATO/APOR Corres. Allis & Oliver, '63–64), for an exchange concerning the amount of emphasis to put on the military side.

87 A. L. Oliver, "The Activities of the Advisory Panel on Operational Research," section 3, for all quotations in this paragraph.

88 "Proposal for Continuation of Support of Operations Research Liaison Activities of Philip M. Morse," October 1964, MIT, MC 75, box 19, folder NATO/APOR IDA Proposal; Ronald M. Murray, Assistant Director, International Programs, Office of the Director of Defense Research and Engineering, to Morse, 16 December 1964 (?), MIT, MC 75, box 19, folder NATO/APOR Corres. McLucas/Oliver. The year this letter was written is uncertain, but it is probably his reply to the October 1964 proposal from Morse to the IDA.

89 Report of Meeting of Group of Experts held on 14 and 15 January, 1963, OECD document DAS/BS/63.5, 4 April 1963, MIT, MC 75, box 20, folder OECD/APOR 1962-3.

90 The next two paragraphs draw on Kirby and Capey, "The Origins and Diffusion of Operational Research in the UK"; Kirby, "Operations Research Trajectories."

91 Kirby, "Operations Research Trajectories," 665.

92 For detailed data see table VIII.4 in *NATO and Science: An Account of the Activities of the NATO Science Committee, 1958–1972* (Brussels: NATO, 1973).

93 A. L. Oliver, "On Operational Research," speech to the NATO Parliamentarians, attached to Oliver to Morse, 28 October 1963, MIT, box 18, folder NATO/APOR 1963–4.

94 "Report by the Exploratory Group" [chaired by Pierre Aigrain], NATO, Document AC/137(DR)D/4(Final), AC/74(DR)D/4(Final), 9 December 1963.

9   Concluding Reflections

1 Roy MacLeod, ed., *Nature and Empire: Science and the Colonial Enterprise, Osiris,* 2nd ser., 15 (2000) has an excellent bibliography of pertinent material.

2 John Lewis Gaddis, *We Now Know: Rethinking Cold War History* (Oxford: Oxford University Press, 1997), 26. There is an ongoing and spirited debate over Gaddis's interpretation of the origins of the Cold War and the place of the Marshall Plan within it. See, for example, Michael Cox and Caroline Kennedy-Pipe, "The Tragedy of American Diplomacy? Rethinking the Marshall Plan," *Journal of Cold War Studies* 7:1 (2005), 97–134.

3 Geir Lundestad, *"Empire" by Integration: The United States and European Integration, 1945–1997* (Oxford: Oxford University Press, 1998), 55, for Eisenhower, 158, for McGeorge Bundy.

4 Lundestad, *"Empire" by Integration,* 22.

5 Ibid., 17.

6 Lily E. Kay, *The Molecular Vision of Life: Caltech, the Rockefeller Foundation, and the Rise of the New Biology* (New York: Oxford University Press, 1993), 217.

7 Daniel J. Kevles, "Principles and Politics in Federal R&D Policy, 1945–1990: An Appreciation of the Bush Report," introduction to *Science—The Endless Frontier,* by Vannevar Bush, 40th anniversary ed., Report NSF 90-8 (Washington, D.C.: National Science Foundation, 1990), ix–xxxiii, xv–xvi.

8 The most notable examples are Alan S. Milward, *The Reconstruction of Western Europe, 1945–1951* (Berkeley: University of California Press, 1984); Milward, *European Rescue of the Nation State,* 2nd ed. (London: Routledge, 2000).

9 When I recently gave a paper at the Niels Bohr Institute on Bohr's relationships with the Ford Foundation, his visit to the CIA, and his willingness to do their bidding (through Stone), his link with the CIA was front-page news in a Copenhagen daily the next day. The journalists were apparently stunned: no one in the audience, including some of Bohr's family, seemed surprised or offended. Perhaps they were just being polite.

10 On de Gaulle, see Robert O. Paxton and Nicholas Wahl, eds., *De Gaulle and the United States: A Centennial Reappraisal* (Oxford: Berg, 1994).

11 Richard Kuisel, *Seducing the French: The Dilemma of Americanization* (Berkeley: University of California Press, 1993), is the classic source for the evolution of U.S.-French cultural relationships, especially after World War II.

12 Charles Maier, "The Marshall Plan and the Division of Europe," *Journal of Cold War Studies* 7:1 (2005), 173.

13 "Science and NATO," LoC, Rabi Papers, box 39, folder 7. The document is undated and unsigned but obviously by Rabi, from around 1964.

14 On the three, see John Krige, "Philanthropy and the National Security State: The Ford Foundation's Support for Physics in Europe in the 1950s," in *American Foundations and Large-Scale Research: Construction and Transfer of Knowledge,* ed. Giuliana Gemelli (Bologna: Clueb 2001), 3–24.

15 Philip Morse, *In at the Beginnings: A Physicist's Life* (Cambridge: MIT Press, 1977), 322, 344.

16 Ronald W. Pruessen, "Cold War Threats and America's Commitment to the European Defense Community: One Corner of a Triangle," *Journal of European Integration History* (Spring 1966), 51–69.

17 Frank Costigliola, "Kennedy, the European Allies and the Failure to Consult," *Political Science Quarterly* 110:1 (1995), 105–123. See also Costigliola, "Kennedy, De Gaulle and the Challenge of Consultation," in Paxton and Wahl, *De Gaulle and the United States,* 169–194.

18 Costigliola, "Kennedy, De Gaulle," 177.

19 John L. Heilbron and Robert W. Seidel, *Lawrence and His Laboratory: A History of the Lawrence Berkeley Laboratory,* vol. 1 (Berkeley: University of California Press, 1989), 238–252, 350–352, and 239, 240, 350, respectively, for the quotations.

20 Jean-Paul Gaudillière, "Paris-New York Roundtrip: Transatlantic Crossings and the Reconstruction of the Biological Sciences in Post-War France," *Studies in History and Philosophy of Biology and Biomedical Science* 33C (2002), 390.

21 Ibid.

22 Jean-Paul Gaudillière, *Inventer la biomédecine: La France, l'Amérique et la production des savoirs du vivant (1945–1965)* (Paris: Éditions la Découverte, 2002), 31.

23 Rob Kroes, "American Empire and Cultural Imperialism: A View from the Receiving End," *Diplomatic History* 23:3 (1999), 461–477.

24 Reinhold Wagnleitner, *Coca-Colonization and the Cold War: The Cultural Mission of the United States in Austria after the Second World War* (Chapel Hill: University of North Carolina Press, 1994), 29–30.

25 Jonathan Zeitlin and Gary Herrigel, eds., *Americanization and Its Limits: Reworking US Technology and Management in Post-War Europe and Japan* (Oxford: Oxford University Press, 2000), 1–50. It is significant that the literature on science and (post)colonialism uses a similar vocabulary to theorize "contact zones" of empire.

26 "Science and NATO."

27 This paragraph owes much to Gaudillière, *Inventer la biomédecine,* 369–381, and to Victoria de Grazia, "Changing Consumption Regimes in Europe, 1930–1970: Comparative Perspectives on the Distribution Problem," in *Getting and Spending: European and American Consumer Societies in the Twentieth Century,* ed. Susan Strasser, Charles McGovern, and Matthias Judt (Cambridge: Cambridge University Press, 1998), 59–83. On the United States as source of credibility, see John Krige, "Distrust and Discovery: The Case of the Heavy Bosons at CERN," *Isis* 92:3 (2001), 517–540.

28 Robert E. Kohler, *Landscapes and Labscapes: Exploring the Lab-Field Border in Biology* (Chicago: University of Chicago Press, 2002), chap. 1. See also Karen Knorr Cetina, "Laboratory Studies: The Cultural Approach to the Study of Science," in *Handbook of Science and Technology Studies,* ed. Sheila Jasanoff, Gerald E. Markle, James C. Patterson, and Trevor Pinch (London: SAGE, 1995), 140–166; Joseph O'Connell, "Metrology: The Creation of Universality by the Circulation of Particulars," *Social Studies of Science* 23:1 (1993), 129–173; M. Norton Wise, ed., *The Values of Precision* (Princeton: Princeton University Press, 1995).

29 Lew Kowarski's interview with Rabi on 6 November 1973, CERN.

Bibliography

Aaserud, Finn. *Redirecting Science: Niels Bohr, Philanthropy and the Rise of Nuclear Physics.* Cambridge: Cambridge University Press, 1990.

Abir-Am, Pnina G. "The Discourse of Physical Power and Biological Knowledge in the 1930s: A Reappraisal of the Rockefeller Foundation's 'Policy' in Molecular Biology." *Social Studies of Science* 12:3 (1982), 341–382.

———. "'New' Trends in the History of Molecular Biology." *Historical Studies in the Physical and Biological Sciences* 26:1 (1995), 167–196.

———. "The Rockefeller Foundation and Refugee Biologists: European and American Careers of Leading RF Grantees from England, France, Germany, and Italy." In Gemelli, *The "Unacceptables,"* 217–240.

———. "The Rockefeller Foundation and the Rise of Molecular Biology." *Nature Reviews Molecular Cell Biology* 3:1 (2002), 65–70.

Althusser, Louis. "Unfinished History." Introduction to Dominique Lecourt, *Proletarian Science? The Case of Lysenko.* London: New Left Books, 1977.

Amaldi, Edoardo. "Notes from Abroad." *Physics Today* 1:1 (1948), 35–37.

———. "Personal Notes on Neutron Work in Rome in the 30s and Post-war European Collaboration in High-Energy Physics." In *Proceedings of the International School of Physics "Enrico Fermi": Course LVII; History of Twentieth-Century Physics,* edited by Charles Wiener, 294–351. New York: Academic Press, 1977.

Appel, Toby. *Shaping Biology: The National Science Foundation and American Biological Research.* Baltimore: Johns Hopkins University Press, 2000.

Armatte, Michel. "Les sciences économiques reconfigurées par la *pax americana.*" In Dahan and Pestre, *Les sciences pour la guerre,* 130–173.

Arndt, Richard T., and David Lee Rubin, eds. *The Fulbright Difference, 1948–1992.* New Brunswick, N.J.: Transaction Publishers, 1992.

Arnove, Robert F. *Philanthropy and Cultural Imperialism: The Foundations at Home and Abroad.* Boston: G. K. Hall, 1980.

Ash, Mitchell G. "Denazifying Scientists and Science." In Judt and Ciesla, *Technology Transfer out of Germany,* 61–79.

————. Review of *Surviving the Swastika: Scientific Research in Nazi Germany*, by Kristie Macrakis. *Isis* 85:4 (1994), 727–729.

Badash, Lawrence. "Science and McCarthyism." *Minerva* 38:1 (2000), 53–80.

Bähr, Johannes, Paul Erker, and Geoffrey Giles. "The Politics of Ambiguity: Reparations, Business Relations, Denazification and the Allied Transfer of Technology." In Judt and Ciesla, *Technology Transfer out of Germany*, 131–144.

Battimelli, Gianni, and Ivana Gambaro. "Da via Panisperma a Frascati: Gli acceleratori mai realizzati." *Quaderno di Storia della Fisica* 1 (1997), 319–333.

Belhoste, Bruno, Amy Dahan Dalmedico, and Antoine Picon, eds. *La formation polytechnicienne, 1794–1994*. Paris: Dunod, 1994.

Berghahn, Volker R. *America and the Intellectual Cold Wars in Europe: Shepard Stone between Philanthropy, Academy, and Diplomacy*. Princeton: Princeton University Press, 2001.

————. "European Elitism, American Money, and Popular Culture." In Moore and Vaudagna, *The American Century in Europe*, 117–130.

————. "Philanthropy and Diplomacy in the 'American Century.'" *Diplomatic History* 23:3 (1999), 393–419.

————. "Shepard Stone and the Ford Foundation." In Gemelli, *Ford Foundation and Europe*, 69–95.

Bernstein, Jeremy. *Hitler's Uranium Club: The Secret Recording at Farm Hall*. New York: AIP Press, 1996.

Beyerchen, Alan. "German Scientists and Research Institutions in Allied Occupation Policy." *History of Education Quarterly* 22:3 (1982), 289–299.

————. *Scientists under Hitler: Politics and the Physics Community in the Third Reich*. New Haven: Yale University Press, 1977.

Beyler, Richard H., and Morris F. Low. "Science Policy in Post-1945 West Germany and Japan." In *Science and Ideology: A Comparative Perspective*, edited by Mark Walker, 197–203. London: Routledge, 2003.

Bird, Kai. *The Chairman: John J. McCloy; The Making of the American Establishment*. New York: Simon and Schuster, 1992.

Bohr, Niels. *Open Letter to the United Nations, June 9th, 1950*. Copenhagen: J. H. Schultz, 1950.

Bordry, Monique, and Pierre Radvanyi, eds. *Oeuvre et engagement de Frédéric Joliot-Curie*. Les Ulis: EDP Sciences, 2001.

Brinkley, Alan. "The Concept of an American Century." In Moore and Vaudagna, *The American Century in Europe*, 7–21.

Brode, Wallace R. "National and International Science." *Department of State Bulletin* 42, 9 May 1960, 735–739.

Bromberg, Joan Lisa. *Fusion: Science, Politics and the Invention of a New Energy Source*. Cambridge: MIT Press, 1982.

Buderi, Robert. *The Invention that Changed the World: How a Small Group of Radar Pioneers Won the Second World War and Launched a Technological Revolution.* New York: Simon and Schuster, 1996.

Burian, Richard M., and Jean Gayon. "The French School of Genetics: From Physiological and Population Genetics to Regulatory Molecular Genetics." *Annual Review of Genetics* 33 (1999), 313–349.

———. "Genetics after World War II: The Laboratories at Gif (La génétique et les laboratoires de Gif)." *Cahiers pour l'Histoire du CNRS* 7 (1990), 25–48. Available online as "The CNRS Laboratories at Gif-sur-Yvette." http://picardp1.ivry.cnrs.fr/~jfpicard/Burian-Gayon.html.

Burian, Richard M., Jean Gayon, and Doris Zallen. "The Singular Fate of Genetics in the History of French Biology, 1900–1940." *Journal of the History of Biology* 21:3 (1998), 357–402.

Burrin, Philippe. *France under the Germans: Collaboration and Compromise.* New York: New Press, 1996.

Bush, Vannevar. *Science—The Endless Frontier: A Report to the President on a Program for Postwar Scientific Research.* Washington, D.C.: National Science Foundation, 1960. (First published July 1945.)

Carson, Cathryn. "New Models for Science in Politics: Heisenberg in West Germany." *Historical Studies in the Physical and Biological Sciences* 30 (1999), 115–171.

———. "Nuclear Energy Development in Postwar West Germany: Struggles over Cooperation in the Federal Republic's First Reactor Station." *History and Technology* 18:3 (2002), 233–270.

Carson, Cathryn, and Michael Gubser. "Science Advising and Science Policy in Post-war Germany: The Example of the Deutscher Forschungsrat." *Minerva* 40:2 (2002), 147–179.

Cassidy, David. "Controlling German Science, I: U.S. and Allied Forces in Germany, 1945–1947." *Historical Studies in the Physical and Biological Sciences* 24:1 (1994), 197–235.

———. "Controlling German Science, II: Bizonal Occupation and the Struggle over West German Science Policy." *Historical Studies in the Physical and Biological Sciences* 26:2 (1996), 197–239.

Caute, David. *Communism and the French Intellectuals, 1914–1969.* New York: Macmillan, 1964.

Chomsky, Noam, Ira Katznelson, Richard.C. Lewontin, David Montgomery, Laura Nader, Richard Ohmann, Ray Siever, Immanuel Wallerstein and Howard Zinn, *The Cold War and the University: Toward an Intellectual History of the Postwar Years.* New York: New Press, 1997.

Collins, Martin J. *Cold War Laboratory: RAND, the Air Force, and the American State, 1945–1950.* Washington, D.C.: Smithsonian Institution Press, 2002.

Costigliola, Frank. "Kennedy, De Gaulle and the Challenge of Consultation." In Paxton and Wahl, *De Gaulle and the United States,* 169–194.

———. "Kennedy, the European Allies and the Failure to Consult." *Political Science Quarterly* 110:1 (1995), 105–123.

Cox, Michael, and Caroline Kennedy-Pipe. "The Tragedy of American Diplomacy? Rethinking the Marshall Plan." *Journal of Cold War Studies* 7:1 (2005), 97–134.

Creager, Angela N. H. "The Industrialization of Radioisotopes by the U.S. Atomic Energy Commission." In *The Science-Industry Nexus: History, Policy, Implications,* edited by Karl Grandin, Nina Wormbs, and Sven Widmalm, 141–167. Sagamore Beach, Mass.: Science History Publications, 2004.

———. "Tracing the Politics of Changing Postwar Research Practices: The Export of 'American' Radioisotopes to European Biologists." *Studies in History and Philosophy of Biological and Biomedical Sciences* 33C (2002), 367–388.

Crease, Robert P. *Making Physics: A Biography of Brookhaven National Laboratory, 1946–1972.* Chicago: University of Chicago Press, 1999.

Dahan, Amy. "Axiomatiser, modéliser, calculer: Les máthematiques, instrument universel et polymorphe d'action." In Dahan and Pestre, *Les sciences pour la guerre,* 49–81.

Dahan, Amy, and Dominique Pestre, eds. *Les sciences pour la guerre, 1940–1960.* Paris: Éditions de l'École des Hautes Études en Sciences Sociales, 2004.

Daston, Lorraine. "Objectivity and the Escape from Perspective." *Social Studies of Science* 22:4 (1992), 597–618.

David, Jean. "Histoire de la génétique à Gif." www.dr4.cnrs.fr/gif–2000/p.5-99-02b.html. Accessed 22 May 2003.

Day, Michael. A. "In Appreciation: I. I. Rabi, The Two Cultures and the Universal Culture of Science." *Physics in Perspective* 6:4 (2004), 428–476.

Defrance, Corine. "La mission du CNRS en Allemagne (1945–1950): Entre exploitation et contrôle du potentiel scientifique allemand." *Revue pour l'Histoire du CNRS* 5 (2001), 54–65.

De Grazia, Victoria. "Changing Consumption Regimes in Europe, 1930–1970: Comparative Perspectives on the Distribution Problem." In Strasser, McGovern, and Judt, *Getting and Spending,* 59–83.

———. *Irresistible Empire: America's Advance through 20th-Century Europe.* Cambridge: Harvard University Press, 2005.

———. "Mass Culture and Sovereignty: The American Challenge to European Cinemas, 1920–1960." *Journal of Modern History* 61:1 (1989), 53–87.

Deichmann, Ute. "Emigration, Isolation, and the Slow Start of Molecular Biology in Germany." *Studies in History and Philosophy of Biological and Biomedical Sciences* 33C (2002), 449–471.

————. "The Expulsion of German-Jewish Chemists and Biochemists and Their Correspondence with Colleagues in Germany after 1945: The Impossibility of Normalization." In *Science in the Third Reich*, edited by Margit Szöllösi-Janze, 243–283. Oxford: Berg, 2001.

De Maria, Michelangelo. "Fermi, un fisico da via Panisperma all'America." *Le Scienze* 2:8 (1999), 1–111.

Dennis, Michael A. " 'Our First Line of Defense': Two University Laboratories in the Postwar American State." *Isis* 85 (1994), 427–455.

————. "Reconstructing Sociotechnical Order: Vannevar Bush and US Science Policy" In Jasanoff, *States of Knowledge*, 225–253.

Doel, Ronald E. "Evaluating Soviet Lunar Science in Cold War America." *Osiris*, 2nd ser., 7 (1992), 44–70.

————. "Scientists as Policymakers, Advisors and Intelligence Agents: Linking Contemporary Diplomatic History with the History of Contemporary Science." In *The Historiography of Contemporary Science and Technology*, edited by Thomas Söderqvist, 215–244. Amsterdam: Harwood Academic Publishers, 1997.

Dosso, Diane. "Louis Rapkine et la mobilisation scientifique de la France libre." Thèse pour l'obtention du Diplôme de Docteur de l'Université Paris 7, Spécialité: Épistémologie et histoire des sciences et des institutions scientifiques. Paris, 1998.

————. "The Rescue of French Scientists: Respective Roles of the Rockefeller Foundation and the Biochemist Louis Rapkine (1904–1948)." In Gemelli, *The "Unacceptables,"* 195–215.

Dulles, John Foster. "Our Cause Will Prevail." *Department of State Bulletin*, 6 January 1958, 19–22. (Also published in *Life*, 23 December 1957.)

Eckert, Michael. "Primacy Doomed to Failure: Heisenberg's Role as Scientific Adviser for Nuclear Policy in the FRG." *Historical Studies in the Physical and Biological Sciences* 21:1 (1990), 29–58.

Edwards, Paul N. *The Closed World: Computers and the Politics of Discourse in Cold War America*. Cambridge: MIT Press, 1996.

Edgerton, David. "British R&D after 1945: A Reinterpretation." In *Science and Technology Policy: An International Perspective*, edited by Carlye Honig, 126–132. London: British Library, 1995.

————. "Research, Development and Competitiveness." In *The Future of UK Industrial Competitiveness and the Role of Industrial Policy*, edited by Kirsty Hughes, 40–54. London: Policy Studies Institute, 1994.

Ellwood, David W. "Italian Modernisation and the Propaganda of the Marshall Plan." In *The Art of Persuasion: Political Communication in Italy from 1945 to the 1990s*, edited by Luciano Cheles and Lucio Sponza, 23–48. Manchester: Manchester University Press, 2001.

———. "The Propaganda of the Marshall Plan in Italy in a Cold War Context." In Scott-Smith and Krabbendam, *The Cultural Cold War in Western Europe*, 225–236.

———. *Rebuilding Europe: Western Europe, America and Postwar Reconstruction.* London: Longman's, 1992.

Epstein, Jason. "The CIA and the Intellectuals." *New York Review of Books,* 20 April 1967, 16–21.

Ezrahi, Yaron. "Science and the Problem of Authority in Democracy." In *Science and Social Structure: A Festschrift for Robert K. Merton,* edited by Thomas F. Gieryn, 43–60. Transactions of the New York Academy of Sciences, ser. 2, vol. 39. New York: New York Academy of Sciences, 1980.

Fisher, Donald. "The Role of Philanthropic Foundations in the Reproduction and Production of Hegemony: Rockefeller Foundations and the Social Sciences." *Sociology* 17:2 (1983), 206–233.

Forman, Paul. "Behind Quantum Electronics: National Security as Basis for Physical Research in the United States, 1940–1960." *Historical Studies in the Physical and Biological Sciences* 18:1 (1987), 149–229.

———. "Scientific Internationalism and the Weimar Physicists: The Ideology and Its Manipulation in Germany after World War I." *Isis* 64 (1973), 151–180.

Fortun, Mike, and Sylvan S. Schweber. "Scientists and the Legacy of World War II: The Case of Operations Research (OR)." *Social Studies of Science* 23:4 (1993), 595–642.

Fosdick, Raymond B. *The Story of the Rockefeller Foundation.* New York: Harper and Brothers, 1952.

Frutkin, Arnold W. *International Cooperation in Space.* Englewood Cliffs, N.J.: Prentice-Hall, 1965.

Gaddis, John Lewis. "The Insecurities of Victory: The United States and the Perception of the Soviet Threat after World War II." In Lacey, *The Truman Presidency*, 235–271.

———. *Surprise, Security, and the American Experience.* Cambridge: Harvard University Press, 2004.

———. "Was the Truman Doctrine a Real Turning Point?" *Foreign Affairs* 52:2 (1974), 386–402.

———. *We Now Know: Rethinking Cold War History.* Oxford: Oxford University Press, 1997.

Galison, Peter. *Image and Logic: A Material Culture of Microphysics.* Chicago: University of Chicago Press, 1997.

———. "The Ontology of the Enemy: Norbert Wiener and the Cybernetic Vision." *Critical Inquiry* 21:1 (1994), 228–266.

———. "Removing Knowledge." *Critical Inquiry* 31:1 (2004), 229–240.

Gaudillière, Jean-Paul. *Inventer la biomédecine: La France, l'Amérique et la production des savoirs du vivant (1945–1965)*. Paris: Éditions la Découverte, 2002.

———. "Molecular Biology in the French Tradition? Redefining Local Traditions and Disciplinary Patterns." *Journal of the History of Biology* 26:3 (1993), 473–498.

———. "Paris-New York Roundtrip: Transatlantic Crossings and the Reconstruction of the Biological Sciences in Post-war France." *Studies in History and Philosophy of Biological and Biomedical Sciences* 33C (2002), 389–417.

Geiger, Roger L. "American Foundations and Academic Social Science, 1945–1960." *Minerva* 28:3 (1988), 315–341.

Gemelli, Giuliana, ed. *American Foundations and Large-Scale Research: Construction and Transfer of Knowledge*. Bologna: Clueb, 2001.

———, ed. *The Ford Foundation and Europe (1950s–1970s): Cross-Fertilization of Learning in Social Science and Management*. Brussels: European Interuniversity Press, 1998.

———. "From Imitation to Competitive Cooperation: The Ford Foundation and Management Education in Western and Eastern Europe (1950's–1970's)." In Gemelli, *Ford Foundation and Europe*, 167–304.

———. "Introduction: Scholars in Adversity and Science Policies (1933–1945)." In Gemelli, *The "Unacceptables,"* 13–34.

———, ed. *The "Unacceptables": American Foundations and Refugee Scholars between the Two Wars and After*. Brussels: Peter Lang, 2000.

———. "Western Alliance and Scientific Diplomacy in the Early 1960s: The Rise and Failure of the Project to Create a European M.I.T." In Moore and Vaudagna, *The American Century in Europe*, 171–192.

Gemelli, Giuliana, and Roy MacLeod, eds. *American Foundations in Europe: Scientific Policies, Cultural Diplomacy and Trans-Atlantic Relations, 1920–1980*. Brussels: P.I.E.-Peter Lang, 2003.

Gemelli, Giuliana, Jean-François Picard, and William H. Schneider, eds. *Managing Medical Research in Europe: The Role of the Rockefeller Foundation (1920s–1950s)*. Bologna: Clueb, 1999.

Ghamari-Tabrizi, Sharon. "Simulating the Unthinkable: Gaming Future War in the 1950s and 1960s." *Social Studies of Science* 30:2 (2000), 163–223.

Gieryn, Thomas F. *Cultural Boundaries of Science: Credibility on the Line*. Chicago: University of Chicago Press, 1998.

Gieryn, Thomas F. "Boundaries of Science." In Jasanoff, Markle, Petersen, and Pinch, *Handbook of Science and Technology Studies*, 393–443.

Gilpin, Robert. *France in the Age of the Scientific State*. Princeton: Princeton University Press, 1968.

Gimbel, John. *The Origins of the Marshall Plan*. Stanford: Stanford University Press, 1976.

————. *Science, Technology, and Reparations: Exploitation and Plunder in Postwar Germany.* Stanford: Stanford University Press, 1990.

Glatt, Carl. "Reparations and the Transfer of Scientific and Industrial Technology from Germany: A Case Study of the Roots of British Industrial Policy and of Aspects of British Occupation Policy in Germany between Post–World War II Reconstruction and the Korean War." 3 vols. Ph.D. diss., European University Institute, Florence, 1994.

Goldstein, Cora Sol. "The Control of Visual Representation: American Art Policy in Occupied Germany, 1945–1949." In Scott-Smith and Krabbendam, *The Cultural Cold War in Western Europe,* 283–299.

Goudsmit, Samuel A. *Alsos.* Los Angeles: Tomash Publishers, 1983.

Graham, Loren R. *Science in Russia and the Soviet Union: A Short History.* Cambridge: Cambridge University Press, 1994.

Grémion, Pierre. *Intelligence de l'anticommunisme: Le Congrès pour la Liberté de la Culture à Paris (1950–1975).* Paris: Fayard, 1995.

————. "The Partnership between the Ford Foundation and the Congress for Cultural Freedom in Europe." In Gemelli, *Ford Foundation and Europe,* 137–164.

Grinevald, J., A. Gsponer, L. Hanouz, and P. Lehmann. *La quadrature du CERN.* Lausanne: Editions d'En bas, 1984.

Hamblin, Jacob Darwin. "The Navy's 'Sophisticated' Pursuit of Science: Undersea Warfare, the Limits of Internationalism, and the Utility of Basic Research, 1945–1956." *Isis* 93 (2002), 1–27.

————. "Science in Isolation: American Marine Geophysics Research, 1950–1968." *Physics in Perspective* 2 (2000), 293–312.

————. "Visions of International Scientific Cooperation: The Case of Oceanic Science, 1920–1955." *Minerva* 38:4 (2000), 393–423.

Haskins, Caryl P. "Technology, Science and American Foreign Policy." *Foreign Affairs* 40 (1962), 224–243.

Hecht, Gabrielle. *The Radiance of France: Nuclear Power and National Identity after World War II.* Cambridge: MIT Press, 1999.

Heilbron, John L. "The First European Cyclotrons." *Rivista di Storia della Scienza* 3 (1986), 1–44.

Heilbron, John L., and Robert W. Seidel. *Lawrence and His Laboratory: A History of the Lawrence Berkeley Laboratory.* Vol. 1. Berkeley: University of California Press, 1989.

Heilbron, John L., Robert W. Seidel, and Bruce Wheaton. "Lawrence and His Laboratory: Nuclear Science at Berkeley." *LBL News Magazine* 6:3 (1981). http://www.lbl.gov/Science-Articles/Research-Review/Magazine/1981/index.html.

Hermann, Armin, John Krige, Ulrike Mersits, and Dominique Pestre. *History of CERN.* Vol. 1, *Launching the European Organisation for Nuclear Research.* Amsterdam: North Holland, 1987.

Hewlett, Richard G., and Jack M. Holl. *Atoms for Peace and War, 1953–1961: Eisenhower and the Atomic Energy Commission.* Berkeley: University of California Press, 1989.

Hitchcock, William I. *France Restored: Cold War Diplomacy and the Quest for Leadership in Europe, 1944–1954.* Chapel Hill: University of North Carolina Press, 1998.

Hoffmann, Stanley. "Why Don't They Like Us?" *American Prospect* 12:20, 19 November 2001.

Hogan, Michael J. *The Marshall Plan: America, Britain, and the Reconstruction of Western Europe, 1947–1952.* Cambridge: Cambridge University Press, 1987.

Hollinger, David. "The Defense of Democracy and Robert K. Merton's Formulation of the Scientific Ethos." *Knowledge and Society: Studies in the Sociology of Culture Past and Present* 4 (1983), 1–15.

———. "Science as a Weapon in *Kulturkämpfe* in the United States during and after World War II." *Isis* 86 (1995), 440–454.

Holmevik, Jan Rune. *Educating the Machine: A Study in the History of Computing and the Construction of the SIMULA Programming Languages.* STS Report 22/94. Trondheim, 1994.

Hounshell, David. "After September 11, 2001: An Essay on Opportunities and Opportunism, Institutions and Institutional Innovation, Technologies and Technological Change." *History and Technology* 19:1 (2003), 39–49.

———. "The Cold War, RAND and the Generation of Knowledge." *Historical Studies in the Physical and Biological Sciences* 27:2 (1997), 237–267.

———. "The Evolution of Industrial Research in the United States." In *Engines of Innovation: U.S. Industrial Research at the End of an Era*, edited by Richard S. Rosenbloom and William J. Spencer, 1–85. Boston: Harvard Business School, 1996.

Hughes, Agatha C., and Thomas P. Hughes, eds. *Systems, Experts, and Computers: The Systems Approach in Management and Engineering; World War II and After.* Cambridge: MIT Press, 2000.

Hughes, Jeff. "Redefining the Context: Oxford and the Wider World of British Physics, 1900–1940." In *Physics in Oxford 1839–1939: Laboratories, Learning, and College Life*, edited by Robert Fox and Graeme Gooday, 267–300. Oxford: Oxford University Press, 2005.

Hughes, Thomas P. *Rescuing Prometheus.* New York: Random House, 2000.

Huxley, Julian. "Soviet Genetics: The Real Issue." *Nature,* 18 June 1949, 935–942.

Iriye, Akira. *Cultural Internationalism and World Order.* Baltimore: Johns Hopkins University Press, 1997.

Jasanoff, Shelia, ed. *States of Knowledge: The Co-Production of Science and the Social Order.* London: Routledge, 2004.

Jasanoff, Shelia, Gerald E. Markle, James C. Petersen, and Trevor Pinch, eds., *Handbook of Science and Technology Studies*. London: SAGE, 1995.

Joerges, Burghard, and Terry Shinn, eds. *Instrumentation between Science, State and Industry*. Dordrecht: Kluwer Academic Publishers, 2000.

Joravsky, David. *The Lysenko Affair*. Cambridge: Harvard University Press, 1970.

Judt, Matthias, and Burghard Ciesla, eds. *Technology Transfer out of Germany after 1945*. Amsterdam: Harwood Academic Publishers, 1996.

Judt, Tony. "Dreams of Empire." *New York Review of Books*, 4 November 2004, 38–41 www.nybooks.com/articles/17518.

———. *Past Imperfect: French Intellectuals, 1944–1956*. Berkeley: University of California Press, 1992.

Jungk, Robert. *The Big Machine*. New York: Scribner, 1968.

Junker, Detlef. "The United States, Germany and Europe in the Twentieth Century." In Moore and Vaudagna, *The American Century in Europe*, 94–113.

Kagan, Robert. *Of Paradise and Power: America and Europe in the New World Order*. New York: Alfred A. Knopf, 2003.

Kaiser, David, "Cold War Requisitions, Scientific Manpower, and the Production of American Physicists after World War II." *Historical Studies in the Physical and Biological Sciences* 33:1 (2002), 131–159.

———, ed. *Pedagogy and the Practice of Science: Historical and Contemporary Perspectives*. Cambridge: MIT Press, 2005.

———. "The Postwar Suburbanization of American Physics." *American Quarterly* 56:4 (2004), 851–888.

Kaplan, Lawrence S. *The Long Entanglement: NATO's First Fifty Years*. Westport, Conn.: Praeger, 1999.

———. *The United States and NATO: The Formative Years*. Lexington: University Press of Kentucky, 1984.

Kay, Lily E. *The Molecular Vision of Life: Caltech, the Rockefeller Foundation, and the Rise of the New Biology*. New York: Oxford University Press, 1993.

———. "Rethinking Institutions: Philanthropy as an Historiographic Problem of Knowledge and Power." *Minerva* 35:3 (1997), 283–293.

Kevles, Daniel J. "Foundations, Universities and Trends in Support for the Physical and Biological Sciences, 1900–1992." *Daedelus* 121:4 (1992), 195–235.

———. "The National Science Foundation and the Debate over Postwar Research Policy, 1942–1945." *Isis* 68 (1977), 5–26.

———. *The Physicists: The History of a Scientific Community in Modern America*. New York: Alfred A. Knopf, 1978.

———. "Principles and Politics in Federal R&D Policy, 1945–1990: An Appreciation of the Bush Report." Introduction to *Science—The Endless Frontier*, by

Vannevar Bush. 40th anniversary ed. Report NSF 90-8. Washington, D.C.: National Science Foundation, 1990.

Killian, J. R. "An International Institute of Science and Technology." *NATO Letter* 10:4 (1962).

Kirby, Maurice W. "Operations Research Trajectories: The Anglo-American Experience from the 1940s to the 1990s." *Operations Research* 48:5 (2000), 661–670.

Kirby, Maurice W., and R. Capey. "The Origins and Diffusion of Operational Research in the UK." *Journal of the Operational Research Society* 49:4 (1998), 307–326.

Kistiakowsky, George B. "Science and Foreign Affairs." *Department of State Bulletin,* 22 February 1960, 276–283.

Kleinman, Daniel J. *Politics on the Endless Frontier: Postwar Research Policy in the United States.* Durham: Duke University Press, 1995.

Kline, Ronald. "Construing 'Technology' as 'Applied Science': Public Rhetoric of Scientists and Engineers in the United States, 1880–1945." *Isis* 86 (1995), 194–221.

Knorr Cetina, Karin. "Laboratory Studies: The Cultural Approach to the Study of Science." In Jasanoff, Markle, Patterson, and Pinch, *Handbook of Science and Technology Studies,* 140–166.

Kohler, Robert E. *Landscapes and Labscapes: Exploring the Lab-Field Border in Biology.* Chicago: University of Chicago Press, 2002.

———. *Lords of the Fly:* Drosophila *Genetics and the Experimental Life.* Chicago: University of Chicago Press, 1994.

———. "The Management of Science: The Experience of Warren Weaver and the Rockefeller Foundation Program in Molecular Biology." *Minerva* 14:3 (1976), 276–306.

———. *Partners in Science: Foundations and Natural Scientists, 1900–1945.* Chicago: University of Chicago Press, 1991.

———. "Warren Weaver and the Rockefeller Foundation Program in Molecular Biology: A Case Study in the Management of Science." In Rheingold, *The Sciences in the American Context,* 249–293.

Kowarski, Lew. "An Account of the Origin and Beginnings of CERN." *CERN Yellow Report CERN 61–10* (10 April 1961).

———. "The Making of CERN: An Experiment in Cooperation." *Bulletin of the Atomic Scientists* 11:10 (1955), 354–357.

Krementsov, Nikolai. *Stalinist Science.* Princeton: Princeton University Press, 1997.

Krige, John. "Atoms for Peace, Scientific Internationalism and Scientific Intelligence." In Krige and Barth, *Global Power Knowledge,* 161–181.

————. "The Birth of EMBO and the Difficult Road to EMBL." *Studies in History and Philosophy of Biological and Biomedical Sciences* 33C (2002), 547–564.

————. "CERN, l'atome piégé par le 'plan Marshall.'" *La Recherche* 379 (October 2004), 64–69.

————. "Distrust and Discovery: The Case of the Heavy Bosons at CERN." *Isis* 92:3 (2001), 517–540.

————. "Felix Bloch and the Creation of a 'Scientific Spirit' at CERN." *Historical Studies in the Physical and Biological Sciences* 32:1 (2002), 57–69.

————. "The Ford Foundation, European Physics and the Cold War." *Historical Studies in the Physical and Biological Sciences* 29:2 (1999), 333–361.

————. "History of Technology after 9/11: Technology, American Power and 'Anti-Americanism.'" *History and Technology* 19:1 (2003), 32–39.

————. "Isidor I. Rabi and the Birth of CERN." *Physics Today* (September 2004), 44–48.

————. "Isidor I. Rabi and CERN." *Physics in Perspective* 7:2 (2005), 150–164.

————. "NATO and the Strengthening of Western Science in the Post-Sputnik Era." *Minerva* 38:1 (2000), 81–108.

————. "Philanthropy and the National Security State: The Ford Foundation's Support for Physics in Europe in the 1950s," in Gemelli, *American Foundations and Large-Scale Research*, 3–24.

————. "The Politics of Phosphorus-32: A Cold War Fable Based on Fact." *Historical Studies in the Physical and Biological Sciences* 36:1 (2005), 71–91. Reprinted in *The Historiography of Science, Medicine and Technology*, edited by Ron Doel and Thomas Söderqvist. London: Routledge, 2006.

————. "La science et la sécurité civile de l'Occident." In Dahan and Pestre, *Les sciences pour la guerre*, 373–401.

————. "Scientists as Policymakers: British Physicists' 'Advice' to Their Government on Membership of CERN (1951/52)." In *Solomon's House Revisited: The Organization and Institutionalization of Science,* edited by Tore Frängsmyr, 270–291. Canton, Mass.: Science History Publications, 1990.

————. "What Is 'Military' Technology? Two Cases of US-European Scientific and Technological Collaboration in the 1950s." In *The United States and the Integration of Europe: Legacies of the Postwar Era,* edited by Francis H. Heller and John R. Gillingham, 307–338. New York: St. Martins Press, 1996.

Krige, John, and Kai-Henrik Barth, eds. *Global Power Knowledge: Science and Technology in International Affairs. Osiris,* 2nd ser., 21 (2006).

Krige, John, and Arturo Russo. *A History of the European Space Agency, 1958–1987.* Vol. 1, *The Story of ESRO and ELDO, 1958–1973.* Report SP-1235. Noordwijk: European Space Agency, 2000.

Kroes, Rob. "American Empire and Cultural Imperialism: A View from the Receiving End." *Diplomatic History* 23:3 (1999), 463–477.

Kuisel, Richard. "Coca-Cola and the Cold War: The French Face Americanisation, 1948–1953." *French Historical Studies* 17:1 (1991), 96–116.

———. *Seducing the French: The Dilemma of Americanization.* Berkeley: University of California Press, 1993.

Lacey, Michael J., ed. *The Truman Presidency.* Washington D.C.: Woodrow Wilson International Center for Scholars and Cambridge University Press, 1989.

Lasby, Clarence B. *Project Paperclip: German Scientists and the Cold War.* New York: Atheneum, 1971.

Lawrence, Ernest O. "High Current Accelerators." In *Proceedings of the International Conference on the Peaceful Uses of Atomic Energy: Held in Geneva, 8 August–20 August 1955,* vol. 16, *Record of the Conference,* 64–68. New York: United Nations, 1956.

Lazar, Marc. "The Cold War Culture of the French and Italian Communist Parties." In Scott-Smith and Krabbendam, *The Cultural Cold War in Western Europe,* 213–224.

Lecourt, Dominique. *Proletarian Science? The Case of Lysenko.* London: New Left Books, 1977.

Leffler, Melvyn P. *A Preponderance of Power: National Security, the Truman Administration, and the Cold War.* Stanford: Stanford University Press, 1992.

Lenoir, Timothy, and Christophe Lecuyer. "Instrument Makers and Discipline Builders: The Case of Nuclear Magnetic Resonance." *Perspectives on Science* 3:3 (1995), 276–345.

Leslie, Stuart W. *The Cold War and American Science: The Military-Industrial-Academic Complex at MIT and Stanford.* New York: Columbia University Press, 1993.

———. "Profit and Loss: The Military and MIT in the Postwar Era." *Historical Studies in the Physical and Biological Sciences* 21:1 (1990), 59–85.

Leslie, Stuart W., and Robert Kargon. "Exporting MIT: Science, Technology and Nation-Building in India and Iran." In Krige and Barth, *Global Power Knowledge,* 110–130.

Lewontin, Richard, and Richard Levins. "The Problem of Lysenkoism." In *The Radicalisation of Science: Ideology of/in the Natural Sciences,* edited by Hilary Rose and Steven Rose, 32–64. London: Macmillan, 1976.

Lexow, Wilton. "The Science Attache Program." *Studies in Intelligence,* 1 April 1966, 21–27. Available on http://www.foia.cia.gov.

L'Héritier, Philippe. "Entretien avec Philippe L'Héritier." http://picardp1.ivry.cnrs.fr/~jfpicard/LHeritier.html.

Lowen, Rebecca S. *Creating the Cold War University: The Transformation of Stanford.* Berkeley: University of California Press, 1997.

———. "'Exploiting a Wonderful Opportunity': The Patronage of Scientific Research at Stanford University, 1937–1965." *Minerva* 30:3 (1992), 391–421.

Lucas, Scott. "Campaigns of Truth: The Psychological Strategy Board and American Ideology, 1951–1953." *International History Review* 18:2 (1996), 279–302.

Ludmann-Obier, Marie-France. "Un aspect de la chasse aux cerveaux: Les transferts de techniciens allemands en France, 1945–1949." *Relations Internationales* 46 (Summer 1986), 195–208.

————. "La mission du CNRS en Allemagne (1945–1950)." *Cahiers pour l'Histoire du CNRS 1939–1989* 3 (1989), 73–84.

Lundestad, Geir. *The American "Empire" and Other Studies of US Foreign Policy in a Comparative Perspective*. Oxford: Oxford University Press, 1990.

————. *"Empire" by Integration: The United States and European Integration, 1945–1997*. Oxford: Oxford University Press, 1998.

————. "'Empire by Invitation' in the American Century." *Diplomatic History* 23:2 (1999), 189–217.

————. "Empire by Invitation? The United States and Western Europe, 1945–1952." *Journal of Peace Research* 23:3 (1986), 263–277.

McCarthy, Kathleen D. "From Cold War to Cultural Development: The International Cultural Activities of the Ford Foundation." *Daedelus* 116:1 (1987), 93–117.

Macdonald, Dwight. *The Ford Foundation: The Men and the Millions*. New Brunswick, N.J.: Transaction Publishers, 1989.

McGovern, James. *Crossbow and Overcast*. New York: William Morrow, 1964.

MacLeod, Roy, ed. *Nature and Empire: Science and the Colonial Enterprise*. *Osiris*, 2nd ser., 15 (2000).

————. "Science and Democracy: Historical Reflections on Present Discontents." *Minerva* 35:4 (1997), 369–384.

Macrakis, Kristie. "The Rockefeller Foundation and German Physics under National Socialism." *Minerva* 27:1 (1989), 33–57.

Maier, Charles S. "Alliance and Autonomy: European Identity and U.S. Foreign Policy Objectives in the Truman Years." In Lacey, *The Truman Presidency*, 273–298.

————. "An American Empire? The Problems of Frontiers and Peace in Twenty-First-Century World Politics." *Harvard Magazine* 105:2 (2002), 28–31. www.harvardmagazine.com/on-line/1102193.html.

————. "The Marshall Plan and the Division of Europe." *Journal of Cold War Studies* 7:1 (2005), 168–174.

————. "The Politics of Productivity: Foundations of American International Economic Policy after World War II." In Charles S. Maier, *In Search of Stability: Explorations in Historical Political Economy*, 121–152. Cambridge: Cambridge University Press, 1987.

Manzione, Joseph. "'Amusing and Amazing and Practical and Military': The Legacy of Scientific Internationalism in American Foreign Policy, 1945–1963." *Diplomatic History* 24:1 (2000), 21–55.

May, Lary, ed. *Recasting America: Culture and Politics in the Age of Cold War.* Chicago: University of Chicago Press, 1989.

Medvedev, Zhores A. *The Rise and Fall of T. D. Lysenko.* New York: Columbia University Press, 1969.

Mendelsohn, Everett. "Robert K. Merton: The Celebration and Defense of Science." *Science in Context* 3:1 (1989), 269–289.

Miller, James E. "Taking Off the Gloves: The United States and the Italian Elections of 1948." *Diplomatic History* 7:1 (1983), 35–55.

Milward, Alan S. *European Rescue of the Nation State.* 2nd ed. London: Routledge, 2000.

———. *The Reconstruction of Western Europe, 1945–1951.* Berkeley: University of California Press, 1984.

Mindell, David A. "Automation's Finest Hour: Radar and System Integration in World War II." In Hughes and Hughes, *Systems, Experts, and Computers,* 27–56.

Mirowski, Philip. "Cyborg Agonistes: Economics Meets Operations Research in Mid-Century." *Social Studies of Science* 29:5 (1999), 685–718.

———. *Machine Dreams: Economics Becomes a Cyborg Science.* Cambridge: Cambridge University Press, 2002.

———. "The Scientific Dimensions of Social Knowledge and Their Distant Echoes in 20th-Century American Philosophy of Science." *Studies in the History and Philosophy of Science, Part A* 35:2 (2004), 283–326.

Moore, R. Laurence, and Maurizio Vaudagna, eds. *The American Century in Europe.* Ithaca: Cornell University Press, 2003.

Morange, Michel. *A History of Molecular Biology.* Cambridge: Harvard University Press, 2000.

———. "L'institut de biologie physico-chimique, de sa fondation à l'entrée dans l'ère moléculaire." *Revue pour l'Histoire du CNRS* 7 (2002), 32–41.

Morrison, Philip. "*Alsos:* The Story of German Science." *Bulletin of the Atomic Scientists* 3:12 (1947), 354, 365.

Morse, Philip M. *In at the Beginnings: A Physicist's Life.* Cambridge: MIT Press, 1977.

———. "Must We Always Be Gadgeteers?" *Physics Today* 3 (December 1950), 4–5.

———. "Report on the First Summer Program on Operations Research at the Massachusetts Institute of Technology, June 16–July 2, 1953." *Journal of the Operations Research Society of America* 1:5 (1953), 303–305.

Murard, Lion, and Patrick Zylberman. "Seeds for French Health Care: Did the Rockefeller Foundation Plant the Seeds between the Two World Wars?" *Studies in History and Philosophy of Biological and Biomedical Sciences* 31:3 (2000), 463–475.

NATO and Science: An Account of the Activities of the NATO Science Committee, 1958–1972. Brussels: NATO Scientific Affairs Division, 1973.

Needell, Allan A. *Science, Cold War and the American State: Lloyd V. Berkner and the Balance of Professional Ideals.* Amsterdam: Harwood Academic Publishers, 2000.

————. " 'Truth Is Our Weapon': Project TROY, Political Warfare and Government-Academic Relations in the National Security State." *Diplomatic History* 17:3 (1993), 399–420.

Nierenberg, William A. "The NATO Science Program." *Bulletin of the Atomic Scientists* 21:5 (1965), 45–48.

Nye, Joseph S., Jr. "Soft Power." *Foreign Policy* 80 (1990), 153–171.

O'Connell, Joseph. "Metrology: The Creation of Universality by the Circulation of Particulars." *Social Studies of Science* 23:1 (1993), 129–173.

Olff-Nathan, Josiane, ed. *La science sous la Troisième Reich.* Paris: Seuil, 1993.

Ozietzki, Maria. "The Ideology of Early Particle Accelerators: An Association between Knowledge and Power." In Renneberg and Walker, *Science, Technology and National Socialism,* 255–270.

Palayret, Jean-Marie. *A University for Europe: Prehistory of the European University Institute in Florence (1948–1976).* Rome: Istituto Poligrafico e Zecca dello Stato, 1996.

Parmar, Inderjeet. " 'To Relate Knowledge and Action': The Impact of the Rockefeller Foundation on Foreign Policy Thinking during America's Rise to Globalism 1939–1945." *Minerva* 40:3 (2002), 235–263.

Paxton, Robert O., and Nicholas Wahl, eds. *De Gaulle and the United States: A Centennial Reappraisal.* Oxford: Berg, 1994.

Pells, Richard. *Not Like Us: How Europeans Have Loved, Hated, and Transformed American Culture since World War II.* New York: Basic Books, 1997.

Pestre, Dominique. Commentary on *Colloque sur l'Histoire du CNRS des 23 et 24 octobre 1989.* www.ivry.cnrs.fr/politiques_de_la_science/sciences_ex_shs_4. Accessed on 22 May 2003.

————. "Louis Néel, le magnétisme et Grenoble: Récit de la création d'un empire physicien dans la province française, 1940–1965." *Cahiers pour l'Histoire du CNRS* 8 (1990), 1–188.

————. "Le renouveau de la recherche à l'École polytechnique et le laboratoire de Louis Leprince-Ringuet, 1936–1965." In Belhoste, Dahan Dalmedico, and Picon, *La formation polytechnicienne,* 333–356.

————. "From Revanche to Competition and Cooperation: Physics Research in Germany and France." Paper presented at the conference "Society in the Mirror of Science: The Politics of Knowledge in Modern France," Berkeley, 30 September–1 October 1988.

Pestre, Dominique, and John Krige. "Some Thoughts on the Early History of CERN." In *Big Science: The Growth of Large-Scale Research,* edited by Peter Galison and Bruce Hevly, 79–98. Stanford: Stanford University Press, 1992.

Picard, Jean-François. "La création du CNRS." *Revue pour l'Histoire du CNRS* 1 (1999), 50–66.

———. "Un demi-siècle de génétique de la levure au CNRS: De la biologie moléculaire à la génomique." *Revue pour l'Histoire du CNRS* 7 (2002), 42–49.

———. *La Fondation Rockefeller et la recherche médicale.* Paris: PUF, 1999.

———. "The Institut national d'hygiène and Public Health in France, 1940–1946." http://picardp1.ivry.cnrs.fr/INH.html, accessed on 2 Feb 2005.

———. *La république des savants: La recherche française et le CNRS.* Paris: Flammarion, 1990.

Picard, Jean-François, and William H. Schneider. "From the Art of Medicine to Biomedical Science in France: Modernization or Americanization." In Gemelli, *American Foundations,* 91–114.

Pickering, Andy. "Cyborg History and the World War II Regime." *Perspectives on Science* 3:1 (1995), 1–48.

———. *The Mangle of Practice: Time, Agency, and Science.* Chicago: University of Chicago Press, 1995.

Pinault, Michel. *Frédéric Joliot-Curie.* Paris: Odile Jacob, 2000.

Porter, Theodore M. *Trust in Numbers: The Pursuit of Objectivity in Science and in Public Life.* Princeton: Princeton University Press, 1995.

Prados, John. *Keepers of the Keys: A History of the National Security Council from Truman to Bush.* New York: William Morrow, 1991.

Pruessen, Ronald W. "Cold War Threats and America's Commitment to the European Defense Community: One Corner of a Triangle." *Journal of European Integration History* (Spring 1996), 51–69.

Rabi, Isidor I. *My Life and Times as a Physicist.* Claremont, Calif.: Claremont College, 1960.

———. *Science: The Center of Culture.* New York: World Publishing, 1970.

Ramunni, Girolamo. "La réorganisation du Centre de la Recherche Scientifique, 7 septembre 1944." *Revue pour l'Histoire du CNRS* 3 (2000), 60–70.

Rasmussen, Nicolas. "Instruments, Scientists, Industrialists and the Specificity of 'Influence': The Case of RCA and Biological Electron Microscopy." In *The Invisible Industrialist: Manufactures and the Production of Scientific Knowledge,* edited by Jean Paul Gaudillière and Ilana Löwy, 173–208. London: Macmillan, 1998.

Rasmussen, Nicolas. *Picture Control: The Electron Microscope and the Transformation of Biology in America, 1940–1960.* Stanford: Stanford University Press, 1997.

Rau, Erik P. "The Adoption of Operations Research in the United States during World War II." In Hughes and Hughes, *Systems, Experts, and Computers,* 57–92.

————. "Combat Scientists: The Emergence of Operations Research in the United States during World War II." Ph.D. diss., University of Pennsylvania, 1999.

————. "Morse, Philip McCord." In *American National Biography* 15:937–940. New York: Oxford University Press, 1999.

Renneberg, Monika, and Mark Walker, eds. *Science, Technology and National Socialism.* Cambridge: Cambridge University Press, 1994.

Rheingold, Nathan, ed. *The Sciences in the American Context: New Perspectives.* Washington, D.C.: Smithsonian Institution Press, 1979.

Rigden, John S. *Rabi: Scientist and Citizen.* Cambridge: Harvard University Press, 2000.

Roberts, Henry L. *Russia and America: Dangers and Prospects.* New York: Harper and Bros. for the Council on Foreign Relations, 1956.

Romero, Federico. "Democracy and Power: The Interactive Nature of the American Century." In Moore and Vaudagna, *The American Century in Europe,* 47–65.

Rees, Mina. "Warren Weaver, July 17, 1894–November 24, 1978." *Biographical Memoirs, National Academy of Sciences* 57 (1987), 492–530.

Rose, Paul Lawrence. *Heisenberg and the Nazi Atomic Bomb Project: A Study in German Culture.* Berkeley: University of California Press, 1998.

Ross, Kristin. *Fast Cars, Clean Bodies: Decolonization and the Reordering of French Culture.* Cambridge: MIT Press, 1998.

Rudolph, John L. *Scientists in the Classroom: The Cold War Reconstruction of American Science Education.* New York: Palgrave, 2002.

Salomon, Jean-Jacques. "The *Internationale* of Science." *Science Studies* 1:1 (1971), 23–42.

Sapolsky, Harvey M. "Academic Science and the Military: The Years since the Second World War." In Rheingold, *The Sciences in the American Context,* 379–399.

————. *Science and the Navy: The History of the Office of Naval Research.* Princeton: Princeton University Press, 1990.

Saunders, Frances Stonor. *Who Paid the Piper? The CIA and the Cultural Cold War.* London: Granta Books, 1999.

Schneider, William H., ed. *Rockefeller Philanthropy and Modern Biomedicine: International Initiatives from World War I to the Cold War.* Bloomington: Indiana University Press, 2002.

————. "War, Philanthropy and the National Institute of Hygiene in France." *Minerva* 41:1 (2003), 1–23.

Schweber, Silvan S. *In the Shadow of the Bomb: Bethe, Oppenheimer, and the Moral Responsibility of the Scientist.* Princeton: Princeton University Press, 2000.

"Science and Foreign Relations: Berkner Report to the U.S. Department of State." *Bulletin of the Atomic Scientists* 6:6 (June 1950), 293–298.

Scott-Smith, Giles. *The Politics of Apolitical Culture: The Congress for Cultural Freedom, the CIA and Post-war American Hegemony.* London: Routledge, 2002.

Scott-Smith, Giles, and Hans Krabbendam, eds. *The Cultural Cold War in Western Europe, 1945–1960.* London: Frank Cass, 2003.

Seitz, Frederick. *On the Frontier: My Life in Science.* New York: American Institute of Physics, 1994.

Shapin, Steven, and Simon Schaffer. *Leviathan and the Air Pump: Hobbes, Boyle and the Experimental Life.* Princeton: Princeton University Press, 1985.

Shryock, Richard Harrison. *Medicine in America: Historical Essays.* Baltimore: Johns Hopkins University Press, 1966.

Siegmund-Schultze, Reinhard. "Rockefeller Support for Mathematicians Fleeing from the Nazi Purge." In Gemelli, *The "Unacceptables,"* 83–106.

Sime, Ruth Lewin. *Lise Meitner: A Life in Physics.* Berkeley: University of California Press, 1996.

Simpson, Christopher, ed. *The Cold War and the University: Toward an Intellectual History of the Postwar Years.* New York: New Press, 1997.

———, ed. *Universities and Empire: Money and Politics in the Social Sciences during the Cold War.* New York: New Press, 1998.

Skolnikoff, Eugene B. *Science, Technology and American Foreign Policy.* Cambridge: MIT Press, 1967.

Smith, Tony. *America's Mission: The United States and the Worldwide Struggle for Democracy.* Princeton: Princeton University Press, 1995.

———. "Making the World Safe for Democracy in the American Century." *Diplomatic History* 23:2 (1999), 173–188.

Snead, David L. *The Gaither Committee, Eisenhower, and the Cold War.* Columbus: Ohio State University Press, 1999.

Solomon, Susan Gross, and Nikolai Krementsov. "Giving and Taking across Borders: The Rockefeller Foundation and Russia, 1919–1928." *Minerva* 39:3 (2001), 265–298.

Solovey, Mark. "Project Camelot and the 1960s Epistemological Revolution: Rethinking the Politics-Patronage–Social Science Nexus." *Social Studies of Science* 31:2 (2001), 171–206.

Strasser, Bruno. "Institutionalizing Molecular Biology in Post-war Europe: A Comparative Study." *Studies in History and Philosophy of Biological and Biomedical Sciences* 33C (2002), 515–546.

Strasser, Bruno J., and Frédéric Joye. "L'atome, l'espace, les molécules: Les pays neutres dans la coopération scientifique européenne (1949–1969)." Unpublished manuscript.

Strasser, Susan, Charles McGovern, and Matthias Judt. *Getting and Spending: European and American Consumer Societies in the Twentieth Century.* Cambridge: Cambridge University Press, 1998.

Sutton, Francis X. "The Ford Foundation: The Early Years." *Daedelus* 116:1 (1987), 41–91.

———. "The Ford Foundation and Europe: Ambitions and Ambivalences." In Gemelli, *Ford Foundation and Europe*, 21–66.

———. "The Ford Foundation's Transatlantic Role and Purposes, 1951–81." *Fernand Braudel Center Review*, 24:1 (2001), 77–104.

Teller, Edward. "Back to the Laboratories." *Bulletin of the Atomic Scientists* 6:3 (1950), 71–72.

Trachtenberg, Marc. *A Constructed Peace: The Making of the European Settlement, 1945–1963.* Princeton: Princeton University Press, 1999.

Van der Bliek, Jan, ed. *AGARD: The History, 1952–1997.* Neuilly-sur-Seine: NATO Research and Technology Organization, 1999.

Vassails, Gérard. "Les monopoles financiers américains contre la science française." *La Pensée* 41 (1952), 121–136.

Von Laue, Max. "The Wartime Activities of German Scientists." *Bulletin of the Atomic Scientists* 4:4 (1948), 103, 104.

Wagnleitner, Reinhold. *Coca-Colonization and the Cold War: The Cultural Mission of the United States in Austria after the Second World War.* Chapel Hill: University of North Carolina Press, 1994.

———. "The Empire of the Fun, or Talkin' Soviet Union Blues: The Sound of Freedom and U.S. Cultural Hegemony in Europe." *Diplomatic History* 23:3 (1999), 499–524.

Walker, Mark. *German National Socialism and the Quest for Nuclear Power.* Cambridge: Cambridge University Press, 1992.

———. "Legends Surrounding the German Atomic Bomb." In *Science, Medicine and Cultural Imperialism*, edited by Teresa Meade and Mark Walker, 178–204. New York: St. Martin's Press, 1991.

———. "The Nazification and Denazification of Physics." In Judt and Ciesla, *Technology Transfer out of Germany*, 49–59.

———. "Physics and Propaganda: Werner Heisenberg's Foreign Lectures under National Socialism." *Historical Studies in the Physical Sciences* 22:2 (1992), 339–389.

Wang, Jessica. *American Science in an Age of Anxiety: Scientists, Anticommunism, and the Cold War.* Chapel Hill: University of North Carolina Press, 1999.

———. "Merton's Shadow: Perspectives on Science and Democracy since 1940." *Historical Studies in the Physical and Biological Sciences* 30:2 (1999), 279–306.

Weart, Spencer R. *Scientists in Power.* Cambridge: Harvard University Press, 1979.

Weaver, Warren. *Scene of Change: A Lifetime in American Science.* New York: Scribner, 1970.

Weindling, Paul. "'Out of the Ghetto': The Rockefeller Foundation and German Medicine after the Second World War." In Schneider, *Rockefeller Philanthropy and Modern Biomedicine*, 208–222.

———. "Public Health and Political Stabilisation: The Rockefeller Foundation in Central and Eastern Europe between the Two World Wars." *Minerva* 31:3 (1993), 253–267.

Weiss, Burghard. "The 'Minerva' Project: The Accelerator Laboratory at the Kaiser Wilhelm Institute/Max Planck Institute of Chemistry: Continuity in Fundamental Research." In Renneberg and Walker, *Science, Technology and National Socialism*, 271–290.

Wheeler, John Archibald. *At Home in the Universe.* New York: American Institute of Physics, 1994.

———. *Geons, Black Holes, and Quantum Foam.* New York: W. W. Norton, 1998.

Wicken, Olav. "Cold War in Space Research: Ionospheric Research and Military Communication in Norwegian Politics." In *Making Sense of Space: The History of Norwegian Space Activities*, edited by John Peter Collett, 41–73. Oslo: Scandinavian University Press, 1995.

———. "Space Science and Technology in the Cold War: The Ionosphere, the Military and Politics in Norway." *History and Technology* 13:3 (1997), 207–229.

Wiener, Charles. "A New Site for the Seminar: The Refugees and American Physics in the Thirties." In *The Intellectual Migration*, edited by D. Fleming and B. Bailyn, 190–233. Cambridge: Harvard University Press, 1966.

Wise, M. Norton, ed. *The Values of Precision.* Princeton: Princeton University Press, 1995.

Zallen, Doris. "Louis Rapkine and the Restoration of French Science after the War." *French Historical Studies* 17:1 (1991), 6–37.

———. "Louis Rapkine and the Rockefeller Foundation: Rebuilding French Science after the War." In Gemelli, *American Foundations*, 157–173.

———. "The Rockefeller Foundation and French Research." *Cahiers pour l'Histoire du CNRS* 5 (1989), 35–58.

Zeitlin, Jonathan, and Gary Herrigel, eds. *Americanization and Its Limits: Reworking US Technology and Management in Post-war Europe and Japan.* Oxford: Oxford University Press, 2000.

Index